The Management of Construction

To

Matthew Lawrence Bennett

and

Andrew Lee Bennett

Two fine sons

Both outstanding managers, though not in the realm of construction

The Management of Construction: A Project Life Cycle Approach

F. Lawrence Bennett, PE, PhD

Emeritus Professor of Engineering Management

University of Alaska, Fairbanks

AMSTERDAM BOSTON HEIDELBERG LONDON NEW YORK OXFORD
PARIS SAN DIEGO SAN FRANCISCO SINGAPORE SYDNEY TOKYO

Butterworth-Heinemann
An imprint of Elsevier
Linacre House, Jordan Hill, Oxford OX2 8DP
200 Wheeler Road, Burlington MA 01803

First published 2003

British Library Cataloguing in Publication Data
Bennett, F. Lawrence
The management of construction: a project life cycle approach
1. Construction industry – Management 2. Project management
I. Title
624'.068

Library of Congress Cataloguing in Publication Data
A catalogue record for this book is available from the Library of Congress

ISBN 0 7506 5254 3

For information on all Butterworth-Heinemann publications visit our website at www.bh.com

Composition by Genesis Typesetting Limited, Rochester, Kent
Printed and bound in Great Britain

Contents

Preface *xiii*
Acknowledgements *xv*

1 Introduction 1
Construction industry segments 1
General building construction 1
Engineered construction 2
Construction industry characteristics 3
Overview of the construction project life cycle 7
Pre-project phase 8
Planning and design phase 8
Contractor selection phase 9
Project mobilisation phase 9
Project operations phase 9
Project closeout and termination phase 10
Discussion questions 10
References 11

2 Pre-project phase 12
Introduction 12
Selection of project delivery system 12
Traditional design–tender–build 12
Design–build 14
Construction manager 16
Project manager 18
Document and construct 18
Separate prime contracts 20
Turnkey 20
Build–own–operate–transfer 21
Joint venture 22
Force account 23
Phased construction 24
Evaluation and comparison of project delivery systems 26

Selection of type of contract 26
Lump sum/fixed price 27
Unit price/measure and value 27
Cost plus 31
Variations of basic cost plus 32
Time and materials 33
Discussion questions 33
References 34
Case study: Project organisation innovation in Scandinavia 36

3 Planning and design phase 40
Introduction 40
The parties and their roles 40
Architect 40
Engineer 41
Geotechnical specialist 41
Other specialists 42
Land surveyor 42
Cost estimator 42
Quantity surveyor 43
Project manager 43
A note on partnering 43
Planning and feasibility study stage 44
Consultant selection 44
The brief 47
Programme development 47
Identification of alternatives 48
Site investigation 49
Constructability analysis 49
Public input 50
Code analysis 50
Preliminary cost estimate 50
Financial feasibility analysis 52
Project recommendation 53
Funding 53
Site selection and land acquisition 54
Design stage 54
Introduction 54
Schematic design 54
Design development 57
Contract document development stage 57
Introduction to the form of the contract 57
Drawings 59
General conditions 60
Special conditions 61
Technical specifications 62
Schedule of quantities 65

Invitation to tender 65
Instructions to tenderers 65
Tender form 66
Agreement 67
Surety bond forms and insurance certificates 67
Summary of planning and design 67
Cost estimates 68
Constructability analysis 68
Involvement of public agencies 69
Discussion questions 69
References 70

4 Contractor selection phase 72
Introduction 72
Methods for contractor selection 72
Pre-qualification/post-qualification 72
Open tender 74
Invited tender 75
Negotiation 75
The contractor's tender decision 76
Project information sources 76
Considerations in deciding to tender 78
Preliminary job planning 80
Method statement 80
Constructability analysis 81
Jobsite visits and checklists 81
Preliminary schedule 82
Pre-tender meetings 83
Cost estimating 83
Levels of detail 87
Rough order of costs 87
Preliminary assessed costs 87
Firm estimate of costs 88
The estimating process 88
Elements of net project cost 90
Labour 90
Materials 91
Equipment 92
Subcontract work 93
Provisional and prime cost allowances 93
Site overheads 93
Company overheads 94
Steps in the process 94
A lump-sum/fixed-price example 96
A unit-price/measure-and-value example 98
Cost-estimating software 101
Value engineering 102
Proposal preparation, submittal and opening 105

Turning the estimate into a tender 105
Submittal and opening process 106
Selecting the successful contractor 107
Criteria 110
Qualifications 114
Notice to proceed and contract agreement 114
Discussion questions 115
References 117
Case study: Is that a late tender? 118

5 Project mobilisation phase 120
Introduction 120
Legal and contractual issues 120
Permits, consents and licences 120
Bonding 122
Insurance 127
Partnering 128
Programming, planning and scheduling 129
Work breakdown structure 130
Bar charts 131
Network schedules 134
General concepts: a modest example 134
Computer applications: a larger example 144
Computer applications: some software features 149
Data input and error checking 149
Basic time calculations 154
Tabular output 154
Resource analysis 154
Cost analysis and cost control 154
Schedule monitoring and updating 156
Graphical output 156
Electronic communication of input and output 156
Operations modelling 156
Budgeting and cost systems 158
From estimate to budget 158
Cash flow projection 160
Organising the worksite 163
Temporary services and facilities 163
Offices 163
Workshops and indoor storage 163
Dry shacks 164
Temporary housing and food service 164
Temporary utilities 164
Sanitary facilities 164
Medical and first aid facilities 164
Access and delivery 165
Storage/laydown areas 165
Security and signage 166

Quarries and borrow areas 166
The site layout plan 166
Buying out the job 168
Material procurement 168
Subcontracting 170
To subcontract or not? 172
Subcontract proposals 172
Subcontract negotiation and award 174
Project staffing 174
Worksite organisation structure 174
Union labour 185
Open shop and merit shop – non-union contracting 187
Special considerations in mobilising for remote projects 187
Discussion questions 188
References 191
Case study: Wilmot pass and Wanganella – logistical planning and personnel accommodation in a remote corner of New Zealand 193
Case study: Temporary quarters for pipeline workers on a far-away project – Trans-Alaska Oil Pipeline construction camps 199

6 Project operations phase 205
Introduction 205
Monitoring and control 205
Schedule updating 206
Cost control 207
Data sources 209
Cost reports 210
Quantity section 211
Cost section 211
Unit cost section 213
Other aspects of financial control 216
Quality management 218
Safety management 222
Environmental management 228
Water drainage and runoff 231
Compacted soil from equipment operations 231
Mud, dust and slurry from tyres 231
Air pollution 232
Contamination from petroleum and other spills 232
Discharge into waters 233
Solid waste disposal 233
Products of demolition and renovation 233
Worker sanitation 234
Endangered species 234
Wildlife protection 234
Noise 235
Archaeology 235

Resource management 235
Personnel supervision and labour productivity 235
Sources of craft personnel, their assignment and supervision 235
Labour productivity 238
Training 239
Working conditions 239
Employee motivation 239
Tardiness and absenteeism 240
Programming and schedule control 240
Material management 240
New technology 240
Length of the working week 240
Changes in the work 240
Project uniqueness 240
Environmental conditions 240
Contractual arrangements 241
Materials management 241
Equipment management 243
Documentation and communication 245
Submittals 245
Variations 246
Measurement, progress payments and retainage 248
Value engineering 251
Other project documentation and document management 252
General 253
Contract documentation 253
Communication records 254
Project status documentation 254
Correspondence 254
Materials management 254
Financial management 255
Electronically-enhanced project communications 257
Selected legal issues 261
Claims process 261
Dispute prevention and resolution 262
Dispute prevention 263
Resolution through court systems 264
Alternative resolution methods 266
Example common issues 270
Differing site conditions 270
Delays 272
Contract termination 275
Discussion questions 277
References 281
Case study: Long-distance construction material management – the Korea–Alaska connection 284

7 Project closeout and termination phase 289
Introduction 289
Completing the work 290

Testing and startup 290
Cleanup 290
Preliminary punch lists 290
Pre-final inspection 291
Final punch list 291
Final inspection 291
Beneficial occupancy 292
Keys 292
Personnel actions 292
Closing the construction office 293
Closing out the project 293
Subcontractor payment 293
Final release or waiver of liens 294
Consent of surety 294
Final quantities 294
Request for final payment 295
Liquidated damages 295
Final payment and release of retainage 295
Final accounting and cost control completion 296
Certificates 296
Certificate for payment 296
Contractor's certificate of completion 297
Certificate of substantial completion 297
Certificate of completion 297
Certificate of occupancy 297
As-built drawings 298
Operating and maintenance manuals 298
Records archiving and transfer 298
Training sessions 299
Warranties, guarantees and defects liability period 299
Post-project analysis, critique and report 300
Owner feedback 300
A closing comment 300
Discussion questions 301
References 301

Index *303*

Preface

An understanding of the construction project life cycle is essential for all who are associated with the construction industry. The materials in this book have been prepared primarily for use as a textbook for students in the design professions, construction management programmes and facility development and management curricula. The topics covered here ought to be of interest to others involved with the project development process as well, including material and equipment suppliers, insurers, owner representatives and members of the legal profession.

As the subtitle indicates, the book is designed around the framework of the project life cycle, commencing with various pre-project decisions and then ranging, in order, through planning and design, contractor selection, project mobilisation, on-site project operations and project closeout and termination. In addition to this very intentional aspect of the book's design, there are other themes that are of special interest to the member of the modern-day construction management team. The reader will find a very international flavour, as we refer to examples of practices, regulations and terminology from around the world, even as we emphasise the commonality of much that happens in construction. The early decades of the twenty-first century will see increasing use of information technology in construction practice and the book devotes several sections to the application of various kinds of software and electronic-based communication methods to project management. Likewise, because environmental impacts due to construction operations have become important concerns of the construction manager, a section is devoted to the causes, mitigation and prevention of such impacts. At several places in the book, the relatively new practice of formalised partnering is discussed. The construction team's responsibility for constructability and value engineering analyses to improve the efficiency of construction operations and the value of the completed facility is a theme that is also found throughout these pages.

A danger in striving for a presentation that will be useful across the international construction community is that some basic construction terminology differs in different parts of the world. For example, in some places, the term *tender* refers to the priced offer prepared and submitted by the contractor, while the term *bid* means the same thing in other locations. A *unit price* contract in some places is a *measure and value* contract in others. For some, a *programme* lays out the project's time dimensions, while a *schedule* does the same for others. A *change* in the work is also known as a *variation*. The student contemplating a construction career in a multitude of locations across the globe will need to learn all of these terms and understand where they are used, but the terminology found in a first reading of a work such as this can lead to considerable confusion. We have tried to minimise confusion by identifying these multiple terms that have identical, or nearly identical, meanings.

The end of each chapter includes discussion questions; in total 103 such questions will be found. Five case studies, drawn from construction practice in Denmark and Sweden, Canada, New Zealand, Alaska and Korea, supplement the text material. The work is fully documented with an extensive reference list at the end of each chapter. An instructor's manual is available for instructors adopting the book for classroom use.

F. Lawrence Bennett PE
Lost Lake, Alaska

Acknowledgements

In listing the many persons who have contributed to this book, I think first of those industry leaders who have influenced my construction career over more than 40 years. Each has taught me about construction practice and has affected the ways in which I think about construction; their impact upon what I have written here is greater than they or I ever imagined. My first real job in construction was with Streeter Associates; 'Big Jim' Norris, Jim Norris Jr and Don 'Brownie' Brown helped lay the foundation for all of my later experiences and explorations. In somewhat chronological order after that, I had the privilege of working with E.L. 'Woody' Lunkenheimer in the home office of United Engineers and Constructors Inc. and with such colourful and influential field supervisors and colleagues as Maurice Rogers, Jim McKee PE, Joe Kane and Harry Alexander at United Engineers and Constructors Inc. I remember also my rich association with S.C. 'Steve' Stephens, Jim Carlson, Howard Tomlinson and Ed Tolerton at Peter Kiewit Sons and Con Frank PE, Bert Bell PE and Galen Johnson PE at Ghemm Company. Alaskans Ross Adkins PE and Mac McBirney, New Zealanders Peter Turner PE and Ernesto Henriod PE and my neighbour Wendell Shiffler, CEO of TRAF Construction, likewise are all part of this book in one way or another. Construction administrators, later turned academics, such as J.C. Gebhard, George Blessis and Ken Hover PE, all at Cornell, provided essential mentoring in the early and later phases of my career. There have been others, both design professionals and owner representatives, but those whom I name above were, and many still are, construction contractors who directed real projects in the field and knew much better than I the importance of the principles outlined in these pages.

In assembling the materials for this textbook, I have relied upon assistance in many forms from a large number of people. I received helpful information for the main body of the text from Jack Wilbur PE of Design Alaska; Tim Anderson of the Carson Group Ltd; Tom Kinney PE of HWA Geosciences; Galen Johnson PE, from Ghemm Company; and Bob Perkins PE of the University of Alaska Fairbanks. Those contributing to the five case studies included Tony Savage, Plant Manager of New Zealand's Manapouri Power Station; Scott W. Bennett at Meridian Energy New Zealand; Evan Stregger PQS, CArb of Costex Management Inc., whose paper appears in Chapter 4; Mike Heatwole and Donelle Thompson of Alyeska Pipeline Service Company; Dermot Cole of the *Fairbanks Daily News-Miner*; and Bert Bell PE and Dick Houx of Ghemm Company. Jon Antevy, e-Builder, Inc., Dan Gimbert, Peachtree Software, Inc. and John Hanson, WinEstimator, Inc. willingly provided samples of their software and helpful advice on its use. Felix Krause contributed the case study in Chapter 2, based on his managerial experience on that project; he also assisted with a review of a portion of the text. Dave Lanning PE, Lanning Engineering, helped with raw materials for the book and then a review of part of the written work. Ernesto Henriod PE contributed valuable ideas and reviews from the inception of this project through the later writing stages. My close friend and colleague Arnim Meyburg reviewed a portion of the text and offered helpful suggestions. Deb Knutsen and the staff of the University of Alaska Fairbanks' Rasmuson Library assisted throughout the project with the magic of interlibrary loan acquisitions.

In addition to providing helpful insight into the process of gathering community input during project planning, my wife Margaret reviewed the entire manuscript with her professional editor's eye and offered many helpful suggestions; beyond all that, she endowed this project with vital moral support throughout the 15 months of the book's development. I acknowledge with love and appreciation her constant and unwavering devotion to this project and all of my other crazy undertakings. Lastly, I express thanks to our sons Matthew and Andrew for their interest in and support of my professional endeavours over many years and I dedicate this one to them with gratitude and love.

While many of the ideas and much of the material here have come from professional colleagues, I accept full responsibility for the book's accuracy and for decisions to include some materials and omit others. If there are mistakes, I will appreciate your communicating them to me.

F. Lawrence Bennett, PE
947 Reindeer Drive
Fairbanks Alaska 99709
flb1@cornell.edu

1

Introduction

The management of construction is an enterprise that involves many people with diverse interests, talents and backgrounds. The owner, the design professional and the contractor comprise the primary triad of parties, but others, such as subcontractors, material suppliers, bankers, insurance and bonding companies, attorneys and public agency officials, are vital elements of the project team whose interrelated roles must be coordinated to assure a successful project. Throughout the project life cycle, from the time the owner first contemplates launching a construction project to that celebrated time, many months or years later, when the completed project is ready for use, the tasks carried out by the various parties vary in type and intensity.

In this book, we consider the roles and responsibilities of the many parties at each phase of the construction project life cycle. The primary focus here is on the construction contractor, who carries the lead responsibility for the on-site installation work and all of the associated planning and followup. It is important, at the same time, to understand how other people and organisations contribute to project success.

Construction industry segments

The construction industry can be broken down into two very broad categories, general building construction and engineered construction. Most construction contractors concentrate on one of these categories, or even on a specialty within one of them. A third category of contractor is the specialty trade contractor, who usually works as a subcontractor for a general, or prime, contractor responsible for the construction of the entire project. We can understand something about the nature of the industry by describing the various types of construction work.

General building construction

Within this very broad category we find projects that include residential, commercial, institutional and industrial buildings. *Residential construction* produces buildings for human habitation, including single-family dwellings, condominiums, multifamily townhouses, flats and apartments and high-rise apartment buildings. Depending on the project's complexity, such work is usually designed by architects, owners or builders themselves, with construction performed by contractors who hire specialty subcontractors as needed; some of this work may be built by owners themselves.

Commercial construction includes retail and wholesale stores, markets and shops, shopping centres, office buildings, warehouses and small manufacturing facilities. Examples of *institutional construction* are medical clinics and hospitals, schools and universities, recreational centres and athletic stadiums, governmental buildings and houses of worship and other religious buildings. Architectural firms usually take the lead in the design of commercial and institutional facilities, with assistance from engineering firms for such specialties as structural and electrical elements. Because this type of work is usually more complex and time consuming than residential construction, owners usually engage general contractors to perform the field construction; subcontractors usually provide specialty services such as plumbing, painting and electrical work.

Often categorised separately from general building construction, *industrial construction* is a special segment of the industry that develops large-scale projects with a high degree of technical complexity. Such endeavours result in facilities that manufacture and process products; examples include steel mills, electric power-generating plants, petroleum refineries, petrochemical processing plants, ore-handling installations and heavy manufacturing factories that produce such products as vehicles, rolling equipment and various kinds of large machinery. The engineer, rather than the architect, usually assumes the lead responsibility for the designs of these kinds of projects. Often the owner selects a single entity to provide both design and construction services under a 'design–build' contract and works closely with the design professional to assure that the owner's special requirements are met.

Engineered construction

This broad category of construction, sometimes called *engineering construction*, is characterised by designs prepared by engineers rather than architects, the provision of facilities usually related to the public infrastructure and thus owned by public-sector entities and funded through bonds, rates or taxes and a high degree of mechanisation and the use of much heavy equipment and plant in the construction process. These projects usually emphasise functionality rather than aesthetics and involve substantial quantities of such field materials as timber, steel, piping, soil, concrete and asphalt. More so than other types of construction, engineered construction is often designed by an owner's in-house staff, such as a provincial highway department or a federal public agency; the Army Corps of Engineers is an example of the latter in the USA. A general contractor is usually engaged to install the work, with subcontractors as needed to contribute specialty services. With these kinds of projects, the exact quantities of some materials can seldom be ascertained in advance; thus these construction contracts are often arranged such that the contractor is paid a pre-agreed-upon unit price (US$ per cubic metre of concrete, for example) for each unit of material actually required.

Two common subcategories of engineered construction are highway construction and heavy construction. *Highway construction* typically requires excavation, embankment construction, paving, installation of bridges and drainage structures and associated lighting and signage. *Heavy construction* projects include dams, tunnels, pipelines, marine structures, water and sewage treatment plans, railroads, rapid transit systems, airports and utility work such as electrical transmission and distribution systems, water lines, sanitary and storm drains, pumping stations and street paving. Utilities, upon completion, are often owned and operated by semi-public entities such as electric associations or water authorities.

Table 1.1 Impact of construction on national economies

Country	Year	% of total value of goods and services or % of gross domestic product	% of total employment	% of total firms[1]
Australia	1997–1998	5.7	7.2	
Austria	1999	6.1	11.0	8.9
Czech Republic	1999	11.3	9.8	13.7
Denmark	1998			10.2
Finland	1999	8.5	8.8	12.5 (1998)
Italy	1998	7.7	9.7	12.3
New Zealand	1998–1999		8.0	12.6
USA	1997	5.0	6.4	11.2
UK	1999	8.2	4.5 (1997)	11.2

[1] A firm is either a single business unit in a single location or a legal entity with one or more business units; country statistics use one definition or the other.

Sources: Organisation for Economic Cooperation and Development (2002); US Census Bureau (2000); Australian Bureau of Statistics (1999).

Construction industry characteristics

Having introduced the broad categories of construction project types, we shall now consider further aspects of the industry by way of some representative statistical data. We want to look at the role that construction plays in the overall economy, the relative proportions of the various construction categories and the character of the industry in terms of the sizes of companies that carry out construction work.

Construction is big business! The industry's significant impact on the world economy can be demonstrated by reviewing construction's proportion of the total value of goods and services, as well as the number of people employed in construction as a proportion of the total workforce and the number of construction firms compared with the total businesses in all industries. Table 1.1 contains representative statistics for several countries. Note that, for the countries listed, the construction industry's contribution to the total value of the economy, measured either as a proportion of value of goods and services or of gross domestic product, ranged between 5.0% and 11.3%. For Australia, for example, the 5.7% for 1997–1998 represents $30 billion (Australian) out of a total of $522 billion. In that country, construction ranked eighth, after manufacturing (13.2%), property and business services (10.8%), ownership of dwellings (9.7%), finance and insurance (6.4%), transport and storage (6.3%), health and community services (6.2%) and retail trade (5.8%).

The construction workforce comprises between 4.5% and 11.0% of the total workforce for the countries listed in Table 1.1. Using Australia as an example once again, the 7.2% (719 000 workers out of a national workforce of 8 555 000 in 1997–1998) places construction fifth in contribution to total employment, after retail trade (14.5%), manufacturing (12.8%), property and business services (10.8%) and health and community services (9.6%) (Australian Bureau of Statistics, 1999).

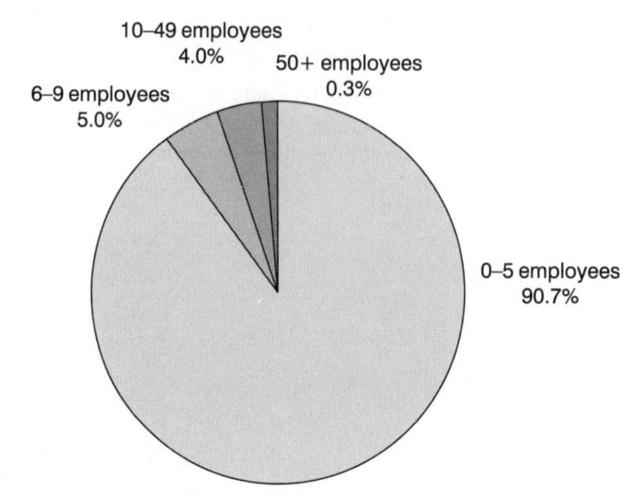

Figure 1.1 Number of New Zealand construction firms by size, February 2001. Source: Statistics New Zealand Te Tari Tatau (2002).

Another measure of an industry's impact on a nation's economy is its number of companies, or business units. For the countries listed in Table 1.1, construction firms represented between 8.9% and 13.7% of the country's firms, in all but one country more than 10%. Italy, for example, reported about 482 000 construction business entities in 1998, out of a total of about 3 905 000 such entities for all industries nationwide. Based on the categorisation in this study, Italian construction had the fourth largest number of firms, after wholesale and retail trade, real estate, renting and business services and manufacturing (Organisation for Economic Cooperation and Development, 2002).

The data that are the basis for the employment and business unit statistics discussed above provide another interesting insight into the nature of the construction industry: the relatively small sizes of construction firms, as measured by their numbers of employees. In the Czech Republic in 1999, for example, the number of employees per construction firm averaged 2.7, while the same statistic for Finland was 3.5 and for the UK, it was 5.3. For every country among the selected sample in Table 1.1, the number of employees per construction firm is significantly lower than the corresponding figure for that country's manufacturing sector; in the Czech Republic, Finland and the UK, the average number of employees per manufacturing firm can be calculated as 9.2, 14.8 and 23.9, respectively. Although there may be some differences among reporting practices in the various countries, the conclusion is clear: construction work is typically performed by organisations with small numbers of employees.

In Figures 1.1 and 1.2, we show some information about the distribution of construction firm sizes for the New Zealand Construction Industry at one point in 2001 (Statistics New Zealand Te Tari Tatau, 2002). At that time, when the country's construction firms averaged 3.1 employees, 90.7% of the firms had five or fewer employees, while only 0.3% of the firms had 50 or more employees, as shown in Figure 1.1. An alternate way to analyse these data is shown in Figure 1.2, which shows the proportion of employees in firms of different sizes. Note that the results are more evenly balanced among the four categories, with 47.3% of all construction employees engaged by firms with five or fewer employees and 18.5% employed by firms with 50 or more employees. Similar statistics are available for construction in the USA, where about 80% of construction firms employ fewer than 10 people (US Department of Commerce, 1997).

Figure 1.2 Number of employees in New Zealand construction firms, by size of firm, February 2001. Source: Statistics New Zealand Te Tari Tatau (2002).

Recall the earlier description of the various construction industry segments, or sectors, in which we suggested two general categories, general building construction and engineered construction, with subcategories within each. The construction activity for a country or other geographical area can be characterised by the value of work performed in each sector during a

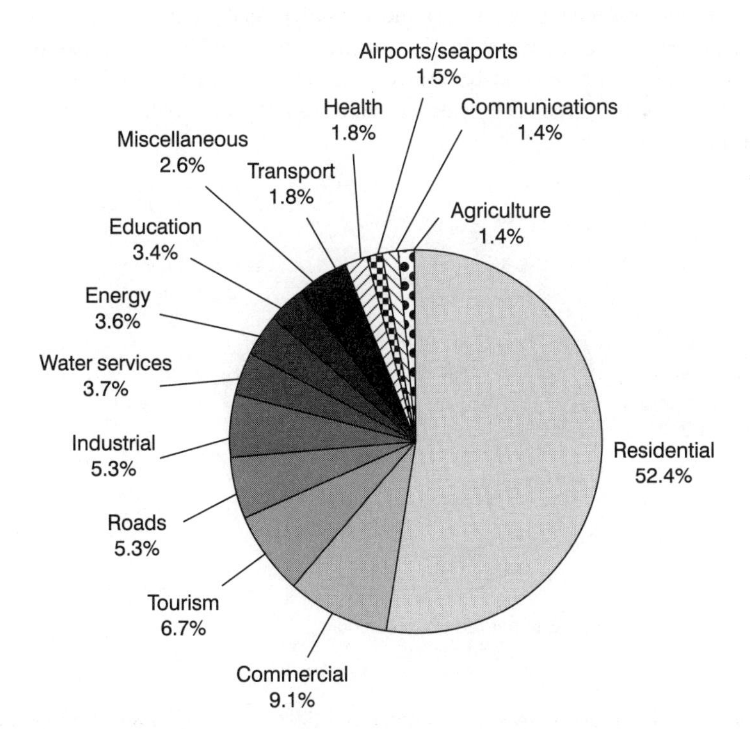

Figure 1.3 Percentage of Irish construction output by sector, 2000. Source: irishconstruction.com (2001).

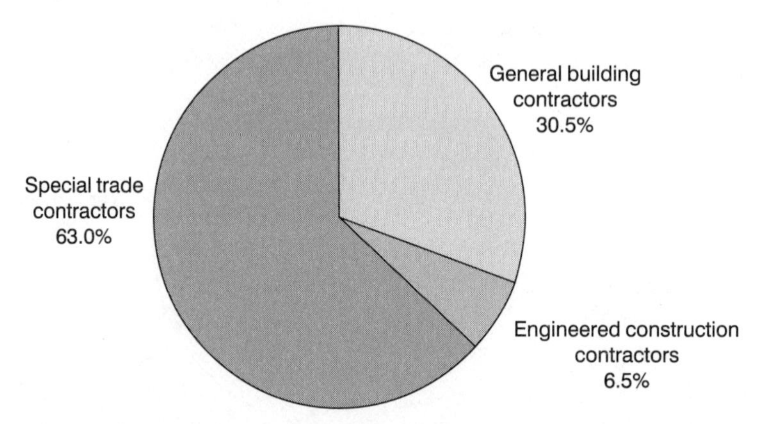

Figure 1.4 Number of US construction firms by type of contractor, 1997 (total = 649 601). Source: US Census Bureau (2000).

certain period. We show one such example in Figure 1.3, in which Irish construction output for 2000 is subdivided into 14 sectors (irishconstruction.com, 2001). This chart shows the major importance of residential construction in this particular country during the time period analysed.

A final interesting characteristic of the construction industry is the large number of specialty trade contractors. Such organisations confine their work to one or two trades, such as electrical work, painting or plumbing and heating. They typically work for general contractors as subcontractors and have no responsibility for the overall project. Figure 1.4 shows the large proportion of specialty trade contractors in US construction in 1997, as compared to general building contractors and engineered construction contractors, while Figure 1.5 indicates that the value of work installed by the three categories was considerably more balanced among the three (US Census Bureau, 2000).

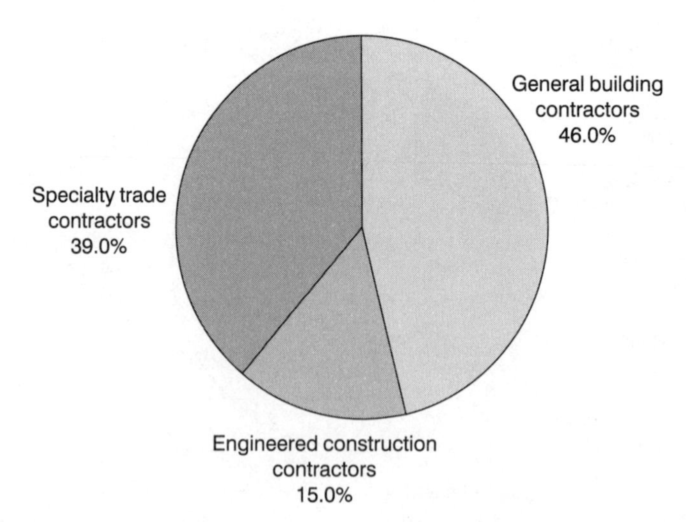

Figure 1.5 Value of US construction by type of contractor, 1997 (total = US$ 865.3 billion). Source: US Census Bureau (2000).

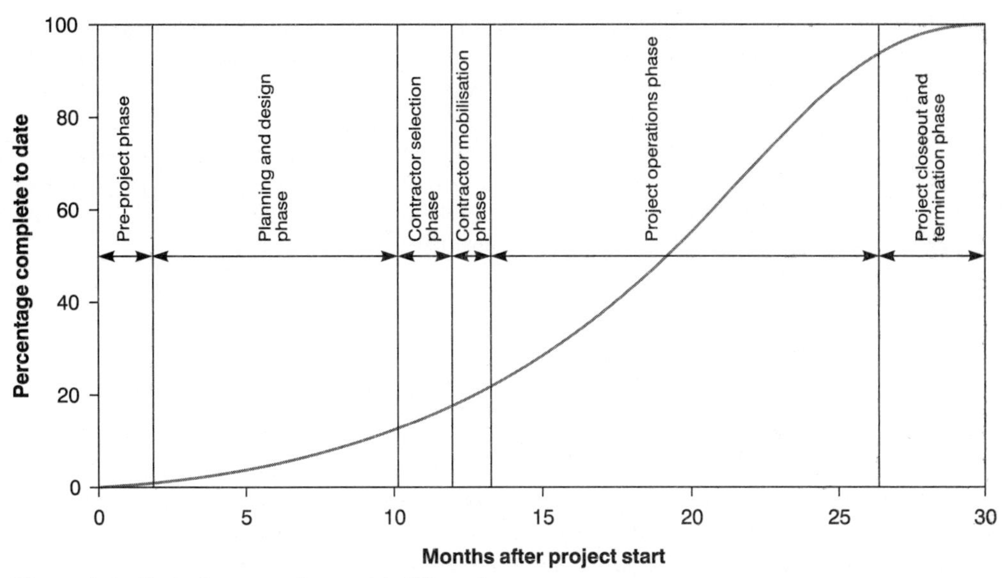

Figure 1.6 Typical construction project life cycle.

Overview of the construction project life cycle

This book is organised around the typical pattern of the flow of the project work from its inception to its closeout and termination. Every project, not just those in the construction industry, goes through a series of identifiable phases, wherein it is 'born', it matures, it carries through to old age and it 'expires'. A software development project manager, for example, might define the following phases in the project's life cycle: initial proposal, process engineering – requirements analysis, process engineering – specifications, design, development, testing, deployment and support (Vision7 Software, 2002). Likewise, a project that results in the development of a new product might contain the following phases: conceptual, technical feasibility, development, commercial validation and production preparation, full-scale production and product support (Babcock and Morse, 2001). Although there may be some overlap in the phases, the work generally flows from the first phase to the last, with the outcome of one phase providing the basis for efforts carried out in the phase that follows.

So it is also with construction projects. In this book, we identify six phases in the construction project life cycle, each with its own purposes and characteristics. First, the owner must make certain pre-project decisions. Then the planning and design of the project is carried out. Next, the contractor is selected, after which the contractor mobilises in order to carry out the field operations. The field work that the lay person often considers to be 'construction' can be considered a separate phase. Lastly, the project must be terminated and brought to a close; because these activities are distinct from the installation work, we separate them into a distinct, final phase. Figure 1.6 is a conceptual diagram that shows the various phases in the typical construction project life cycle. It also shows the amount of funds the owner might commit by the end of each phase, as a percentage of the total project budget.

To attempt to understand the management of construction by organising the study on the basis of the project life cycle may be somewhat arbitrary, because there is admittedly some overlap between phases and thus some duplication in the presentation. However, this deliberate design of the book will provide a logical basis for tracking the project's activities and understanding the roles of the people responsible for those activities, from the time the owner first conceives the idea for a construction project until that point when the contractor has vacated the site for the final time.

Structured in this way, each chapter provides a description of one of the project's phases. The result should be an understanding not only of the importance of each phase individually but also of the way they interrelate to form an integrated whole project. We begin with a brief overview of each of the six phases of the construction project life cycle.

Pre-project phase

A construction project begins with an idea, a perceived need, a desire to improve or add to productive capacity or the wish for more efficient provision of some public service. Whether the idea will be converted into a completed project will be decided during the planning and design phase. However, prior to that, among the first things the owner must do is to decide what sort of *project delivery system* will be used. How will the various parties be related? Will the owner engage a design professional to prepare plans and specifications and then contract separately with a construction contractor? Or, will a single entity be responsible for the entire project? Other possible options include several separate specialty contractors, each related by contract with the owner, the use of a construction manager as an advisor to the owner, the use of the owner's own construction forces and the phasing of the project such that individual portions of the field work are commenced prior to the completion of all design work.

The other primary decision required by the owner early in the project relates to the *type of contract* to be used with the contractor. Will the contractor be paid a specified fixed price, regardless of the actual quantities used in the project and regardless of the contractor's actual costs? Will the quantities of materials be measured and the contractor paid on the basis of those quantities and pre-agreed-upon unit prices for each material? Or, will the contractor be reimbursed for its actual costs, plus a fee, perhaps with an agreed-upon upper limit? The owner will also need to decide the basis upon which the design professional will be paid. Often these decisions are not made without consultation and advice. Depending upon the owner's expertise and experience in administering construction contracts, the owner may engage a professional engineer, an architect or a project manager during this pre-project phase to advise on these important decisions.

Planning and design phase

The project is fully defined and made ready for contractor selection and deployment during the planning and design phase. It is convenient to divide this phase into three stages. The goal of the first stage is to define the project's objectives, consider alternative ways to attain those objectives and ascertain whether the project is financially feasible. In this process of planning and feasibility study, a project brief will be developed, more details will be set forth in a programme statement, various sites may be investigated, public input may be sought, a preliminary cost estimate will be prepared, funding sources will be identified and a final decision on whether to proceed with the project will be rendered.

In the second stage, the design professional will use the results of the planning efforts to develop schematic diagrams showing the relationships among the various project components, followed by detailed design of the structural, electrical and other systems. This latter activity is the classical hard core engineering familiar to students in the design professions, in which various engineering principles are used to estimate loads and other requirements, select materials, determine component sizes and configurations and assure that each element is proper in relation to other elements. The output from this design development effort is used in the final stage, wherein contract documents are prepared for use in contractor selection and installation work at the construction site. The design professional prepares not only the detailed construction drawings but also written contract conditions containing legal requirements, technical specifications stipulating the materials and the manner in which they shall be installed and a set of other documents related to the process of selecting the contractor and finalising the contract with the successful tenderer.

Contractor selection phase

In anticipation of selecting a contractor, the owner must decide whether an open invitation will be issued to all possible tenderers or whether only certain contractors will be invited to submit offers and whether any sort of pre-qualification process will be invoked to limit the number of tenders. On the other side, contractors will have to consider a number of factors in deciding whether they will make the effort to assemble a proposal for a particular project. If a contractor finds the prospective project attractive, two major tasks will be required. First, a series of planning steps will be carried out, including studies of various methods and equipment that would be employed and the development of a preliminary project programme setting forth an approximate time schedule for each major activity. Second, a priced proposal will be prepared, including the direct costs of labour, materials, plant and subcontractors, various overhead charges and a sufficient added amount for profit. The last step in this phase is the submittal, opening and evaluation of tenders, the selection of the successful contractor and the finalisation of the construction contract.

Project mobilisation phase

After the contractor is selected, a number of activities must be completed before installation work can begin at the project site. Various bonds, licences and insurances must be secured. A detailed programme for the construction activities must be prepared. The cost estimate must be converted to a project budget and the system for tracking actual project costs must be established. The worksite must be organised, with provisions for temporary buildings and services, access and delivery, storage areas and site security. The process of obtaining materials and equipment to be incorporated into the project must be initiated and arrangements for labour, the other essential resource, must be organised. With the completion of this phase, it is finally time to begin the actual field construction.

Project operations phase

In presenting the contractor's activities on the construction site, we will suggest, perhaps too simply, that the responsibilities involve three basic areas: monitoring and control, resource management and documentation and communication. Five aspects of monitoring and controlling the work are important. Actual schedule progress must be compared against the project programme to determine whether the project is on schedule; if it is not, actions must be undertaken to try to bring the programme back into conformance. Likewise, the cost status must be checked to establish how actual performance compares with the budget. An equally important part of monitoring and control is quality management, to assure that the work complies with the technical requirements set forth in the contract documents. In addition, the contractor has an important role to play in managing the work safely and in a way that minimises adverse environmental impacts.

In managing the project's resources, the contractor will, first, be concerned with assigning and supervising personnel and assuring that the labour effort is sufficiently productive to meet schedule, cost and quality goals. In addition, materials and plant must be managed so that these same goals are met. Because construction projects require large amounts of paperwork, a special effort is required to manage this documentation effectively. Examples include the various special drawings and samples that must be submitted to the owner or design professional for approval prior to installation, the frequent need to respond to requests for changes in the project after the on-site work has begun and the all-important process for periodically assessing the value of work completed and requesting payment for this work. Various on-line and other electronic means are available to assist contractors with document management and project communications.

Project closeout and termination phase

Finally, as the project nears completion, a number of special activities must take place before the contractor's responsibilities can be considered complete. There are the various testing and startup tasks, the final cleanup, various inspections and remedial work that may result from them and the process of closing the construction office and terminating the staff's employment. In addition, a myriad of special paperwork is required, including approvals and certifications that allow the contractor to receive final payment, a set of as-built drawings that include all changes made to the original design, operating manuals, warranties and a final report. The contractor will also be responsible for transferring and archiving project records and will conduct some sort of project critique and evaluation; operator training may also be part of the contractor's contractual responsibilities.

We begin, in Chapter 2, with a description of the various options available for project delivery systems and the types of contracts that can be used to bind the owner and contractor in a legal agreement.

Discussion questions

1 Draw a chart that depicts the various categories of construction described in the text. For each category, name an example project in your area.
2 Seek statistics on the construction industry in your country or region, similar to those given in the chapter, for proportion of the economy and firm size. Identify any significant differences between your numbers and those in the text and suggest possible reasons for the differences. A likely source of this information will be the World Wide Web.

3 Try to find statistics on the amount of construction carried out in your region, broken down by type of construction or type of contractor. Compare this information with that given in the text and discuss possible reasons for similarities and differences.
4 Another way to characterise an area's construction activity is to look at the number of building consents (also called building permits or building approvals) issued by public authorities. For two or more countries or regions in which you are interested, obtain such statistics for the past several years. Graph your findings and comment on the results.
5 Each phase in the construction project life cycle utilises the talents of various members of the project team. For each of the six phases described in the text, indicate the people with primary roles. In addition, indicate the role you might play, if any, in each phase at some point in your career.

References

Australian Bureau of Statistics. 1999. *Australian System of National Accounts. 1997–98.* Publication 5204.0.

Babcock, D.L. and L.C. Morse. 2001. *Managing Engineering and Technology*, 3rd edn. Prentice-Hall.

irishconstruction.com. 2001. *Business & Information Online*. http://www.irishconstruction.com/pksstats.html.

Organisation for Economic Cooperation and Development. 2002. *Structural Statistics for Industry and Services, Core Data, 1992–1999.*

Statistics New Zealand Te Tari Tatau. 2002. *New Zealand Business Demographic as at February 2001*. http://www.stats.govt.nz/.

US Census Bureau. 2000. *1997 Census of Construction Industries*. http://www.census.gov/.

US Department of Commerce. 1997. *County Business Patterns.*

Vision7 Software. 2002. *Software Development*. http://www.vision7.com.

2

Pre-project phase

Introduction

Prior to the commencement of the construction project, even before the selection of the designer and the accomplishment of any planning activities, the project owner faces two important decisions regarding the relationships among the various parties and the basis upon which the contractor will be paid. Because these two issues can be dealt with somewhat independently, we describe each in two separate sections of this chapter. In the first, we review the various project delivery systems that form the basis for the project's contractual relationships and dictate the span and duration of responsibility of each party. In the second, we identify different types of contracts that are used to measure how the construction contractor will be paid for completed construction work.

Selection of project delivery system

What Dorsey (1997) describes as the 'eternal triangle' of construction consists of the owner, designer and construction organisation. All project delivery systems include these three as participants, with others often part of the project team as well. Their relationships vary according to the different systems and ownerships. This section describes several ways in which the various parties involved in the construction contract can be related. They range from the traditional method utilising a designer separate from the construction organisation, whose work is usually completed prior to the engagement of a constructor, to various methods in which a single entity is responsible for execution of the entire project.

Traditional design–tender–build

We call this approach to construction project delivery 'traditional' because it has been the approach of choice for owners of most construction projects during many centuries. With this method, the owner contracts with a design organisation to perform preliminary planning, carry out design work and prepare contract documents (all of which will be described in some detail in Chapter 3). Following the completion of this phase, a construction organisation is selected,

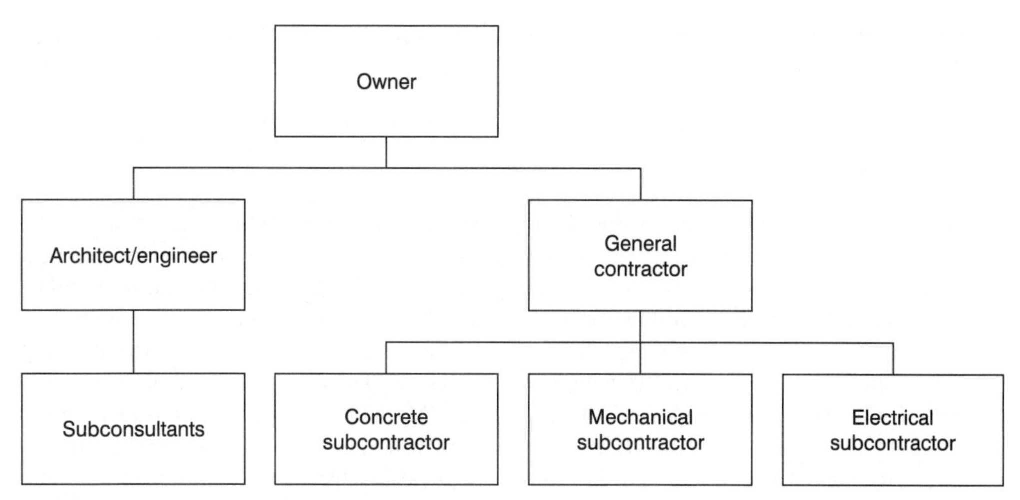

Figure 2.1 Traditional design–tender–build relationship chart.

based upon the owner's criteria, and the owner enters into a contract with the successful contractor for the assembly of the project elements in the field. In this method, the contract for the design work is separate from that for the construction work.

Figure 2.1 shows an organisation chart that might apply to such an arrangement. Note that the owner is under contractual obligation to two parties: the design professional to prepare the design and the general contractor to carry out the field installation. Each of these two parties typically has its own contracts with various subconsultants and subcontractors.

The relationships shown in Figure 2.1 and elsewhere in this chapter are quite simplified. For example, we have not shown any of the 'subsubcontractors' that might be engaged by subcontractors. The mechanical subcontractor might contract with an insulation contractor and an instrumentation contractor for the specialty works that are part of the mechanical subcontract but not within the particular mechanical subcontractor's expertise. Furthermore, it is usually the case that subcontractors will contract with various material suppliers, also not shown on our relationship charts.

The various lines connecting the entities in Figure 2.1 and the other figures in this section represent contractual relationships; a contract would be in place between any two parties connected by such lines prior to the commencement of any obligations by either party. It is important to indicate that non-contractual types of relationships may also exist between parties involved in a construction project. For example, the architect/engineer in Figure 2.1 may have some oversight or inspection responsibilities in connection with the general contractor's work. Furthermore, despite a lack of contractual relationship between the two parties, the law may permit the contractor or a subcontractor to sue the designer for certain alleged design deficiencies or other acts or conditions that may have impacted the contractor's operations. For simplicity, we have not tried to draw such relationships on these charts.

As described above, an important characteristic of the design–tender–build type of delivery system is that the design and construction organisations are separate. Another especially important feature is that all of the project's design work must be completed prior to solicitation of tenders and no construction work can begin until a successful tenderer has been selected. Thus, the term design–tender–build implies a strict, and sometimes time-consuming, project schedule sequence – designing, followed by tendering, followed by constructing. Future chapters describe these steps in detail.

Design–build

The distinguishing characteristic of the design–build, or design–construct, method is that the owner executes a single contract with an organisation that becomes responsible for both the design and the construction of the project. This approach closely resembles the 'master builder' whose tradition goes back to Biblical times (Tenah, 2001). If you needed a project built, you needed only to contact a single expert, a master builder, whose expertise, experience and contacts would assure a successfully completed project. In their *Construction 21: Reinventing Construction* report, the Ministry of Manpower and Ministry of National Development of Singapore (1999) note that one of the primary reasons for low productivity in the construction industry is the lack of integration of activities across the project life cycle. Indeed the traditional design–tender–build approach described above, with its organisational separation of design and construction, has great potential for such lack of integration. *Construction 21: Reinventing Construction* urges that the design–build method, and others, be implemented as a means of improving productivity.

The Design–Build Institute of America (1994) lists potential benefits from the design–build method as follows.

- Singular responsibility. There is a single point of responsibility for quality, cost and schedule adherence, avoiding 'buck passing' and 'finger pointing'.
- Quality. The greater responsibility implicit in this method provides motivation for high-quality and proper performance.
- Cost savings. The single entity with whom the owner contracts can work together as a team to evaluate alternative methods and materials efficiently and accurately.
- Time savings. Design and construction can be overlapped, and bidding time after design is eliminated, thus offering the possibility of substantially reduced project duration.
- Potential for reduced administrative burden. After the contract is agreed upon, the owner will have relatively little investment in coordinating and arbitrating between designer and contractor, since they are a single entity.
- Early knowledge of firm costs. The single design–construction entity is responsible for both design and cost estimates at an early stage, thus allowing early establishment of financing and reduced exposure to cost escalation.
- Risk management. Cost, schedule and quality can be clearly defined and appropriately balanced. The design–build organisation will manage many of the risks that the owner might otherwise be responsible for.
- Balanced award criteria. The owner can give credit in the award process for such considerations as design quality, functional efficiency and team experience, as well as lowest first cost.

To be fair, one should also recognise some disadvantages of the design–build method. These include the following.

- Importance of the project brief. Without a clear statement defining the owner's needs, the design–build organisation cannot understand the required scope prior to contracting with the owner.

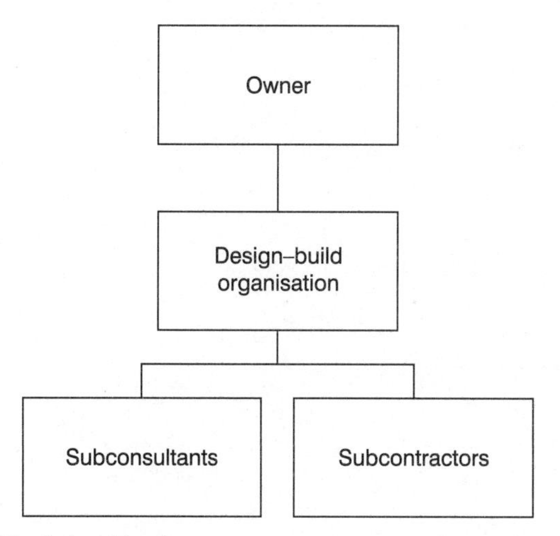

Figure 2.2 Design–build relationship chart.

- Difficulty of establishing a price for the work. It is difficult to establish a price for the work until the design is complete, which it is not when the design–build organisation is selected.
- Costly tendering. Owners must expect to pay for the efforts by design–build organisations to assemble their tenders. These efforts often include preliminary design work, necessary to determine prices.
- Short tender periods. Design professionals often are subjected to inordinate pressure to complete their tenders, including any necessary preliminary design, in a few weeks.
- Potential low quality. Oversight of the on-site construction activities is left to the owner. Because both designer and contractor are one organisation, the designer cannot fairly represent the owner on site.
- Less control over subcontractor and consultant selection. The owner is farther away, contractually, from the process of engaging subcontractors and subconsultants, whereas in the traditional design–tender–build process often the owner has the role of approving nominated subs.
- Generally less control by the owner over both project definition and execution than design–tender–build projects.

A study of design–build projects in Hong Kong found that the proportion of public sector contracts of the design–build type increased from 0.9% to 9.9% between 1992 and 1999. The report lists single-point responsibility, sole liability, better project management, better time control and better cost control as potential advantages of this method. It also notes that limitations include design and quality management, restricted variations, heavy client burden and lack of experienced design–build firms (Chan et al., 2001).

In Figure 2.2, we show a simplified relationship chart for a design–build project. Note that the owner has a contractual relationship with a single organisation that is responsible for both design and construction. This organisation, in turn, engages subconsultants and subcontractors to assist with design and construction.

The Singapore *Construction 21: Reinventing Construction* report (Ministry of Manpower and Ministry of National Development of Singapore, 1999) cites an example of time savings with the design–build method, as follows:

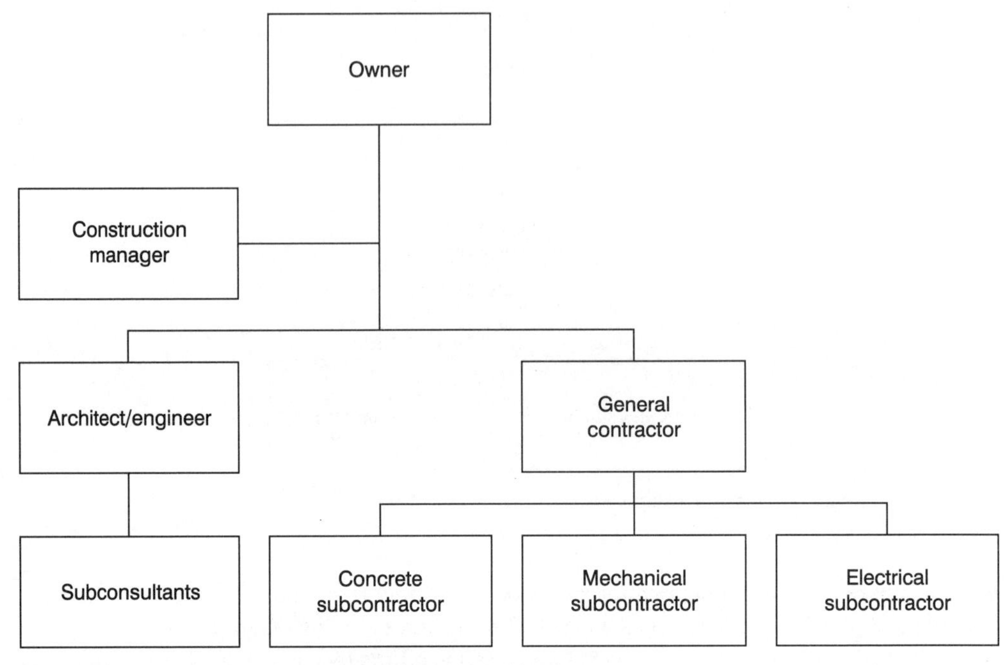

Figure 2.3 Agency construction manager relationship chart.

Millenia Tower (42 floors) and Centennial Tower (37 floors) were both constructed by Dragages Et Travaux Publics (Singapore) Pte Ltd. Both buildings were built 2 years apart, and have a floor area of about 2,000 m^2 each. Millenia Tower was initially designed by the owner's team but modified by the contractor, whereas Centennial Tower was a D&B project (by the private sector). Millenia Tower was built in 32 months with 6 days per floor for the structure, whereas Centennial was built in only 23 months with 4 days per floor for the structure, 30% faster.

Construction manager

The owner may engage a construction manager to provide professional construction management services. The construction manager organisation provides advice to the owner regarding construction matters, including cost, schedule, safety, the construction process and other considerations; such advice may be offered throughout the project life cycle or at selected portions thereof. Two types of construction management have evolved (Rubin et al., 1999; Clough and Sears, 1994); they are depicted in Figures 2.3 and 2.4 respectively. In the 'agency' type of construction management arrangement, the construction manager acts as advisor to the owner for a fee and the owner engages separate contractor and designer organisations. Here, the construction manager acts as an extension of the owner's staff and assumes little risk except for that involved in fulfilling its advisory responsibilities. Note that Figure 2.3 shows the construction manager in a position out of the direct line between the owner and contractor.

In contrast, the 'at-risk' construction manager occupies a contractual position between the owner and the execution contractors, as shown in Figure 2.4. It is not a coincidence that Figure

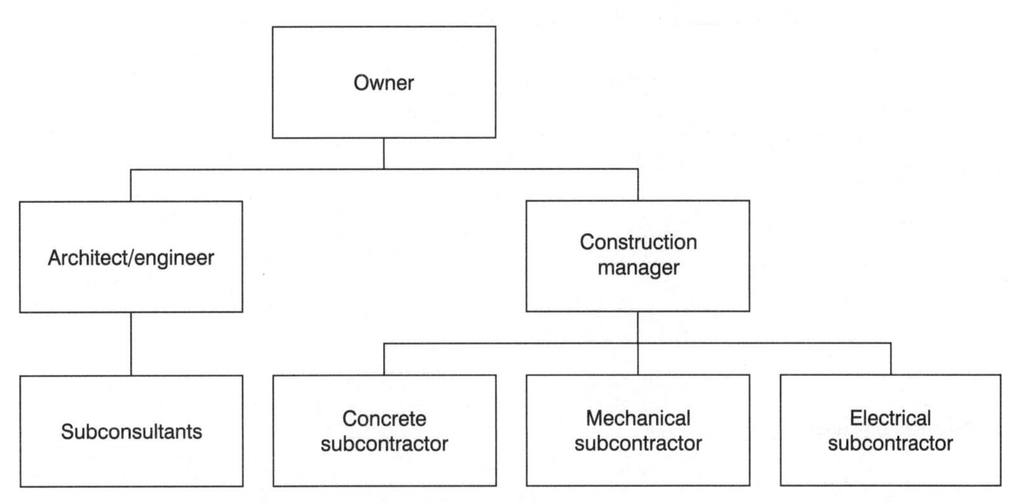

Figure 2.4 At-risk construction manager relationship chart.

2.4 is very similar to Figure 2.1, for the construction manager replaces the general contractor in this arrangement and thus holds the various trade contracts. In addition to the general contractor roles as described in the section on the traditional design–tender–build approach, the at-risk construction manager provides expert advice to the owner on all matters related to the construction, usually beginning well before the field work begins. Often, the contract between the owner and the construction manager is based on a guaranteed maximum contract price, as described later in this chapter. Whereas the general contractor of Figure 2.1 often performs some of the field construction work itself, most, if not all, of the actual construction in at-risk construction management is performed by subcontractors.

Whether 'agency' or 'at risk', construction management contracts have certain inherent advantages and disadvantages. By engaging an expert advisor early in the process, the owner can achieve a project with a near optimal balance of time, cost and quality features, due to the opportunities for review of design alternatives as they are developed. Materials and equipment with long delivery times can be identified and ordered early in the process. There may be less exposure to contract claims and litigation. A disadvantage is a lack of consistency in contractual arrangements among different projects. Furthermore, even though it is the owner's advisor, the construction manager may tend to emphasise the traditional contractors' interests in cost and time savings, to the detriment of high-quality construction. Even though the 'at-risk' construction manager enters into contracts with subcontractors, these contracts are often in the name of the owner; thus, these subcontractors may look to the owner for payment, unlike in the Figure 2.1 type of general contractor–subcontractor arrangement.

A lawsuit that arose from a construction management arrangement helps clarify the differences between 'agency' and 'at-risk' contracts (Simon, 1989). In Owen Steel Co. versus George A. Fuller Co. (563 F. Supp. 298 [S.D.N.Y. 1983]), a structural steel and metal decking subcontractor sued Fuller, the project's construction manager, as well as the owner, for approximately US$ 1 million allegedly due. The construction manager argued successfully that it was in an agency relationship with the owner and thus was not obligated for the owner's responsibility, based on the New York State law that precludes agents from liability for sums owed to third parties by the principal. Thus, the agent, Fuller, was not assuming any liability

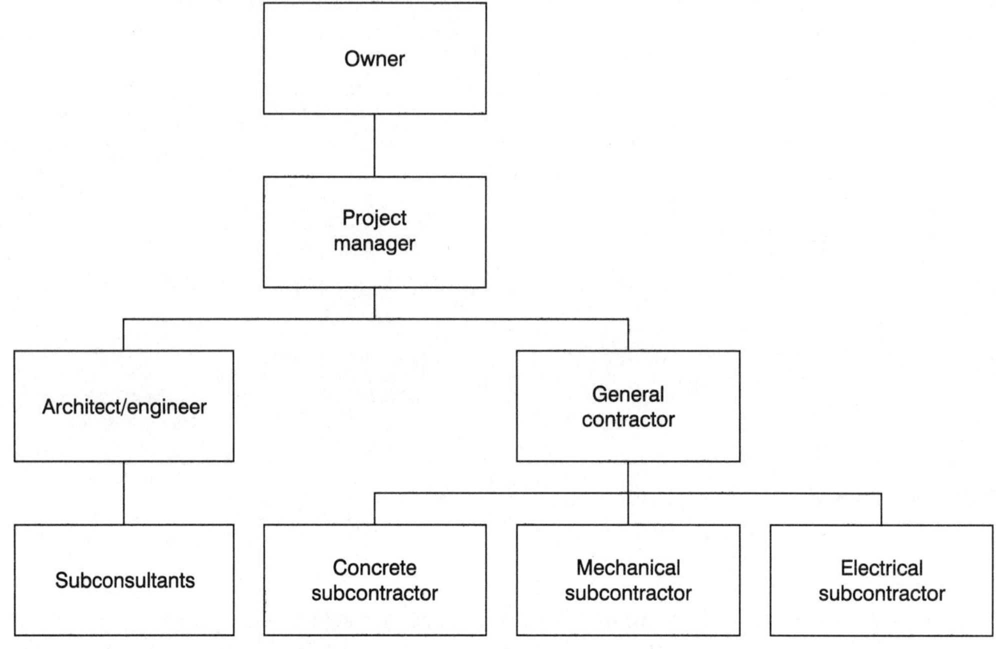

Figure 2.5 Project manager relationship chart.

under the contract and was in no way responsible for any lack of payment that existed between the subcontractor and the owner.

Project manager

Sometimes the owner decides to turn the entire project over to an independent manager. As one example, a school district may have little or no in-house expertise on capital project development; rather than employ a large temporary staff to oversee planning, design and construction, they might use the 'project manager' approach. Figure 2.5 shows one possible form of the contractual relationships. Note that the figure simply modifies Figure 2.1 by adding a project manager between the owner and the architect/engineer and general contractor. Thus the project manager manages the project on the owner's behalf. This arrangement implies that the project manager contracts with the designer and the general contractor. Below the level of project manager on the chart, other arrangements are possible. For example, the project manager might decide to engage a single design–build organisation or might employ a construction manager of the type described earlier. The distinguishing characteristic of the project manager form is the assumption of the responsibility for the entire project by a separate firm, on behalf of the owner.

Document and construct

This is a variation of the design–build arrangement, wherein the owner engages a design team to develop a project concept, overall schematic layouts, performance specifications and other preliminary design details sufficient to select a contractor or construction manager through a

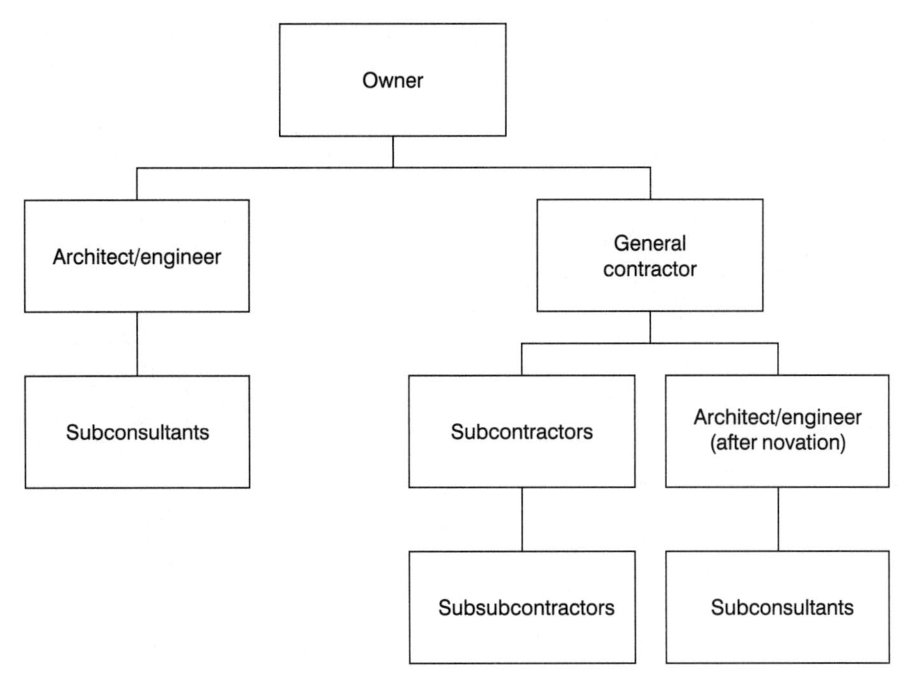

Figure 2.6 Document and construct relationship chart.

tender process. After the contractor is selected, the design consultant's contract with the owner is transferred to the contractor through a process called *novation*. The contractor then becomes responsible to the owner for completion of design, as well as construction, with the understanding that the design work will be completed by the originally selected design organisation. Advantages of this method include a greater control of planning and specification by the owner, as well as the advantages listed for the design–build method. In addition to the disadvantages listed for design–build, the design firm bears the uncertainty of not knowing to whom they will eventually be contracted for completion of the design. In Figure 2.6, we show one possible document and construct arrangement, with the architect/engineer assuming a role as a 'subcontractor' to the general contractor after the construction contract is signed. Another arrangement would be to add the use of a project manager with the role described in the previous section.

In the USA, the term *bridging* refers to the process similar to that described as document and construct in the previous paragraph. The bridging process involves two design firms. The first is under contract with the owner and its responsibilities extend part way through the design process. The resulting documents define the parts of the project that the owner desires to control; they also provide sufficient definition to allow the selection of a construction organisation. The documents leave latitude for the contractor to seek alternative means of construction and thus achieve economies in construction technology. After the contractor is selected, a second design firm is appointed by the contractor (with approval by the owner). This design firm becomes a subcontractor to the contractor and has responsibility for final construction drawings and specifications. Construction does not begin until construction drawings are complete and all parties agree that the owner's intentions will be fulfilled. It is clear that *document and construct* and *bridging* are nearly synonymous; the significant difference is that bridging utilises two

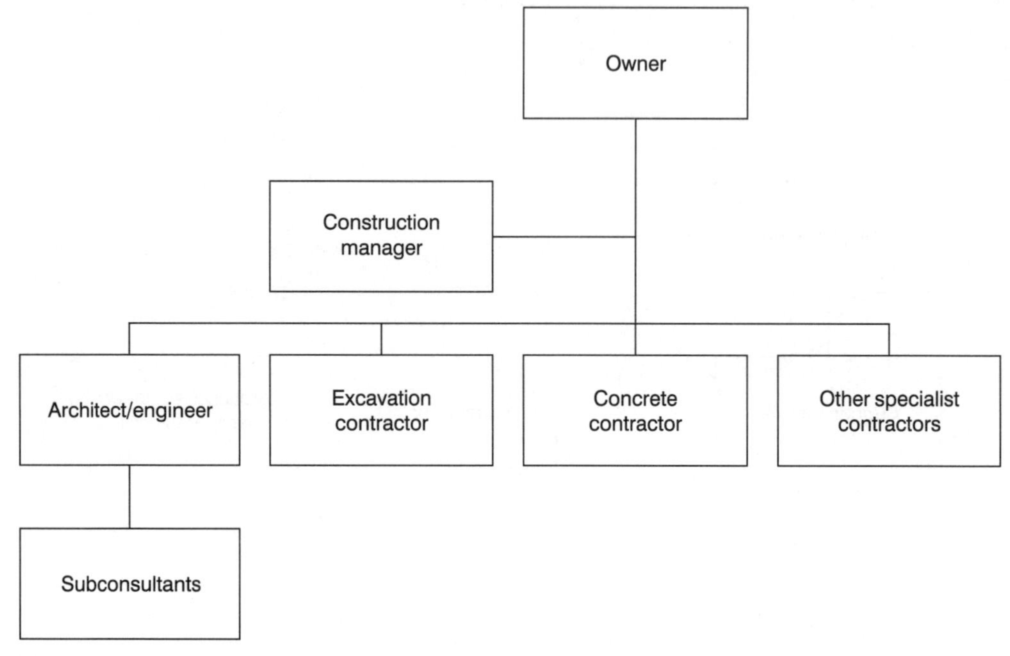

Figure 2.7 Separate prime contracts relationship chart.

separate design firms, in contrast to the single design firm under 'document and construct' and the attendant need for contract novation (Project Delivery Strategy, 2002; Bridging, 2002).

Separate prime contracts

Figure 2.7 shows an example of an arrangement in which the owner contracts directly with individual specialty contractors, each of whom can be considered as a 'prime' contractor, because there is no single general contractor to coordinate their work. In this particular example, we include an 'agency' construction manager, who will assist the owner in this coordination, but the chart makes it clear that the construction manager is not related contractually to the several prime contractors. In the absence of a construction manager, the owner's in-house staff would assume responsibility for coordination of the several prime contractors. Note that there must be enough separate prime contracts to complete the entirety of the work, because the owner will not have any construction forces of its own. This feature differs from the general contractor–subcontractor arrangement, in which the general contractor is responsible for the entire construction project and thus will install any works not performed by subcontractors. It is likely that the various specialist prime contractors in Figure 2.7 would have their own specialist subcontractors, although they are not shown in the figure.

Turnkey

A turnkey contract is one in which the owner and contractor agree on a fixed contract sum for a contract under which the contractor will take responsibility for the entire project. Such agreements are often designated as EPC contracts, because of the prime responsibilities for engineering – providing basic and detailed design, procurement – supplying parts and other goods required for the project and construction – erecting and commissioning the project (Majmudar, 2001). The well-known and much-respected Federation Internationale des Ingenieurs-Conseils (FIDIC) designates such projects as EPC turnkey projects (Fédération Internationale des Ingénieurs-Conseils, 1999). In this type of project, it is a construction organisation that usually enters into the contract with the owner (called the Employer by FIDIC and often the Principal in some countries).

At first glance, this form looks much like the design–build form discussed earlier. However, the scope of the contractor's responsibility is typically broader than basic design, procurement and construction, as it includes such services as project financing and land procurement and other tasks not generally within the design–build scope. The owner provides a brief describing desired outcomes, performance criteria and standards, the parties agree on a fixed price and the contractor proceeds, with no participation by the owner in the performance of the work (McMullan, 1994). When the project is complete, the owner inspects the work, and, if it meets the requirements, hands over the cheque, 'turns the key', enters the facility and commences operation (Bennett, 1996; Clough and Sears 1994). If the services of the contractor have included land acquisition, the owner literally purchases the entire completed project from the contractor at the end of the project. Projects ranging from process and power plants to public housing projects have been brought to completion successfully using the turnkey method. Dorsey (1997) suggests that turnkey contracting is rather like purchasing a new automobile, but with such important differences as (1) the owner is bound by contract to buy the project upon completion, with serious legal consequences for rejecting the completed project, (2) performance specifications are often used to establish materials, equipment and general quality and (3) owner-generated changes will alter the price once the project is underway.

In the UK, a recent turnkey example is the completion of the world's first commercial digital terrestrial television network for the British Broadcasting Corporation and ONdigital. Crown Castle International was responsible for design specification, network planning, selection, purchase, installation, operation and maintenance of transmitter electronics, antennas, power and other components, specification and procurement of a signal distribution network and operation and management of the transmitter and transmission network (Crown Castle International, 2001). Note that the operation and management of these facilities may be somewhat atypical compared to projects whose owners assume operating responsibility after completion of construction. The importance to the owner of a fixed final price, and often a certain completion date, makes this method attractive. The owner expects to pay a premium for the certainties provided in such a contract and for the wider risks assumed by the contractor, such as unexpected ground conditions.

Build–own–operate–transfer

The build–own–operate–transfer (BOOT) type of project has evolved as a means of involving the private sector in the development of the public infrastructure. The concept, in use in some parts of the world for some centuries, requires the private sector to finance, design, build, operate and manage the facility and then transfer the asset to the government free of charge after a specified concession period (Tiong, 1992). The term BOT, for build–operate–transfer, was first coined by the Turkish Prime Minister in 1984 in connection with the privatisation of that country's public-sector projects. The terms BOOT and BOT are used synonymously, while terms like DBO (design–build–operate) and BOO (build–own–operate) imply construction and operation but no transfer (Carson Group, 2000). In Australia, for example, the transfer option is omitted (McMullan, 1995).

Examples of projects that have used the BOOT approach include power stations, toll roads, parking structures, tunnels, bridges and water supply and sewage treatment plants. Consider a parking structure as an example. A municipality may identify a need for such a facility. A project sponsor is selected and the design and construction are carried out under the sponsor's overall direction. Upon completion of the structure, the sponsor is responsible for operation and maintenance; income to cover the cost of construction plus ongoing periodic costs is derived from parking fees. At the end of a period of, say, 20 years, the sponsor transfers the facility to the municipality free of charge, with a guarantee that it is in satisfactory condition. Typical 'concession periods' range from about 10 years, not including project development time, for some power station projects to as long as 55 years, including approximately 7 years for construction, for the Eurotunnel project (Tiong, 1992).

It is apparent that such an approach requires a complex organisational structure and carries considerable long-term risk for the project sponsor, while minimising such risk for the governmental owner. In a typical BOOT project, the parties are likely to include the following (Tiong, 1992; McMullan, 1995):

- client – usually a governmental agency;
- constructor – often responsible for both design and construction, although the general design will usually be dictated by the client and may be carried out by the client;
- operation and maintenance contractor;
- offtakers – entities that agree to purchase the outputs of the project, such as water or electricity, usually a governmental agency, not always 'the client' as listed above;
- suppliers of materials, equipment and so on for both the construction and long-term operation of the facility;
- lenders and investors;
- sponsor – a consortium of interested groups, usually including the constructor, operator and financial institution, that prepares the proposal and, if successful, contracts with the client to carry out the design, financing, construction and operation (note that the 'sponsor' thus includes at least three of entities named earlier in this list).

A BOOT type of project can utilise any of several of the organisational relationships described earlier in this chapter. For example, the sponsor may employ a project manager to oversee the entire design–finance–build process, a construction manager to advise on construction and/or a design–build firm as a single entity to accomplish those two parts of the scheme.

Joint venture

Adding further complication to our consideration of project organisational relationships is the joint venture. Such an arrangement is 'a voluntary association of two or more parties formed to conduct a single project with a limited duration' (Bennett, 1996). Joint venture agreements are formed between construction firms or between design firms and construction firms; they do not include owners. In a sense they are special purpose partnerships, because of their separate, temporary nature.

The usual purpose of such an arrangement is to spread the risks inherent in large projects and to pool resources in a way that permits the joint venture to execute a project that would be beyond the capabilities of one of the parties individually. Each party is called a coventurer, or joint adventurer; they begin with an agreement between themselves, whose purpose is to seek the contract for a project. If they are successful, the joint venture itself contracts with the project owner, with one of the coventurers stipulated as the sponsoring coventurer. The temporary combination of two or more companies allows them to pool construction equipment, personnel, office facilities, financial means and other resources. Each coventurer has an active role in performing the work and shares in any profits or losses, in accordance with their agreement (Clough and Sears, 1994).

Other reasons for forming joint ventures include the ability to meet mandated minority business enterprise requirements, if one of the coventurers is so classified, and to allow a contractor a chance to participate in a different type of work or a new geographical area (Tenah, 2001). During the construction of the Trans-Alaska Oil Pipeline in the 1970s, for example, many of the pipeline's sections were built by joint ventures consisting of a company specialising in pipeline construction, with little or no experience in cold-regions construction, and a company with experience in the Far North but little pipeline background. They successfully pooled their talents, equipment, experience and financial strengths to carry out this US$ 9 billion project on a year-round basis over 3 years.

An interesting case study analysis of the use of joint ventures in developing countries suggests that these international construction joint ventures (ICJVs) can be appropriate market penetration strategies (Lim and Liu, 2001). One such project was the £414 million Pergau Hydroelectric Project in Malaysia. The joint venture partners were the local Malaysian company Kerjaya Binaan, Balfour Beatty of the UK and Cementation, also from the UK. The resulting joint venture was called Kerjaya Balfour Beatty Cementation (KBBC); it was organised as an independent operation in Malaysia and key management personnel were seconded from the coventurers for the duration of the 67-month project.

Note that joint ventures can be applied to most types of project delivery systems. The traditional design–tender–build type of project might attract a joint venture contractor. Or, a design–build project might be carried out by a temporary organisation consisting of an engineering design firm and a construction contractor; the temporary nature of the organisation is the key for the 'joint venture' designation. Furthermore, a construction manager or project manager could be part of the project organisation in which a joint venture was responsible for field construction. And finally we should note that nothing precludes the use of subcontractors, subsubcontractors, subconsultants and other parties within the overall joint venture scheme.

The case study at the end of this chapter describes an innovative approach to a joint venture organisation utilised for a very large construction project in Scandinavia.

Force account

The so-called force account approach to construction project organisation is very simple – the project owner acts as the prime contractor and carries out the work with its own forces by providing field supervision, materials, equipment and labour (Clough and Sears, 1994). This method is usually confined to relatively small, uncomplicated projects that are built for the owner's use rather than sold to another party upon completion. Examples might include a highway department that uses its own forces to perform a culvert replacement or a minor re-alignment and an educational institution that uses its own building services crews to renovate classrooms. Such owners usually have their own guidelines for deciding the project magnitude and complexity beyond which they choose to contract with separate construction organisations. They must consider the tradeoffs among the extra time and expense of the formal contracting process and the possible construction cost savings, potential time savings and improved quality resulting from these competitive or negotiated contracts. Often, such small-scale projects are also designed in house by the owner's staff.

The force account method has been used in many settings around the world. One interesting example is the large-scale implementation of force account road building in Kenya. In 1993–1994, the Minor Roads Laborers Program employed 11 125 casual workers and 1397 permanent staff. This programme has been cited by the World Bank for its positive effects on reducing poverty in the region, in addition to improving the transportation infrastructure. The programme

> employed primarily low-income residents, more than half of whom earned wages for the first time; created a means to market surplus food crop and livestock production; increased cash earnings from both farm and non-farm sources; and contributed to reducing the gap between upper- and lower-income groups. (World Bank Group, 2000)

Phased construction

Although more a scheduling approach than a delivery system, we discuss phased construction, or 'fast-tracking', here because several of the delivery systems are better suited than others to this method of planning and carrying out the project schedule. The basic idea is that some portions of the design work occur concurrently with field construction, thus achieving an overall savings in total project duration. Recall the traditional design–tender–build approach in which the entire project is fully designed prior to the calling for tenders for the construction phase. Under this method, construction does not begin until the end of the tendering and contractor selection efforts, which necessarily must follow completion of design.

Contrast this approach with a scheme under which enough design work is completed to begin field work. An example might be a building project that is designed and built in several parts. After the foundations are designed, foundation construction begins. Design for the balance of the structure proceeds concurrently with that first construction effort. The structural frame design is completed while foundation installation work is underway, after which the structural frame is erected. The process continues, with design and construction occurring in overlapped phases. Figure 2.8 is a bar chart that compares the time schedule for a building project designed and constructed using the traditional design–tender–build approach with that for a similar project using phased construction. For this hypothetical project, the total design time, 12 months, is the same under both approaches; similarly, the duration of construction, from start to finish is the same, 26 months. The 12-month difference in total project duration results from being able to start the foundation work at the beginning of the fifth month under the phased approach, whereas that work would not start until the beginning of the seventeenth month using the traditional method. The phased-construction approach would be well suited to the separate prime contracts delivery system, under which the owner, or the owner's project manager, coordinates the several contractors and assures that the design schedule proceeds sufficiently to provide documents for the various construction phases. In Figure 2.8, we include a tender period for each construction phase, as required by this separate-prime-contracts delivery system. The design–build delivery system would also be compatible with a phased-construction approach; in

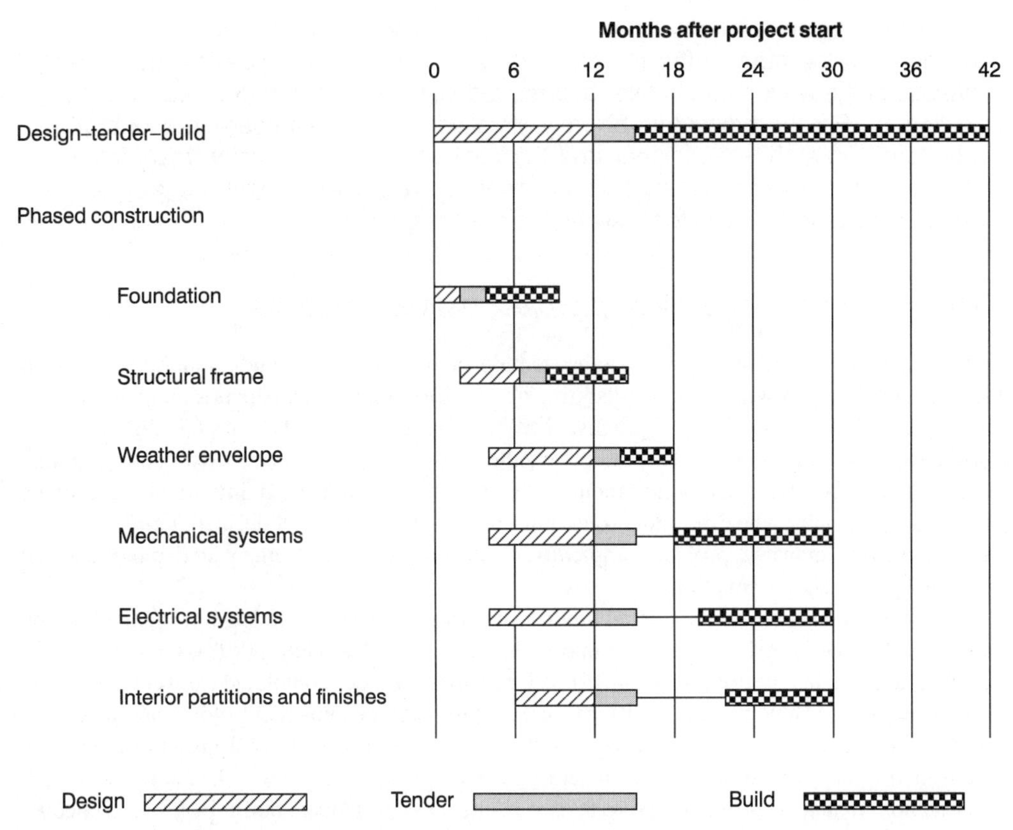

Figure 2.8 Time schedule comparison. Traditional design–tender–build versus phased construction.

this case, the design–build organisation would oversee the coordination of all phases of design and construction and it is likely that fewer formal tendering periods would be needed.

An obvious risk inherent in the phased-construction/fast-track approach is that the early design work must make assumptions about the outcomes of later design work; that early work may be incorporated into the finished construction before completion of all design and changes may be needed if those early assumptions were not accurate. A simple but basic example relates to foundation design. If the building is completely designed prior to tendering (that is, if it is not 'fast-tracked'), the foundation design will be based on actual design loads of the structure, weather envelop, equipment, furnishings and any other building components, as well as live loads. However, to save time, under phased construction the foundation design will be completed before all of these loads have been determined and some assumptions will be required. If the foundations are not sufficient for the loads in the final structural, mechanical and other design, extra costs will ensue from the required foundation strengthening or replacement.

Clough and Sears (1994) offer some helpful cautions and comments about phased construction:

> . . . fast-tracking has received harsh criticism because it emphasizes time rather than quality and can sometimes take longer than the usual, sequential process when

applied to complex projects. It has been pointed out that the final construction cost is unknown at the start of the fast-tracked construction and if bids for subsequent phases of the work come in over budget, redesign options to reduce cost are very limited . . . For the process to be truly effective . . . the best possible cooperative efforts of the architect-engineer and the contractor or construction manager are required. It is also necessary that the owner have a well-thought-out program of requirements and be able to make immediate design decisions.

Evaluation and comparison of project delivery systems

In Table 2.1 we show a summary of salient features and some of the benefits and limitations of the several delivery systems described above. The choice for an owner depends on many factors. One source (Project Delivery Systems, 2001) suggests several key issues that must be considered, including speed of execution, schedule delays, change orders, cost control, quality control, owner risk, pre-construction input, level of team communication, importance of a single point of responsibility, the extent that roles and responsibilities are well defined, familiarity with the construction industry, project complexity, level of owner involvement and importance of minimising adversarial situations.

Many investigators have conducted studies of various delivery systems and their effect on project performance criteria. For example, Konchar and Sanvido (1998) compared cost, schedule and quality performance of 351 US building projects completed between 1990 and 1996 using design–tender–build (33% of the 351 projects), design–build (44%) and an at-risk construction manager (23%). The design–build projects had a median cost growth (increase in cost from original estimate) of 2.17%, whereas the median cost growth was 3.37% for the at-risk construction manager projects and 4.83% for the design–tender–build projects. Schedule growth had median values of 0.0% for both design–build and at-risk construction manager projects and 4.44% for design–tender–build. A multivariate analysis found that design–build projects had unit costs averaging 4.5% less than those for at-risk construction manager projects and 6% less than design–tender–build, while the unit costs of at-risk construction manager projects averaged 1.5% less than those for design–tender–build. The multivariate study also compared delivery systems in terms of construction speed (defined as project area per as-built construction duration), finding that the design–build method resulted in projects 23% faster, on average, than at-risk construction manager projects and 33% faster than design–tender–build projects. At-risk construction manager projects averaged 13% faster than design–tender–build projects. A similar study of 332 projects in the UK (Bennett et al., 1996) found that design–build projects had unit costs that averaged 13% less than unit costs for design–tender–build projects and construction speeds that averaged 12% faster.

Selection of type of contract

Having explained the several types of contractual relationships that can exist among the parties to a construction contract, we now turn to consideration of the types of contract for construction services. We wish to separate these two issues, as they are separate decisions that must be made by the owner, even though some types of contracts are better suited for certain types of project delivery systems. This section, then, explains how the contractor is chosen and the basis for payment for each of the contract types. In a later section, we consider in detail the process for selecting and paying the contractor.

Lump sum/fixed price

Perhaps the simplest type of contract is the one under which the contractor is paid a pre-agreed fixed amount for the project, based on a contract for a specified amount of in-place finished construction work. This amount is paid without regard for the actual costs experienced by the contractor. For such a contract, the project must be completely defined by the contract documents prior to receipt of proposals, selection of the contractor and formation of the contract. The price of the lump sum, or 'stipulated sum', contract must include all direct costs of labour, materials, equipment and subcontractors, as well as such indirect costs as field supervision, field office, equipment maintenance and the like, plus general company overhead, plus profit. The contractor in essence guarantees the completed project for the stipulated price.

This type of contract is suitable for such projects as buildings, which can be completely designed and whose quantities are thus definable, at the beginning of the project. For a project involving large quantities of earthwork, it is unlikely that those quantities can be accurately predicted in advance and thus another type of contract would be more appropriate. Two advantages to the owner of lump-sum contracts are the fact that the total cost of the project is known before construction begins and the lack of a need to monitor and approve the contractor's costs. On the other hand, the owner bears the risk of poor quality from a contractor trying to maximise profit within the fixed sum; other potential disadvantages to the owner include the high cost and long time required for contractors to prepare tenders; although such costs are borne by the contractor, ultimately they are built into the successful contractor's total price. In addition, the flexibility of this contract form is limited; any variation from the original plans and specifications requires a change order, a process that can be time consuming and expensive and may even lead to contract disputes.

From the standpoint of the contractor, a fixed-price job has the potential for a satisfactory profit, if the project is well managed and costs are controlled; it is the contractor who wields this control. Also, the need for detailed cost records for the owner's use does not exist; the contractor does not need to 'prove' its costs in order to be paid. (The contractor will still need to keep cost records in order to compare actual performance against estimated costs, but these records are for the contractor's use only.) Another potential advantage to the contractor is the prospect of profit from changes in the work after the contract is signed. However, a poorly managed or unlucky job can lead to problems, because payments are not adjusted if the contractor gets into difficulties. Also, each contractor must prepare a detailed cost estimate and prepare a tender, even though only one such tender will lead to a contract.

The term 'fixed price' is often used synonymously with lump sum, but this may be somewhat of a misnomer, as the 'guaranteed' fixed price may change during the project's lifetime. We shall deal later in some detail with the process involved in changing such a contract. Suffice it to say here that changes may occur because of two primary reasons: changes in the conditions at the site, such as differing soil conditions and changes in owner desires, ranging from minor cosmetic changes (wall coverings in a building, for example) to complete changes in scope. Unless the contract provides that the contractor is to cover the costs of such changes within the lump sum amount, a change order will likely lead to a change in the total contract price.

Unit price/measure and value

The New Zealand *Conditions of Contract for Building and Civil Engineering Construction* (Standards New Zealand Paerewa Aotearoa, 1998) describe a *measure-and-value contract*:

Table 2.1 Project delivery system options

System	Features	Advantages	Limitations
Traditional design–tender–build	Separation of design and construction responsibilities Completion of design prior to selection of contractor	Certainty of price Clarity of roles No coordination risk to owner Easy to accomplish changes during design	No opportunity for phased construction Fixed price established late in process Owner administers all design and construction contracts No contractor input to design
Design–build	Single organisation responsible for design and construction	Single point of responsibility Constructability input during design Fixed price early in process Opportunity for phased construction	Difficulty of formulating price prior to design Lack of oversight by designer Costly tendering process Less control by owner
Construction manager	Professional manager to advise owner and designer on construction aspects May be agency type (advisory role only) or at-risk type (more responsibility for on-site performance)	Construction expertise available during design phase Construction manager provides advice to owner during construction phase Under 'at risk', some risk is removed from owner	Increased overhead costs Owner may take on greater risks under 'agency' type, especially if multiple primes are used Owner relinquishes some control
Project manager	Professional manager to advise owner and designer on all aspects of project	Owner relies on project manager for coordination of most aspects of project Potential for rapid project start-up and prosecution	Owner relinquishes considerable control Increased overhead costs
Document and construct	Early design performed under contract to owner Later design performed (possibly by same designer) under contract to contractor	Fixed-price contract and complete documentation before construction begins Centralised responsibility Constructability considered during design	Designer may not control whom it ultimately works for New and unfamiliar method Limitations similar to design–build

Table 2.1 Continued

System	Features	Advantages	Limitations
Separate prime contracts	Owner contracts with individual specialty contractors	High degree of control by owner Savings in cost of engaging general contractor Potential for effective phased construction	Requires owner construction expertise General contractor risks assumed by owner Less clear relationship between designer and on-site activities
Turnkey	Single organisation responsible for all aspects of project including, but not limited to, design and construction	Owner relies on turnkey contractor for entire project Potential cost and time savings	Requires clear and detailed scope and needs statement at beginning of project Owner relinquishes almost all control
Build–own–operate–transfer	Single organisation responsible for designing, building and operating facility for a certain time period, after which it is transferred to owner	Owner transfers most risk to project sponsor Design tends to recognise long-term cost impacts Constructability considered during design	Major risks to project sponsor Complexity due to large number of parties Long lead time Large up-front costs
Joint venture	Two contractors in a temporary partnership to build a single project	Takes advantage of strengths of each coventurer Allows combined expertise to build large projects Allows each contractor to gain experience in new area or with new type of work	Coordination challenges Like a partnership, requires very clear agreement between the coventurers
Force account	Construction project carried out by owner's own forces	Avoids time and expense of tendering Owner can exercise more direct control	Workforce may lack needed skills Owner may lack needed managerial expertise No price competition

> The Principal shall pay the Contractor for the measured quality as determined by the Engineer of each item carried out at the rate set out in the Schedule of Prices [and in the case of Variations in accordance with Section 9].

In some other parts of the world, such a contract might be designated as a *unit-price contract*. In contrast to the lump-sum contract, this method determines the amount the contractor will be paid as the project proceeds by requiring that the actual quantities of finished product be measured and then multiplied by pre-agreed per-unit prices. Contractors provide tenders based on estimated quantities provided by the owner, so that each tenderer's price is based on a common set of quantities. Thus, prior to the work, the tender prices are based on *estimated* quantities, whereas during and after the work, the payment is based on *actual* quantities. In a later section we shall describe a process for developing a unit-price tender. Suffice it to say here that a contractor's unit prices must cover not only direct costs but also indirect costs, overheads, contingencies and profit; the contractor is paid only for items on the quantity list and paid at the pre-agreed unit prices.

This approach is well suited to projects whose quantities are not well defined in advance of the start of field work. An example might be pile driving, where piles are to be driven 'to refusal'. The owner might provide information that a total of 60 such piles are to be driven, with an estimated depth of 12 metres each. Thus, each prospective contractor would provide a unit price to drive the estimated 720 linear metres of piling. If the successful tenderer has submitted a unit price of US$ 150 per linear metre and if the actual quantity placed is 720 linear metres, the contractor will be paid 720 × US$ 150 = US$ 108 000 for this item. If, however, the actual quantity turns out to be 685 linear metres, the payment would be US$ 102 750, whereas, if the actual quantity is 740 linear metres, the payment for this item would be US$ 111 000. Total payment to the contractor will be a collection of items in addition to the piling, such as excavation, backfill, concrete, paving and the like.

If the actual quantity varies substantially (say, more than 10% in some jurisdictions) from the estimated quantity, many conditions of contract provide for a means of adjusting the unit prices appropriately. The New Zealand *Conditions of Contract for Building and Civil Engineering Construction* (Standards New Zealand Paerewa Aotearoa, 1998) provide as follows:

> Any quantities given in the Schedule of Prices are provided for the purpose of evaluating tenders and may be taken as a reasonable assessment of the quantities involved in the Contract Works. Where the actual quantity of any single item differs from that given in the Schedule of Prices to such an extent as to make the scheduled price for that or any other item unreasonable then the change in quantity shall be treated as if it was a Variation.

In this document, a *Variation* has special meaning, and a strict process that seeks to be fair to both owner and contractor is set forth to provide recompense for the contractor.

For the owner, this approach has the advantage of removing some risk to the contractor, thus leading to somewhat smaller contingencies. Also, pre-tender documents may be prepared in less detail than under the lump-sum contract, although they must be in sufficient detail to allow tenderers to assess the overall magnitude and complexity of the work. However, under the unit-price/measure-and-value method, the final project cost cannot be known with certainty until the project is complete. Furthermore, an effort is required by the owner or owner's representative to 'track' actual quantities by some means of measuring – counting truckloads, weighing steel and the like. From the contractor's viewpoint, some of the risk in the bidding process is removed, because payment is based on actual quantities rather than on a lump sum. Also, there is a possibility of unanticipated profit if the actual quantities are greater than forecast (but not as great as would require a renegotiation or Variation); in this case, the fixed costs should have been covered by the original quantities. On the other hand, lower-than-expected profits will result if actual quantities are lower than estimated. Another disadvantage to the contractor is the effort required to keep track of quantities. For a US perspective on unit price contracts, see Bennett (1996), Bockrath (1999) and Clough and Sears (1994).

Cost plus

A third basic method by which the contractor may be paid is the *cost-plus method*, under which the owner pays the contractor's costs related to the project plus a fee that covers profit and non-reimbursable overhead costs. This approach seems simple, straightforward and desirable, at first glance. However, there are some disadvantages, as we shall see. Two types of cost-plus contracts are used: (1) cost plus a percentage of costs, under which the fee is an agreed-upon percentage of the 'costs' and (2) cost plus fixed fee, wherein the fee does not depend on the contractor's costs. Of particular importance in this type of contract is the need to define 'costs'; Halpin and Woodhead (1998) point out that 'if the owner is not careful, he may be surprised to find out he has agreed to pay for the contractor's new computer'.

In Chapter 4, we shall describe methods used by owners to select contractors and we shall find that cost-plus contracts are often, though not exclusively, negotiated rather than subjected to a tender process. Clough and Sears (1994) list four important considerations to be taken into account by owners and contractors when negotiating such contracts, as follows:

- a definite and mutually agreeable subcontract letting procedure;
- a clearly understood agreement concerning the determination and payment of the contractor's fee;
- an understanding regarding the accounting methods to be followed;
- a list of job costs that will be reimbursable.

Sears and Clough (1994) further note that two categories of expense can be particularly difficult to define and manage. One is contractor's general overhead. Are the costs of preparing payroll and working drawings, of engineering and other office functions, and those general administrative costs involving home office personnel reimbursable or are they to be covered by the fee? Clear definition is essential. The other troublesome category of reimbursable cost is that related to construction equipment; if the contractor owns equipment used on the job, some sort of charge rate must be agreed upon.

The American Institute of Architects (1997, 2001), in its standard forms of agreement between owner and contractor for cost-plus-fee contracts, sets forth a detailed list of the labour, subcontract, materials (both incorporated into the work and used on a temporary basis), equipment and miscellaneous costs that will be reimbursed as well as those costs not to be reimbursed. The general idea is that costs directly related to the work will be reimbursed, while those of a truly general, company-wide nature, will not. But often there will be confusion and disagreement, even in the most carefully drawn contract, as to whether certain 'general' overhead costs are to be reimbursed.

In selecting a cost-plus type of contract, the owner will want to recognise that a cost-plus-percentage contract may lead to overspending by the contractor, because there is little incentive to be efficient and economical. Under this system, the greater the 'costs', the more will be paid as a fee. There is less incentive to overspend under the cost-plus-fixed-fee type of contract and the contractor may be motivated to complete the project quickly in order to save on non-reimbursable overhead expenses and recover the fee quickly. On the other hand, the contractor may tend to 'skimp' on quality, because the fee will be the same no matter how low the costs are. A cost-plus-fixed-fee contract requires the project size and scope to be defined reasonably firmly, so that a fair fixed fee can be defined. Another disadvantage to the owner under any cost-plus arrangement is the lack of definition of total project cost until the project has been completed. But the owner can begin the construction early in the design phase using a phased-construction approach, with a contract that simply defines reimbursable costs and the fee arrangement.

It is to the contractor's advantage to shift the risk of cost increases to the owner and still be assured of recovering reimbursable costs and being paid some fee. However, financial rewards for innovative construction methods leading to cost savings are for the owner's benefit, not the contractor's.

Some of the disadvantages of cost-plus contracts noted above can be mitigated to some degree with variations of the basic cost-plus approach. These variations are discussed in the following section.

Variations of basic cost plus

We mention only the essence of several variations of cost-plus contracts, enough to make it clear that many creative approaches exist. To overcome the objection that the project's cost is not known until after completion, some owners utilise a guaranteed- (or warranted-) maximum-cost contract. This type of contract can be well suited to a project whose scope is well defined and seems particularly suited to turnkey projects. The contractor warrants that the project will be constructed in accordance with the project documents and the cost to the owner will not exceed some total maximum value. Note well that it must be clearly stated whether this maximum is the total 'costs' (not including the fee) or the total that the owner pays, including the fee. Usually it is the latter and the contractor pays for any excess above the maximum.

The use of various penalty and bonus provisions is common with cost-plus contracts. A target estimate of costs is established, with the understanding that (1) savings accruing from actual costs less than the target will be shared between owner and contractor and (2) if actual costs exceed the target, the extra costs will be shared as well. It also possible to have a bonus incentive of type 1 without the type 2 penalty, although in this case a guaranteed maximum price may also be included in the contract. We should also note that incentive clauses may be related not only to costs but also to schedule and quality performance. Also, some contracts are written with variable fee percentage provisions, the actual fee being determined by a formula that considers cost performance relative to the target, as well, sometimes, as schedule performance. There are many possibilities!

In Australia, the use of 'trade contracts' is another variation on the cost-plus theme. Here, a general contractor in a cost-plus contract may be required to secure fixed price subcontracts. As explained by McMullan (1994):

> The Principal enters into a cost-plus contract with the prime contractor, the work is then contracted out by the prime contractor on a fixed price basis, the prime contractor being entitled to cost-plus reimbursement by the Principal for those trade contract prices. Effectively, therefore, the Principal has the benefit of fixed price contracting.

As an example of some of the features described in the previous paragraphs, consider a cost-plus-fixed-fee contract for a building project, where the estimated cost is £15 000 000 and the 'fixed' fee is an additional £1 800 000, so that the best estimate of the total cost to the owner is over £16 800 000. The contract might stipulate a guaranteed maximum price of £17 500 000 as well as a target for costs of £15 250 000. (Note that this target need not necessarily be the £15 000 000 cost estimate.) Any savings in costs below £15 250 000 might be shared, with 25% going to the contractor as a bonus, while 25% (or some other percentage) of the excess costs above £15 250 000 might be assessed as a penalty. In any case, the owner would never have to pay more than £17 500 000. To the reader is left the exercise of calculating payments under various actual scenarios. Note that changes in the work can lead to changes in the agreed-upon values in this example.

Time and materials

A time-and-materials contract is often used on small projects, perhaps a maintenance effort, a small building or a series of small projects. It has elements of both the unit-price and cost-plus approach. The owner pays the contractor based on effort expended, but there is no 'fee' as such. Materials are paid at their actual cost, while labour and equipment inputs are reimbursed at pre-agreed rates. An important element of this method is that these labour and equipment rates must include all indirect and overhead expenses, profit and contingency, in lieu of the payment of any extra 'fee'.

The contract includes a list of hourly payment rates – for carpenters, millwrights, labourers, 10 m^3 dump trucks, front loaders and the like. If carpenters are paid US$ 17.50 per hour, the hourly rate billed for their services might be US$ 42.00 to include all indirect payroll expenses and a portion of the many other overheads, plus profit and contingency. Then, as the basis for a payment request for a given period, the contractor presents material invoices, payroll records with hours by category and similar records for equipment. Subcontract payments would normally be reimbursed at actual cost. If the request is approved, the contractor receives payment based on 'time and materials' – the time for each labour and equipment category multiplied by its respective rate, plus materials and subcontracts at cost.

This method is often used for design services, for which it is usually difficult to determine the total expected effort in advance, thus making a fixed-price design contract impractical.

Having reviewed two important decisions the owner must make during the pre-project phase, we now turn to the planning and design phase, where we describe the several parties and their activities and then explain the development of construction contract documents.

Discussion questions

1 An owner is contemplating the design and construction of a high-rise apartment building in your region's capital city. Identify potential appropriate project delivery systems and construction contract options. Which might be preferred? Why?
2 Answer question 1 if the project is an ore-processing facility in a remote equatorial region.
3 Which type of project delivery system might be best suited to a phased construction schedule? Why?
4 Clarify the differences between the construction manager and the project manager delivery systems. Under what circumstances would one be preferable to the other, from the owner's point of view?

5 Would the design professional tend to prefer the agency or the at-risk construction manager project delivery system? Why? Answer these two questions again from the point of view of the owner and from the point of view of the construction manager.
6 List several potential risks that will be assumed by the sponsor of a build–own–operate–transfer project. Identify those risks on your list that would not fall upon a design–build organisation if such a project were transferred to the owner upon completion of construction.
7 Visit a local building site and try to determine the number and types of subcontracts that are in active use. How might the number of subcontracts vary with different types of construction?
8 Is it better that a design–build organisation be led by a designer or a contractor? Why?
9 Draw a relationship chart for a project that uses an agency construction manager and multiple prime contractors.
10 List several projects, real or hypothetical, for which a joint venture might be well suited to carry out the construction work. For each, suggest the special strengths that each joint venture partner would contribute to the project.
11 Suppose you are invited to consider performing a pile-driving contract, where the steel piles are to be driven to refusal. Subsurface soil data is incomplete and not trustworthy. What type of construction contract would you prefer? Why?
12 Distinguish between a time-and-materials contract and a cost-plus contract.
13 The text described a cost-plus-fixed-fee construction contract for which the estimated sum of all reimbursable costs was £15 000 000 and the fixed fee was £1 800 000. The guaranteed maximum price (costs plus fee) was £17 500 000. Any savings in costs below the target of £15 250 000 was to be shared, with the contractor receiving 25% of the savings. Suppose that any costs above £15 250 000 were to be shared equally between owner and contractor.

   Determine the amount the contractor will be paid if the actual reimbursable costs are (a) £14 850 000, (b) £15 000 000, (c) £15 250 000, (d) £15 650 000, (e) £16 050 000, (f) £17 000 000, (g) £17 500 000 and (h) £17 800 000.
14 Interview a local general contractor to determine the proportion of its contracts that are lump sum, unit price, cost plus and time and materials. Try to find out whether this contractor prefers one over the others.
15 Can a cost-plus contract ever be favourable to both owner and contractor? If so, under what circumstances?
16 The section on design–build includes the following statement: ' . . . one of the primary reasons for low productivity in the construction industry is the lack of integration of activities across the project life cycle.' Based on what you know so far, list several means by which better integration could be achieved across the entire project development cycle.

References

American Institute of Architects. 1997. *Standard Form of Agreement between Owner and Contractor – Cost of the Work Plus a Fee, with a Negotiated Guaranteed Maximum Price (GMP)*. AIA Document A111.

American Institute of Architects. 2001. *Standard Form of Agreement between Owner and Contractor where the basis of payment is the Cost of the Work Plus a Fee without Guaranteed Maximum Price (GMP)*. AIA Document A114.

Bennett, F.L. 1996. *The Management of Engineering: Human, Quality, Organizational, Legal, and Ethical Aspects of Professional Practice*. John Wiley.

Bennett, J., E. Pothecary and G. Robinson. 1996. *Designing and Building a World-Class Industry*. University of Reading, UK.

Bockrath, J.T. 1999. *Contracts and the Legal Environment for Engineers and Architects*, 6th edn. McGraw-Hill.

Bridging. 2002. *3D/International, Inc.* http://www.3di.com.

Carson Group. 2000. *Project Delivery Options*. Class handout for ENCI 491 Construction Management, University of Canterbury, Christchurch, New Zealand.

Chan, A.P.C, D. Scott and E.W.M. Lam. 2001. Study of design–build projects in Hong Kong. In *Proceedings, Third International Conference on Construction Project Management*, Singapore, 29–30 March, pp. 366–376.

Clough, R.H. and G.A. Sears. 1994. *Construction Contracting*, 6th edn. John Wiley.

Crown Castle International. 2001. *Crown Castle Case Studies*. http://www.crowncastle.com.

Design–Build Institute of America. 1994. *The Design–Build Process: Utilizing Competitive Selection*.

Dorsey, R.W. 1997. *Project Delivery Systems for Building Construction*. Associated General Contractors of America.

Fédération Internationale des Ingénieurs-Conseils. 1999. *Conditions of Contract for EPC Turnkey Projects*. 1st edn.

Halpin, D.W. and R.W. Woodhead. 1998. *Construction Management*, 2nd edn. John Wiley.

Konchar, M. and V. Sanvido. 1998. Comparison of U.S. project delivery systems. *Journal of Construction Engineering and Management*, **124**, 435–444.

Lim, E.C. and Y. Liu. 2001. International construction joint venture (ICJV) as a market penetration strategy. In *Proceedings, Third International Conference on Construction Project Management*, Singapore, 29–30 March, pp. 377–389.

Majmudar. 2001. *Engineering and Procurement and Construction Contracts in India*. http://www.majmudarindia.com.

McMullan, J. 1994. Developments in construction management. In *National Construction and Management Conference*, Sydney, Australia, February 18. Reprinted as Project delivery systems. *Electronic Construction Law Journal*, http://www.mcmullan.net/eclj.

McMullan, J. 1995. BOT contracts. In *International Seminar on Property and Investment Laws*, Beijing, People's Republic of China, May 23. Reprinted as Build own operate transfer (BOOT) projects. *Electronic Construction Law Journal*, http://www.mcmullan.net/eclj.

Ministry of Manpower and Ministry of National Development of Singapore. 1999. *Construction 21: Reinventing Construction*.

Project Delivery Strategy. 2002. *3D/International, Inc.* http://www.3di.com.

Project Delivery Systems. 2001. *Genesis Builders and Planners*. http://www.onbuilding.com/cdelivery.html.

Rubin, R.A., V. Fairweather and S.D. Guy. 1999. *Construction Claims: Prevention and Resolution*, 3rd edn. John Wiley.

Simon, M.S. 1989. *Construction Claims and Liability*. John Wiley.

Standards New Zealand Paerewa Aotearoa. 1998. *Conditions of Contract for Building and Civil Engineering Construction*. New Zealand Standard NZS: 3910: 1998.

Tenah, K.A. 2001. Project delivery systems for construction: an overview. *Cost Engineering*, **43**, 30–36.

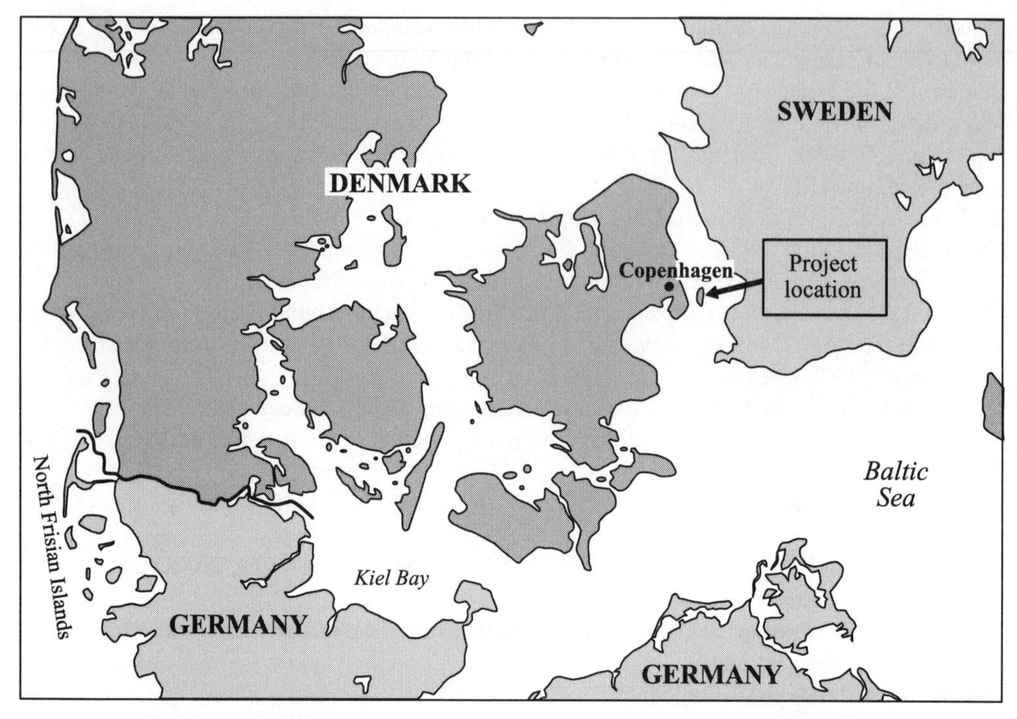

Figure 2.9 Öresund bridge project location map. Courtesy of Hochtief AG, Essen, Germany.

Tiong, R. 1992. *The Structuring of Build–Operate–Transfer Construction Projects*. Construction Management Monograph Series. Nanyang Technological University Centre for Advanced Construction Studies.

World Bank Group. 2000. *Force Account Model. PovertyNet; Safety Nets: Public Works*. http://www.worldbank.org/poverty/safety/pworks/force.htm.

Case study: Project organisation innovation in Scandinavia

by Felix Krause, Dipl Ing, MEng, Section Head, Hochtief AG

Between 1995 and 2000, Sundlink Contractors, a joint venture consisting of Skanska AB, Hochtief AG, Holgaard and Schults A/S and Monberg and Thorsen A/S, designed and built a major transportation project linking Copenhagen, Denmark and Malmo, in South-western Sweden. The Öresund project included a high cable-stayed bridge, east and west approach bridges, an artificial island and a tunnel, a total distance of 15.8 km. The project location is shown in Figure 2.9, while the various elements are depicted in Figure 2.10. This was a project with impressive technical features: 2000 tonnes of cables, 82 000 tonnes of structural steel, 320 000 m^3 of concrete and 7.8 km total bridge length, including a 490 m span with 55 m navigation clearance. Likewise, the construction methodology involved considerable innovation and massive amounts of logistical planning and coordination. For example, 49 steel girders and associated concrete decks were produced at Cadiz, Spain and transported to the site by giant barges towed by tugboats, with two girders per barge load; a production facility in

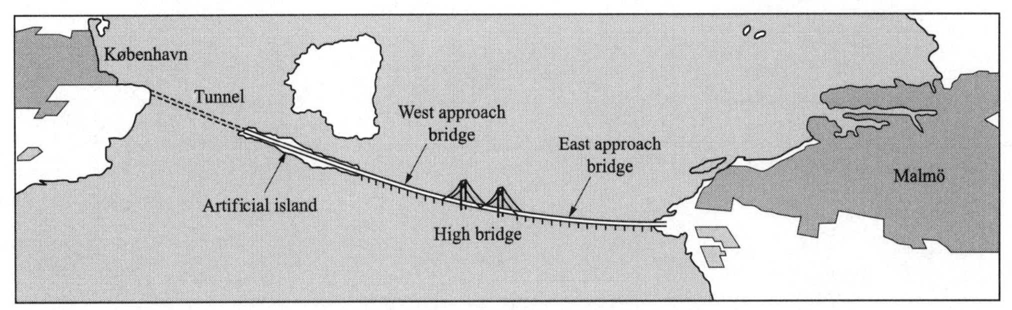

Figure 2.10 Öresund bridge project elements. Courtesy of Hochtief AG, Essen, Germany.

Malmo Harbour pre-fabricated the massive piers and caissons; and a floating crane, Svanen, with an 8700 tonne load capacity, transported bridge components and lifted them into place.

Equally notable was the pioneering approach to the project's management organisation. The same joint venture partners had been involved in Denmark's Great Belt East Bridge project in 1991–1995. A traditional top-down management structure was utilised and large claims, numerous non-conformances, a lack of trust and a tense relationship between builder and client resulted. Intent on improving on their previous performance, the partners felt that this new US$ 1 billion challenge required a new approach to project organisation and a more cooperative, trustful client–contractor relationship. They chose a decentralised and transparent organisation structure whose primary goal was to allow the client, Øresundkonsortiet, owned jointly by the governments of Denmark and Sweden, to interact directly with all layers of the contractor's organisation, all the way to the forepersons on the jobsite (Sundlink Contractors, 1996).

The primary parties involved in the project are shown in the organisation chart in Figure 2.11. Some of the features of this new form of organisation are as follows.

- The project management team assumes an advisory role and develops strategies and transparent guidelines for the different project teams (for example, Offshore Department, Site Support and Logistics and so on) together with an inter-disciplinary task force.
- Each project team sets its own goals, schedules and budgets and bears full responsibility for them.
- The on-site work force of a project team has a direct input in setting up self-defined targets, working schedules, bonus payments and so on.
- The project teams are decentralised operational units that are not supervised by the project management (as long as their self-controlling mechanisms function well).
- The success of each project team is monitored and assessed quarterly by an interdisciplinary committee. The achievements are compared to the previously self-defined goals and forecasts of the project teams. The client is fully integrated in this process and can give recommendations directly to department managers.
- Service groups comprising contract administration, technical, quality assurance and development, accounting and finance groups serve as central specialist units that can be addressed and consulted any time by any team or individual. There are no

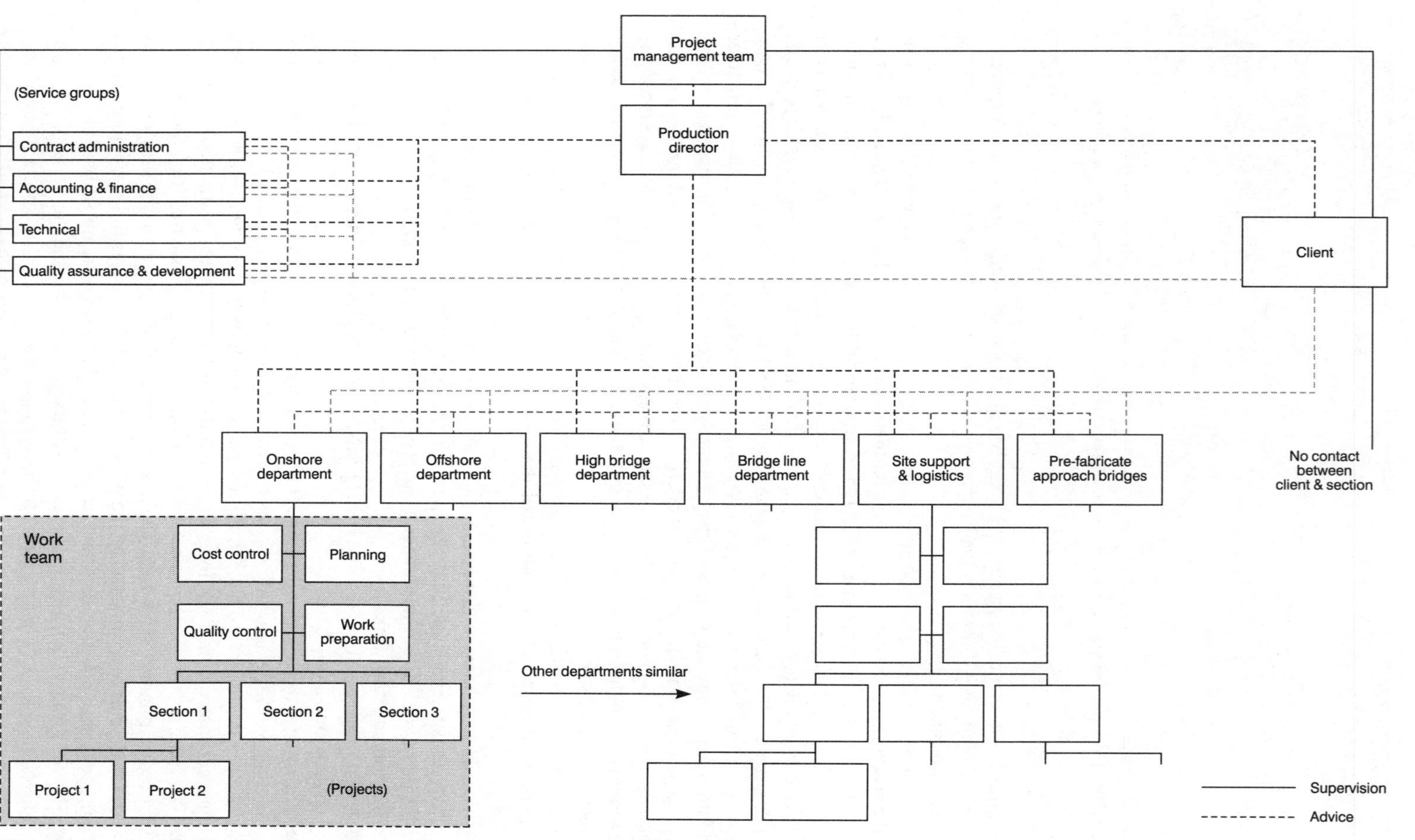

Figure 2.11 Öresund bridge project organisation chart. Courtesy of Hochtief AG, Essen, Germany.

Figure 2.12 Öresund bridge under construction, November 1999. Looking east toward Malmo, Sweden. Courtesy of Hochtief AG, Essen, Germany.

hierarchical dependencies between the service groups and the project teams. Their recommendations are not binding.

- The work team, down to the individual, is committed to report frankly all irregularities or quality deficiencies to the project management team and to the client. In case a work team does not follow this policy, either the work team leader is dismissed or the entire work team is restructured and placed under close surveillance.
- Subcontractors are encouraged to join this self-controlled, frank and interactive construction management system.
- The client commits to refrain from allocating blame and helps to overcome deficiencies and solves encountered problems together with the project teams.
- All trust-building measures between contractor and client are strongly encouraged by both sides.

'All's well that end's well' and this project ended well. The implementation of the system described above was fully successful and will be used as a model for future large-scale construction projects. The project was executed below the projected budget and finished ahead of the time schedule. No claims were filed and the relationship between client and contractor was superb. Figure 2.12 shows an overview of the project's status in November 1999.

Reference

Sundlink Contractors. 1996. *Project Operation System Operation Manual*, 2nd edn. The Öresund Link Contract No. 3 – Bridges. Document No. 01-F.6302-AF0006.2.

3

Planning and design phase

Introduction

In this chapter we describe the significant activities that take place in the project development process prior to the selection of the organisations that will assemble the various project elements in the field. After describing many of the parties involved in the planning and design phase, we divide this phase into three stages. In the first, or planning and feasibility study stage, the various parties define and clarify the project's purpose and scope, conduct feasibility studies, select and acquire land and investigate site conditions and consider options, in a preliminary way, for how the work might be assembled. In the second, or design stage, schematic design lays out the relationships among the project elements, while design development provides the detailed calculations and drawings that specify the sizes and locations of the structural members, earthworks, mechanical systems and all other parts that make up the project. In the contract document development stage, all of the documents required to enter into a construction contract are put together, including the detailed working drawings, technical specifications and legal conditions of the contract. When the planning and design phase is complete, the owner or the owner's representative is ready to select the construction organisation.

As we have already described in Chapter 2, many project delivery options are available, some involving various advisors to the owner and different approaches to the sequencing of the design and construction work. Under a design–build contract, for example, a single organisation is responsible for both design and construction and phased construction may be employed, whereby not all of the design work is completed before construction is begun. This organisation may also be involved in the planning and feasibility study stage. For most projects, the stages described in this chapter must all be carried out, although not always in the strict order presented here. For simplicity, the order given here assumes that all design will be completed prior to contractor selection.

The parties and their roles

Architect

The term 'design professional' is often used to refer to the architect and engineer as they perform their planning, design and construction liaison tasks on a construction project. It is common to use the words architect-engineer, architectural-engineering firm or A/E, for the party engaged carrying out these tasks. The role played by each of these two professionals varies significantly depending upon the type of project. If it is a building or other facility project, the architect is likely to lead the planning and design team; indeed, the owner wishing to construct a new church, residence, school, hotel or apartment building will often first contact an architect about launching the project. On the other hand, if the project involves building a highway, bridge, dam, pipeline or industrial plant, the architect will have a very small role, if any. Professional licensing regulations may stipulate the particular profession responsible for carrying the lead role in certain types of projects.

On a facilities project, the architect takes the lead responsibility for developing the owner's programme, first into a generic or conceptual design, then into more detailed floor plans and finally into a complete set of drawings and other documents sufficient to have the project assembled in the field. The architect is concerned with building space use, appearance, relationships among users and spaces and finishes, as well as the overall coordination of all parties to the planning and design process. In addition, the architect will probably be in charge of the process to select the contractor and may be involved, during construction, in quality control inspections and other activities on behalf of the owner. When the engineer is in charge of a highway, industrial, heavy or utility project, the architect's role, if any, will generally be confined to the visual appearance and finish details of any buildings and other structures. The Royal Incorporation of Architects of Scotland provides a helpful client's advice document at http://www.rias.org.uk (Royal Incorporation of Architects in Scotland, 2001).

Engineer

For non-facility projects, the overall role of the engineer is much the same as that provided by the architect on facilities projects. The engineer takes charge of the planning and design phase, coordinating the various parties and their activities, often including the contractor selection and project inspection tasks. Depending on the project, the engineer (really a number of specialties) may be involved with utilities studies, structural design, heating, ventilating and air conditioning (HVAC) systems, data communications systems, roadway layout, pavement design and earthwork and foundation systems. For a facilities project, the engineer typically fills a supporting role, providing planning and design services in geotechnical, structural, mechanical and electrical aspects of the project.

Geotechnical specialist

Soil and foundation work is integral to most projects, whether facility projects led by architects or other projects coordinated by engineers. Often a specialist will be involved, especially early in the planning and design phase, as a subconsultant to the design professional. Note the position of subconsultants in our various relationship charts in Chapter 2. If a building is being planned, the geotechnical specialist might be engaged to advise on dealing with earthquake susceptibility, soft sediments, alternative foundation solutions, unstable slopes and flooding potential. If the design professional is planning and designing a roadway, geotechnical expertise might contribute studies of environmental impacts, frost-heave potential, erosion control, embankment and base course material source evaluation and slope stability.

Other specialists

The nature of the project obviously determines the other specialists that might be engaged during the planning and design stage. In the case of an educational facility, an educational specialist or facilitator might be engaged to assist in developing the programme of the project. This multifaceted task includes gathering information from various user groups, seeking means to resolve any conflicts and drafting a programme that clarifies expectations and sets forth the basis upon which the architectural staff can plan the various spaces and their relationships. Planning and design of court houses and airports usually requires special expertise from security consultants. If an auditorium were part of a project, an acoustical consultant might be engaged. Specialists in kitchen planning and design, library layout or laboratory design are other examples of the kinds of temporary experts that might assist on building projects. In the case of an industrial manufacturing plant, a specialist in process layout and work flow might join the team. Traffic consultants would likely assist in the planning of a new highway. Hydrologic specialists may work with geotechnical engineers to evaluate flood risks. A complete list of these other consultants is impossible and unnecessary; those given above indicate the wide range of specialties involved in typical projects. Rarely would these experts be part of the permanent staff in the design professional's office. Thus, they join the project team temporarily as subconsultants.

Land surveyor

As various sites are being investigated and especially after one has been selected, land surveyors may be engaged. For building projects, this work includes investigation of property records, easements and rights of way, establishment or re-establishment of property corners and boundaries, determination of land contours and slopes and location of existing improvements, natural features and obstructions. A pipeline or highway project can involve much preliminary route layout as various alternative routes are investigated. This effort includes on-the-ground, aerial/photogrammetric and satellite-based global positioning system methods that lead to route layouts, various kinds of maps and geographic information system databases for use in project planning. Surveyors may also be involved in the construction phase, but they are equally important as part of the planning and design team.

Cost estimator

If funding is limited (which it usually is!), the owner will desire an estimate of the final cost beginning early in the project's life. Of course, until final construction contracts are arranged, this estimate can only be an approximation. The cost estimator is responsible for compiling this information. As we shall see in Chapter 4, cost estimates can be prepared at various levels of detail. In the planning stage, little information is available about the physical properties of the project elements and thus the estimate of cost has little detail. As more design is developed, the cost estimator provides more detailed cost numbers.

The degree of accuracy improves as more detail is developed. Early in the planning process, when little more than the total project area is known (and that perhaps only preliminarily), the cost estimate is probably accurate to within ±30%. Part-way through the design process, cost estimates ought to be accurate to within ±20%; when final construction documents are ready for the tender process, an estimate of accuracy of ±10% can be expected. Although we imply that the cost estimator is a separate member of the team, often the architectural or engineering designer performs this task in association with the various design activities; the term *engineer's estimate* is often used for that estimate prepared at the completion of design development.

Quantity surveyor

The quantity surveyor prepares schedules of quantities of the various project elements. In some cases, this work is performed in conjunction with that of the cost estimator. In many places in the world, a schedule of quantities is prepared for most types of construction contracts, including lump sum and unit price; in other places the quantity surveyor prepares schedules of quantities for unit-price contracts, while the quantity take-off work for lump-sum contracts is left to individual tenderers. We shall consider this matter again later in this chapter during our discussion of construction document development. Quantity surveys are conducted at various points during the planning and design phase, not just during construction document development. Depending on the stage at which a quantity survey is performed, the quantities might include such detailed information as the number of cubic metres of a certain type of concrete and the number of linear metres of a certain size of communication cable; with less detail available, they might indicate the areas of different types of interior walls or even simply the volume or area of a facility devoted to certain types of use.

Project manager

In Chapter 2, we described several types of project delivery systems that utilised parties specifically vested to act on behalf of the owner in managing the project. When the project manager form is used, this person and their staff become integral members of the team. It is likely that the project manager will be engaged early in the planning process and will serve throughout the project life cycle. Sometimes the party carrying this role is designated as programme manager, the distinction apparently being that the latter has even broader responsibilities. A project manager might be engaged somewhat later in the process, after some of the preliminary planning and feasibility analysis is completed, especially if those early efforts were conducted within the owner's organisation. The programme manager will likely act on the owner's behalf for the *entire* project and may serve in this role on multiple simultaneous projects. We should also note that if a construction manager is part of the project team, it is important to add that expertise early in the planning and design phase.

A note on partnering

In a later section, we shall describe an oft-used technique under which the various parties involved in building a project pledge to cooperate as fully as possible, beyond that required as part of the formal construction contract. This 'partnering' philosophy and process lead to a project partnering charter – a 'moral contract' of sorts – in which expectations, understandings and promises are set forth. There is considerable evidence that this structured and formalised approach tends to reduce the incidence of conflicts and disputes, leads to greater project success in terms of schedule, cost and quality and results in more satisfied owners. Usually the partnering process is initiated after the construction contractor is selected.

As we consider the involvement of the parties involved in the planning and design phase, the notion of partnering ought to be emphasised. Even though there is not likely to be a formal partnering agreement at this point in the life cycle, the parties can contribute to project success by adopting a philosophy that emphasises cooperation, clear and frequent communication, active efforts to control costs and avoid delays and rapid resolution of problems and disputes. The owner and all other parties must be willing to trust each other and accept their proportionate share of risks.

Planning and feasibility study stage

Two key things should be accomplished early in the planning and feasibility study stage. First, there must be a clear understanding of the project's objectives, purposes, scope and nature by both the client/owner and organisation responsible for carrying out the work (or at least those members of the project team identified by that time); a brief or other defining document is essential to this process. Second, a relationship between the client/owner and the project delivery organisation must be established, with clearly defined roles and responsibilities.

Consultant selection

Among the first tasks to face the owner when initiating the project will be to select the primary design professional. The selection procedure outlined below can also apply to the selection of subconsultants and others on the project team, although often in a somewhat modified and simplified form.

In order to obtain proposals from prospective consultants, the owner will sometimes begin by soliciting preliminary expressions of interest and qualification. Such solicitations may be made available in the local newspaper, in professional publications and newsletters and by direct mailings to previously engaged firms. They will describe the proposed project in a general way and invite responses by a fixed deadline. Response may require completion of a questionnaire outlining the consultant's qualifications. The Fédération Internationale des Ingénieurs-Conseils (FIDIC) provides a consulting firm registration form (Hicks and Mueller, 1996) that can be filed with the Fédération or used to respond to requests for expressions of interest. This form may be useful to owners as they prepare their own response requirements. For federal government projects in the USA, two questionnaire forms are required (SF 254 and SF 255), one containing general information and the other specifically related to the proposed project. These forms are excellent guides to assist consultants in defining their firms' capabilities and describing the company's background and the expertise of its personnel (Hicks and Mueller, 1996).

After the prospective consultants express their interest, the owner selects the best suited for further consideration. These three to five firms are invited to prepare proposals. These requests for proposals contain at the least the following (Fédération Internationale des Ingénieurs-Conseils, 2001a): statement of work, areas of expertise required, time schedule, type of contract proposed, project budget, submittal date, information to be included in the proposal and expected selection date. Preparation of effective proposals is essential to the long-term health of the consulting organisation; many hundreds of hours can be expended on a single proposal. To be fair to all parties, the owner should inform proposers of the criteria that will be used to evaluate proposals.

Most design professionals believe strongly that selection should be based on qualifications rather than price (Bachner, 1991). *Qualifications-based selection* (QBS) uses criteria other than price to select the consultant, with the fee decided after selection; in this case, the scope of work and project definition are usually left somewhat undefined until after the consultant is selected. *Fee-bidding* is the term used to refer to the process in which design professionals are required to submit a proposed fee with their proposals; in this case, the owner must have prepared a clear project definition and scope statement in order for the consultant to arrive at a fee proposal. While it is recognised that all parties must eventually have an interest in the cost of design services, the fee is proposed and decided earlier in the fee bidding process than under QBS. In the USA, the federal government relies exclusively on QBS for selection of design professionals, based on legislation passed in 1972 known as the Brooks Law; most US states and many other countries use the approach for both public and private construction projects. An engineer employed by New Zealand's Opus International Consultants reported that tenders for a certain project were evaluated by the 'Brooks Law method', using methodology, management skills, technical skills, track record and relevant experience as the criteria.

Assuming a QBS procedure, the proposals would include such elements as the following (based in part on Fédération Internationale des Ingénieurs-Conseils, 2001a):

- past experience with projects of a similar nature;
- details of organisation, project control and financial control;
- size, expertise and responsibilities of staff, especially the qualifications of the proposed project design manager;
- type of organisation and managerial method for executing the work;
- quality assurance organisation;
- knowledge of local conditions;
- local resources;
- project methodology;
- availability of resources;
- schedule;
- an indication of how well the proposer understands the project.

Firms prepare their proposals using these or similar elements as their outlines, knowing that the evaluation will be based on those criteria. The proposals are submitted and evaluated, usually by a committee from the owner's organisation, sometimes with the help of the programme or project manager, if one has been selected by that time. Each member of the assessment committee performs an independent evaluation; a form similar to that in Table 3.1 might be used by each member, after which the results are tallied and the top candidates identified. If Table 3.1 represents the composite rating by the committee, there is a fair indication that Apex Consultants and Bergen Associates, and perhaps Design Ltd, should be dropped from further consideration. The top two or three candidates then become the short-listed firms and interviews are held with each. Following the interviews, a top-ranked firm is identified (not necessarily Engineering International, Inc., depending upon the interviews) and negotiations are undertaken to try to reach an agreement for services. During these negotiations, the issue of an appropriate fee is considered. If these negotiations fail to reach an agreement, negotiations are held with the next-ranked firm, and so forth, until a suitable understanding is reached.

Table 3.1 Consultant selection matrix

Criteria	**Weighting %**	**Firm**				
		Apex Consultants	**Bergen Associates**	**Colp Engineering**	**Design Ltd**	**Engineering International, Inc.**
Past experience with projects of a similar nature	10	5	3	8	8	9
Details of organisation, project control, financial control	8	6	6	9	7	7
Size, expertise and responsibilities of staff	20	4	5	7	6	8
Type of organisation and managerial method for executing the work	15	6	6	8	6	8
Quality assurance organisation	5	5	7	6	7	7
Knowledge of local conditions	7	8	3	6	10	9
Local resources	5	8	2	7	9	8
Project methodology	20	3	6	7	8	8
Availability of resources	10	4	5	8	6	7
Weighted score	100	4.89	5.04	7.39	7.16	7.94

Rate each element between 0 and 10 (10 = high).

Sometimes owners require candidate firms to furnish a proposed fee along with their qualifications. At least two approaches are possible for this fee-bidding procedure mentioned earlier. Under the single-envelope system, proposers furnish all information, including the proposed fee, in a single package. The proposed fee then becomes an additional criterion to be evaluated, with its relevant weighting. As the term implies, the double-envelope system requires proposals to be separated into a package for the technical proposal and another for the fee. Under this system, the technical proposals are usually evaluated first, after which negotiations with the highest-ranked firm are held. During these discussions, that firm's fee envelope is opened and the fee becomes a part of the negotiations. If the parties reach an agreement on all matters, all other second envelopes are returned unopened to their respective firms.

For a very large, complicated or otherwise significant project, a design competition is sometimes held. Top-ranked firms are invited to furnish proposed solutions to, say, an intricate harbour crossing, to a specified level of detail. Evaluation of these solutions becomes another step in the selection process. Owners must weigh the advantages of potentially innovative design solutions against the expense and time required to conduct the competition.

Many variants of the approach described above are possible, but the steps outlined form the basis for most consultant selection procedures, especially when public funds are involved. Private owners often use less formal methods and base their selections on friendship, loyalty and previous experience, as well as cost.

Like construction contracts, payments for professional design services can be determined in several ways. Clough and Sears (1994) list the following:

- percentage of construction cost
- multiple of salary cost
- multiple of salary cost plus non-salary expense
- fixed lump-sum fee
- total expense plus professional fee
- hourly or per diem charge.

It is common practice to pay the designer in instalments as stages of the services are completed. Just as in the case of construction contracts, a number of standard design consultant agreement forms are available from organisations such as FIDIC (Fédération Internationale des Ingénieurs-Conseils, 2001b) and the American Institute of Architects (1997a).

The brief

The *brief* is a statement that specifies the scope of the project. It defines the objectives to be achieved and lays out in a general way what the final product will accomplish. The brief can be prepared by the owner as a means of clarifying the need for the project, even before the project manager or design professional are engaged; if used in this way, it will be the basis upon which those prospective consultants prepare their proposals. On the other hand, it can also be prepared with the help of the project manager or design professional after they are engaged. These consultants, with experience on similar projects, can assist the owner in identifying and clarifying needs and setting forth the project's scope. A well-prepared brief, consisting of only a few pages, is essential in forming the basis for all that follows throughout the project life cycle. The development of the programme will logically follow from this initial defining statement.

Programme development

A construction project is built to achieve certain objectives, as defined in the brief. Those objectives will be achieved if the elements of the programme are satisfied. Thus, after the brief, a more comprehensive statement of those elements must be developed, elements that will be translated into the physical aspects of the completed project. It is likely that several parties will be involved in programme development. Consider a project to construct a new elementary school. Those contributing to the *programme* might well include teachers, parents, local and district administrators, community members including near neighbours, project manager and design professional. For this kind of project, a facilitator may be engaged to 'ask the right questions' and elicit ideas that will be considered as part of the programme. Among hundreds of questions that might be considered are the following (Bennett, personal communication):

- How will this project contribute to the community's educational goals of excellence, diversity, individuals' rights and responsibilities and relationship to the ecosystem?
- What use will the wider community make of this facility?
- What is the facility expected to provide for people with special needs?
- Are there priorities for space allocation among various programme areas: communication, science, arts, gifted/talented, recreation and others?
- Will this facility require special storage areas, utility spaces, administrative offices and counselling rooms?
- If potential sites have been identified, are there concerns among various interest groups about traffic flow, green space, bicycle and ski trails and visual impact?
- To what extent should this school provide for anticipated changes in educational methods, including introduction of various technologies?

The result of this process will be a written document, subject perhaps to approval by the Board of Education or similar authority, describing in words what the completed project is expected to accomplish, including educational goals, community use, land use impacts and neighbourhood harmony.

Educational facilities are perhaps extreme examples of projects subject to community involvement. At the other extreme, a family-owned manufacturing facility may incorporate the ideas of one or a few individuals. Even here, a process that includes first a brief and then a programme will provide the foundation upon which the design professional can begin to identify various alternative means of meeting the project's objectives. The modern trend is toward increasing involvement of the public in all types of projects, both public and private, especially in matters of physical, sociological and environmental impact.

Identification of alternatives

There may be several feasible ways to achieve the objectives set forth in the programme. If a bridge is required to carry highway traffic and utility conduits across a watercourse, what will be its location and alignment? Will it be a suspension bridge, a cable stayed bridge or a plate girder bridge? Will the structural materials be primarily steel, concrete, timber or a composite? Will it consist of a single or double deck?

Our elementary school will need a site and several may be under consideration. For each, there are probably several possible layouts, or 'footprints', the school can occupy on the site. There will be alternative exterior access and traffic flow patterns, utility access and playground locations. Inside the building (or, perhaps, building*s*, as an alternative), will there be a single or multiple storeys? What room layouts are feasible that can lead to pleasing use of exterior light, satisfactory pedestrian flow, energy efficiency and successful interactions among the various educational components? Quickly the various combinations become numerous and the design professional's task is to identify all the possibilities and then select the most feasible alternatives for further study.

For each selected alternative, the planning process proceeds through several steps. For a facility project such as our proposed elementary school, the architect will begin to develop general concepts, perhaps using proximity or bubble diagrams to show relationships among the facility's functions, leading to preliminary floor plans and renderings. The engineer will support this process with general investigations of utilities, structural components, sitework and communication systems. For a highway or bridge project, the engineer will lead the effort to develop preliminary layouts of each alternative, their alignments and general shape. The result of this part of the planning process will be a collection of preliminary drawings and descriptions for alternatives whose costs can be estimated, whose general timelines can be forecast and whose construction techniques can be studied, with the goal of identifying financially and technically feasible options, if any, and then selecting the preferred option.

A master plan is a document that lays outs the various physical elements of the project in their proposed locations on the project site together with general descriptions of the project features. The identified alternatives provide a structured means of studying the options available for supplying the elements in the master plan in order to meet the project's objectives.

It should be noted that the contract with the design professional is often subdivided and agreed to in phases, in order to define the scope and services and attendant fees more accurately. For example, a contract with a defined scope and cost may be issued for the development of a master plan in order to examine alternatives and elicit community input. Then, once the scope of the project is well defined, a second contract can be written that allows better clarification and control of cost and outcome.

Site investigation

In parallel with other preliminary planning, investigation of the site is conducted at varying levels of detail. When several alternatives are still under consideration, the various potential sites will be studied from the standpoint of general soil conditions, topography, access and cost. As the options are narrowed, further detail will be developed on the preferred option or options. The geotechnical specialist will provide major input at this step. Soil conditions that influence foundation types must be identified. Environmental aspects of the site, including potential underground contaminants, wetland issues and endangered species must be studied. Site investigation also includes options for preserving existing vegetation and existing improvements, if any, various ways for accessing the site and interfacing with off-site traffic flow, flow of people and vehicles on site, availability of utilities and various security matters. In addition, property surveys may be needed to establish corners and boundaries and investigation into ownership records may be required to establish who owns title to the land and in what form. Additional investigation may be needed later; at this planning phase, the site studies are oriented toward providing sufficient information to assist with decisions about feasibility and selection.

Constructability analysis

Throughout planning and design, it is essential to consider whether proposed alternatives can be built and whether they can be built efficiently. The term 'constructability' is used for this evaluation, which is a continuing process, perhaps more active during the design stage to be described later in this chapter. DeWitt (1999) describes constructability analysis (also called constructability review) as

> . . . a process that utilizes experienced construction personnel with extensive construction knowledge early in the design stages of projects to ensure that the projects are buildable, while also being cost effective, bidable, and maintainable.

During the planning and feasibility study stage, even though the development to that point is mainly conceptual, each alternative will be studied for such things as ease of construction, impact on the project schedule, effects that different materials might have on procurement and installation, safety considerations and various coordination issues among personnel, equipment and materials. The design professional may perform this analysis with its own forces, or it may engage a consultant with special knowledge of construction procedures. If a construction manager is a part of the project team, constructability will be a major responsibility. We shall have occasion to consider constructability again in our discussion of the design stage.

Public input

We have already suggested that members of the general public may be involved in the development of the project programme, prior to the identification and study of alternative approaches to meeting the programme objectives. The public will likely have other opportunities as well to be part of the project. At various points in the planning and design process, public hearings and workshops may be held. An invitation for the public to attend such a meeting is reproduced as Figure 3.1. Although this project is small by most standards, the process is representative, as the Alaska Department of Transportation and Public Facilities solicited comments on various aspects of this highway pullout or 'wayside' project during the planning phase. Although hearings are more likely to be held for publicly funded projects such as this one, legal regulations may stipulate that the public be invited to comment on issues such as land use and environmental impact for any project, whether public or private. The increasing use of the Internet makes this technology a convenient means for gathering public comments. In a later section we shall describe project websites and their use not only to enhance communications among members of the project team but also for providing information about the project to the general public and obtaining public feedback.

Code analysis

Building codes, fire codes, plumbing codes, electrical codes and a number of other codes and regulations may apply to the project being planned. During the planning stage, it is important that all alternatives under consideration can be made to comply with these documents. Thus, a code analysis will be required. Are there restrictions on the number of storeys the building can contain? What earthquake loadings apply to our bridge design and can the materials under consideration be used in this earthquake zone? How many elevators and stairways are required in our elementary school? Will a sprinkler system be required for fire suppression? Do zoning regulations allow an oil pipeline in this neighbourhood? What lot setback requirements are in effect for a manufacturing facility in this area? The answers to questions such as these are available in codified form and through discussions with enforcement personnel if clarification is needed. The answers will impact on the various alternatives as they emerge from the planning process and will thus influence the estimated costs of each.

Preliminary cost estimate

In order to determine whether an alternative under consideration is financially feasible, an estimate of its cost is needed. During the planning process, such an estimate cannot be made

Online Public Notice

State of Alaska Online

Home Page | Go Back

Submitted by	Date Modified	Ak Admin Journal	Attachments	Public (Web edit)
Pamela Lord/NR/DOTPF on 10/31/2001 at 11:40 AM	11/14/2001 12:57:19 PM	[not printed]	No files attached	

Public Workshop: Birch Lake Wayside Project

Category: Public Notices

Department: Transportation & Public Facilities

Publish Date: 10/31/2001

Location: Fairbanks
Region: Northern

Body of Notice:
BIRCH LAKE WAYSIDE PROJECT
PUBLIC WORKSHOP

The Alaska Department of Transportation and Public Facilities (ADOT&PF) plans to upgrade the existing turnout at Birch Lake, Mile 306 of the Richardson Highway. The project entails: installing trash receptacles, picnic tables, vaulted toilets, informational signs, and landscaping features. Improvements will affect public access to the lake. We will be holding a public workshop at:

Noel Wien Library
1215 Cowles Street
Thursday, November 15, 2001
4:00 to 7:00 p.m.

Interested individuals are welcome to come and meet with ADOT&PF personnel involved in this project.

For more information about this project, please contact:

Stephen D. Henry, P.E.
Design Engineering Manager
Department of Transportation and Public Facilities
2301 Peger Road
Fairbanks, Alaska 99709-5316
(907) 451-2230
Email: steve_henry@dot.state.ak.us

Figure 3.1 Public hearing notice for a project in the planning stage.

Table 3.2 Sample preliminary cost estimate for high-rise hotel with steel frame, precast concrete slabs and walls

Item number	Item	Quantity	Unit	Unit cost (US$)	Extension (US$)
1	Hotel floor space	8050	m^2	1550	12 477 500
2	Furnishings	230	Room	2300	529 000
3	Swimming pool	1	Lump sum	105 000	105 000
4	Parking	150	Vehicle	1100	165 000
5	Contingency	10%			1 327 650
6	Overhead and profit	20%			2 920 830
	Total				17 524 980

with a large degree of precision, as we indicated in our discussion of the role of the cost estimator. Usually the estimator uses some broad measures of cost to develop the estimate. For example, the cost of a multi-storey concrete parking structure may be estimated based on the number of vehicle spaces multiplied by a historic construction cost per vehicle space for the locality; the costs of any extraordinary features such as a toll station or extra-long approach ramps would be added to the estimate. In the case of a single-storey apartment complex built of timber, an estimate based the cost per square metre of floor area might be prepared. Extras such as a swimming pool, parking areas and public common rooms would be estimated separately. Sources of this per-vehicle space or per-square metre cost information include the planner's own records of historic data from previous projects and various published databases, either in book or electronic form. Example cost data sources are *Rawlinson's Australian Construction Handbook* (2001) in Australia and New Zealand and *Means Square Foot Cost Data 2002 Book* (2002) in the USA. Table 3.2 shows a sample preliminary cost estimate for one alternative of a hypothetical proposed hotel project.

Financial feasibility analysis

Not only must a project meet certain technical and performance requirements, it must also be financially viable. The funds proposed to be invested in the project must show the potential to generate an economic return to those investing in the project that is at least equal to that available to them from other similarly risky investments. This concept of 'return on investment' is appropriate for most projects, whether in the private or public sector. For a private-sector manufacturing facility, the company making the investment of funds expects to generate sufficient cash flows from operating the facility to pay for the construction and the ongoing operating expenses and, in addition, have an attractive interest rate of return. In the case of a publicly funded roadway or school facility, those ratepayers who provide the funding expect that the benefits, either in money terms or in non-quantifiable measures, will be at least equal to the funds invested in the project. Admittedly, the benefits derived from a school building project are more difficult to estimate than are its costs. However, some consideration must be given to whether the benefits, both financial and otherwise (the 'otherwise' being more important in the case of a school!) are at least equal to the monetary costs.

All cash flows throughout the project's development and operation life cycle must be considered in making such analyses. The term *life cycle costing* refers to economic studies that include all such cash flows. The cost of maintaining and periodically upgrading a highway and bridge project must be estimated, along with its design and construction costs, to determine whether it is likely to be feasible. Estimates of the cost of operating and maintaining a manufacturing plant, as well as the income its operation is expected to generate, will be essential in determining financial viability.

The cost estimates over the life cycle for each alternative form the basis for these feasibility studies. Typically, some proposed options will be discarded at this stage because they are not financially viable. Those remaining will be compared with each other, considering both financial and other factors, to arrive at a recommended option.

Project recommendation

Having studied the various options from the standpoint of costs, economic benefits, ease of construction, schedule impacts, as well as the all-important matter of alignment with user objectives and the project programme, the design professional is then in a position to recommend an option. Sometimes the recommendation will be not to proceed; 'do nothing' is always an option! Whatever the recommendation, the design professional will prepare a report for the owner describing the planning process, identifying the alternatives and explaining the rationale for the selected option. If the recommendation is to proceed, the report will suggest the appropriate steps to be taken to launch the design stage. The choice of whether to implement the recommendation is the owner's, of course.

Funding

The point in time at which funding approval is sought for a project varies with the nature of the project. Sometimes the money has been allocated before much of the planning process has been carried out. Occasionally a complete design is prepared, ready for contractor selection, prior to availability of funds. More often, however, the process outlined above leads to a recommendation that the project proceed and that it be funded. In this case, before further design efforts are undertaken, assurance of available funding is sought. A municipality may need to seek voter approval for a sewage treatment facility whose preliminary planning has been completed. A Hospital Foundation Board may be asked to approve an extension to the hospital at this stage. A University Board of Trustees may give its approval for the expenditure of funds. The Board of Directors of a corporation may likewise become involved in funding approval for a new processing plant at this point in the process. Later approvals may also be required, such as approval of the selection of the prime construction contractor for a certain fixed price, but the approvals that follow the planning stage set funds for the project aside, with the expectation that the work will be completed if it can be done in accordance with the plans developed thus far.

Site selection and land acquisition

Until funding is approved, there is no need to acquire land for the project (assuming the site is not already in possession of the owner). As explained earlier, various sites will likely have been investigated, each with alternative layouts specific to the site. In our site investigation section, we listed such considerations as soil conditions, topography, access and environmental impacts as important site characteristics. Note that 'site' will have different meanings depending on whether the project is a building, a roadway, an airport or some other development. In roadway construction, for example, right-of-way specialists are important members of the project team.

When funding is approved, it may be for a project at a specific site or the approval may allow for alternative sites, depending upon further investigation and negotiation. In any case, ultimately the project site, if not already owned, must be acquired, through purchase, lease or other arrangement. Sometimes public projects utilise the process of eminent domain by which private land may be acquired even from reluctant sellers under certain limited conditions.

The site acquisition step completes the usual process of planning and feasibility study and allows the design stage to begin. For a typical project, approximately 10% of the planning and design effort has been completed at the end of the planning and feasibility study stage.

Design stage

Introduction

As the discussion in the previous sections implies, the planning stage usually involves considerable back-and-forth deliberation of several alternatives, as options are modified and refined, in an attempt to find the 'best' solution to the stated programme objectives. Feedback is an important part of this process, as the various parties evaluate the alternatives, suggest changes and reach tentative decisions. Although some of this methodology continues into the design stage, for the most part the basic design decisions have been made and the task is now to refine and amplify the programme in a series of steps that result in documents that will be used for the construction process. The two parts of the design stage are *schematic design* and *design development*, after which the *contract document* (or construction document) development stage results in fully completed documents. The process is a continuum, with some overlaps and considerable similarity among the topics and tasks. We begin by describing schematic design.

Schematic design

The design professional is responsible for producing a set of preliminary drawings as well as a written report, for approval by the owner. On a building project, the architect will take the lead in developing a design concept with plan, elevation and sectional views that meet the space and programme requirements. Engineers will develop concepts of how the various systems will fit into the facility: foundation systems, cooling and heating systems and data communication systems. These preliminary drawings may be freehand sketches, but they should illustrate the project's character and emphasise such aspects as harmony with the surrounding area and any improvements, architectural style, exterior appearance, planning and zoning requirements and overall structural concepts. Civil engineers will become involved in site analysis and layout, based in part on legal and topographic surveys prepared earlier, with some updating following site selection; features include topography, soil conditions, parking, access, utilities, setback requirements, water features, drainage, required easements and existing structures. Surveyors will continue their previous efforts with title searches, photogrammetry and on-site surveying and mapping. Special consultants may play a role in schematic design; one example might be an acoustical expert who would begin to advise on layout and basic features of a large auditorium.

As we noted earlier, most decisions regarding alternatives will have been made during the planning stage. Thus, an important characteristic of schematic design is that it emphasises the expansion of a single set of concepts rather than multiple options. One building agency of a US state lists the following schematic design drawings required after approval of the design concept (State of Mississippi, 2000):

- the basic design approach drawn at an agreed-upon scale;
- site location in relationship to the existing environment;
- relationship to master plans;
- circulation;
- organisation of building functions;
- functional/aesthetic aspects of the design concepts under study;
- graphic description of critical details;
- visual and functional relationship;
- compatibility of the surrounding environment.

That same agency requires the design professional to prepare narrative descriptions of the following building systems at the completion of schematic design:

- structure
- foundations
- floor grade and systems
- roof
- exterior/interior walls and partitions
- interior finishes
- sight lines
- stairs and elevators
- specialty items
- mechanical systems
- built-in equipment
- site construction.

A cost estimate will be prepared as a part of the schematic design effort. The State of Mississippi *Planning and Construction Manual* (State of Mississippi, 2000) requires that such estimates be 'based on adjusted square foot or cubic foot cost of similar construction in the area of the project or on a system cost study of the project', with appropriate cost escalation factors.

Figure 3.2 shows a project development schedule for a major engineering laboratory and classroom building at Cornell University in Ithaca, New York, USA (Cornell University, 2001). Of interest are the several steps in the process, including schematic design, and the special-programme features of this project, including laboratory safety concerns, the need to demolish

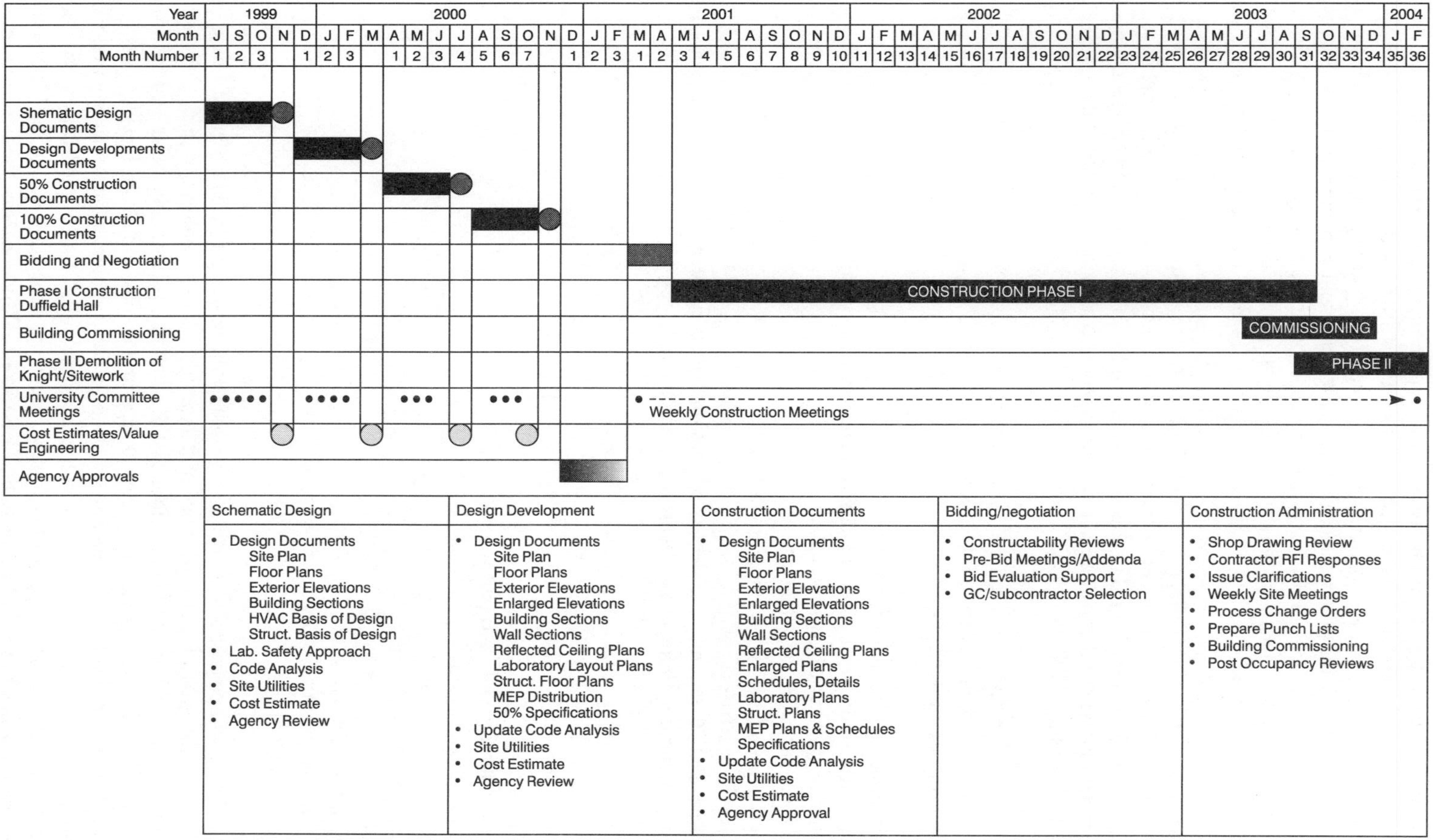

Year	1999					2000												2001												2002												2003												2004	
Month	J	S	O	N	D	J	F	M	A	M	J	J	A	S	O	N	D	J	F	M	A	M	J	J	A	S	O	N	D	J	F	M	A	M	J	J	A	S	O	N	D	J	F	M	A	M	J	J	A	S	O	N	D	J	F
Month Number	1	2	3		1	2	3		1	2	3	4	5	6	7		1	2	3	1	2	3	4	5	6	7	8	9	10	11	12	13	14	15	16	17	18	19	20	21	22	23	24	25	26	27	28	29	30	31	32	33	34	35	36

Shematic Design Documents

Design Developments Documents

50% Construction Documents

100% Construction Documents

Bidding and Negotiation

Phase I Construction Duffield Hall — CONSTRUCTION PHASE I

Building Commissioning — COMMISSIONING

Phase II Demolition of Knight/Sitework — PHASE II

University Committee Meetings — Weekly Construction Meetings

Cost Estimates/Value Engineering

Agency Approvals

Schematic Design	Design Development	Construction Documents	Bidding/negotiation	Construction Administration
• Design Documents Site Plan Floor Plans Exterior Elevations Building Sections HVAC Basis of Design Struct. Basis of Design • Lab. Safety Approach • Code Analysis • Site Utilities • Cost Estimate • Agency Review	• Design Documents Site Plan Floor Plans Exterior Elevations Enlarged Elevations Building Sections Wall Sections Reflected Ceiling Plans Laboratory Layout Plans Struct. Floor Plans MEP Distribution 50% Specifications • Update Code Analysis • Site Utilities • Cost Estimate • Agency Review	• Design Documents Site Plan Floor Plans Exterior Elevations Enlarged Elevations Building Sections Wall Sections Reflected Ceiling Plans Enlarged Plans Schedules, Details Laboratory Plans Struct. Plans MEP Plans & Schedules Specifications • Update Code Analysis • Site Utilities • Cost Estimate • Agency Approval	• Constructability Reviews • Pre-Bid Meetings/Addenda • Bid Evaluation Support • GC/subcontractor Selection	• Shop Drawing Review • Contractor RFI Responses • Issue Clarifications • Weekly Site Meetings • Process Change Orders • Prepare Punch Lists • Building Commissioning • Post Occupancy Reviews

Figure 3.2 Project development schedule for Duffield Hall, Cornell University, with tasks in each development stage.

an existing building as part of Phase II and the involvement of university committees throughout the project's life cycle.

Typically, 35% of the total planning and design effort has been expended upon completion of schematic design.

Design development

Design development activities flow naturally from, and are based on, documents produced during schematic design. In Figure 3.2, note the progression from schematic design to design development. Work continues on the site plan, floor plans, exterior elevations and building sections, with detailed work on wall sections, structural floor plans, reflected ceilings and laboratory layouts. Mechanical, electrical and plumbing distribution design receives major emphasis. Code analysis work is updated and site utility, cost estimate and agency review efforts are expanded and refined. In addition, work on construction specifications is begun.

Fivecat Studio (2001) is so bold as to state the following:

> All of the major decisions for the project will be made during design development. The sketches prepared during schematic design will be thoroughly detailed and developed into a complete set of design drawings. All building materials, fixtures and finishes will be selected . . . When [design development] is complete, the project will be fully developed . . .

By comparison with schematic design, design development is specific and detailed. By the end of this step, most major decisions about the entire project are resolved. The design professional's submittal to the owner will generally consist of a written report and a set of detailed drawings. In addition, a more refined cost estimate will be required, based on preliminary quantities of the various elements of the finished project (cubic metres of concrete, square metres of paving, linear metres of cabling and kilograms of steel), rather than overall area or volume measures as in earlier estimates. The report will include a first try at technical specifications, particularly proposed materials, to be refined later during development of contract documents.

With the completion of design development, approximately 65% of the planning and design effort has been expended and the last step prior to contractor selection, the development of construction contract documents, can begin.

Contract document development stage

Introduction to the form of the contract

During this final stage of the design phase, all of the previous effort is transformed into documents that will form the basis for the construction contract. When this stage is completed, the contractor, or contractors, can be selected, after which the work involved in procuring and assembling the physical parts can begin. The term *contract* can refer either to the formal agreement signed by the parties, usually the owner and the contractor, or to this agreement plus a multitude of other documents that are referenced by the agreement. Thus, when we talk about the *contract documents*, we include three types: (1) the *technical specifications*, which describe the project, its materials and its methods or performance requirements, (2) the so-called *up-front* documents that describe the relationships and responsibilities of the parties (including at least the agreement, general and special (or supplementary) conditions, invitation to tender, instructions to tenderers, form for submitting the tender and bond documents) and (3) the *drawings*, which are scaled and dimensioned graphic depictions of the project features, usually with notes that explain related details. The technical specifications and the up-front documents are typically part of a project manual distributed to prospective tenderers for use in the contractor selection process and to guide the successful tenderer during on-site construction. It is the designer's responsibility to produce all of these documents or to assemble them if they are furnished by the owner, during this, the contract document development stage.

The bases for many of these documents have been developed over many years by various national and international organisations; these so-called 'standard' forms and conditions are used in both private and public construction projects. On the international level, FIDIC, to which we referred briefly in Chapter 2, provides many resources useful for designers preparing contract documents for construction projects. FIDIC was founded in 1913 by the national consulting engineering organisations of Belgium, France and Switzerland (Fédération Internationale des Ingénieurs-Conseils, 2001b); today it consists of the member associations of 67 countries, representing some 560 000 professional engineers. Among the member associations are the Association of Consulting Engineers in the UK, the Consulting Engineers Association of India and Camara Argentina de Consultores. The Federation publishes standard forms of contract documents used in the procurement of building and engineering works. Of special interest here are three sets of documents: (1) *Conditions of Contract for Construction*, used when the owner or the owner's designer provides the design, which is then built by a construction organisation, (2) *Conditions of Contract for Plant and Design–Build* and (3) *Conditions of Contract for EPC/Turnkey Projects* (Fédération Internationale des Ingénieurs-Conseils, 2000). Each set includes general conditions, guidance for preparing particular (special) conditions, example bond security forms, letter of tender forms, a contract agreement and a dispute adjudication agreement, all of which will be discussed in a general way in subsequent sections herein.

In addition to the internationally recognised standard documents promulgated by FIDIC, many countries have developed such contract materials. Standards New Zealand, the trading arm of the New Zealand Standards Council, a crown entity established in 1938, publishes *Conditions of Contract for Building and Civil Engineering Construction* (Standards New Zealand Paerewa Aotearoa, 1998). This publication, developed and updated by a broad-based construction industry committee, is utilised throughout New Zealand for most construction projects, both public and private. It includes general conditions of contract; schedules to the general conditions including suggested special conditions, an agreement, surety bond forms and insurance certificates; conditions related to the tendering process; a series of guidelines for utilising the document; and a general aid to evaluating the costs of changes in the work. Individual owners have the choice of using this document or some other, but in reality it is used almost universally within New Zealand. A quotation from the foreword is instructive (Standards New Zealand Paerewa Aotearoa, 1998):

> A major advantage of using a standard conditions of contract is that its provisions are well known and understood. It does not need to be searched for project specific changes buried within every time a tender is priced or construction is being managed.

Construction industry associations in the USA have produced a large number of standard contract documents. For building construction, the American Institute of Architects (AIA) and the Associated General Contractors (AGC) have been active in this area for many years. While choices of document forms may be important, such efforts can also lead to unnecessary duplication. In recent years, AIA and AGC have collaborated to produce common documents for construction management types of delivery systems. For public works, bridges and power stations in the USA, documents promulgated by the Engineers Joint Contract Documents Committee are often used; this committee was established by several prominent professional engineering and construction societies (Dorsey, 1997). Furthermore, the American Association of State Highway and Transportation Officials (AASHTO) furnishes a standard format for highway construction specifications that is used by most states in the USA (Fisk, 2003).

Instead of using 'standard' documents such as those described above, many construction projects use non-standard documents developed by individual design or owner organisations. For example, a consulting firm might use its own documents for its projects, tailoring them as needed for any particular project. Such materials are sometimes referred to as *master* documents (Robbins, 1996). At the other extreme, the contract documents might be produced anew for an especially complex project with many irregularities; it is unlikely, however, that a completely new document would be invented for even such a project.

Having introduced the sources from which construction contract documents can be obtained and having mentioned the names of many of those documents, we now define and describe the salient features of each of the especially important types.

Drawings

Drawings have already been produced as part of the schematic design and design development efforts. However, the task during the contract document development stage is to provide even more detail, such that the project can be tendered and built with these drawings as the primary guidance. Such documents prepared by the design professional are often called *design drawings* when they are prepared for tendering purposes. The same drawings are then referred to as *contract drawings* after a construction organisation has been engaged, because they become officially part of the contract by reference.

Standard classifications of drawings and prescribed orders in which they appear have evolved over time. In the case of building construction, the following sets are typical (Bennett, 1996):

- general layout and civil works such as roads and parking areas, drainage and landscaping;
- architectural, showing all dimensions and locations of all features in the building;
- structural, with details of all major elements including connections and fastenings;
- mechanical, including plumbing, heating, ventilation, air conditioning and special mechanical equipment;
- electrical, with light fixtures, motors, conduit and cable, instrumentation, communication system components, junction boxes and other such details.

The sheets in each of these five sets are numbered separately and sequentially, with a special capital-letter prefix; for example, architectural drawings might be numbered A1, A2, and so on, while structural drawings would begin with S1. For other types of construction, such as highway or utilities, the makeup and format can vary considerably. Computer-aided drafting plays a large role in the preparation of these final drawings. After their production, rolls or booklets of these drawings (also called *plans*) are assembled and provided to prospective contractors along with the other documents described below.

General conditions

We now turn to the written materials that, together with the drawings, completely describe the design professional's intent with respect to the project. Historically, the term *specifications* has referred only to the requirements for the project's technical details. However, as Clough and Sears (1994) point out,

> it has become customary to include the bidding and contract documents together with the technical specifications, the entire aggregation being variously referred to as the project manual, project handbook, construction documents book, or most commonly, simply as the specifications or 'specs'.

One essential document in the project manual is the general conditions.

The *general conditions*, sometimes called the *general provisions*, set forth the rights and responsibilities of the owner and contractor and also of the surety bond provider, the authority and responsibility of the design professional and the requirements governing the various parties' business and legal relationships. These conditions are 'general' in the sense that they can apply to any project of a certain type; sometimes they are referred to as the 'boilerplate', based on the somewhat erroneous notion that they are simply attached to the other documents with no further thought. As explained in the introduction to this section, many organisations have produced standard documents and these certainly include general conditions. A review of several such sets indicates that they provide much similarity of subject matter, despite various headings and topics. The New Zealand *Conditions of Contract for Building and Civil Engineering Construction* (Standards New Zealand Paerewa Aotearoa, 1998) include 15 sections, including such topics as the contract, general obligations, engineer's powers and duties, time for completion, payments and disputes. One helpful outline of the numerous provisions in the typical general conditions is provided by Robbins (1996) as follows:

- definitions and abbreviations
- bidding requirements
- contract and subcontract procedure
- scope of the work
- control of the work
- legal and public relations
  - damage claims
  - laws, ordinances and regulations
  - responsibility for work
  - explosives
  - sanitary provisions
  - public safety and convenience
  - accident prevention
  - property damage
  - public utilities
  - abatement of soil erosion, water pollution and air pollution
- prosecution and progress
  - commencement and prosecution of the work
  - time of completion
  - suspension of work

  - unavoidable delays
  - annulment and default of contract
  - liquidated damages
  - extension of time
- measurement and payment
  - measurement of qualities
  - scope of payment
  - change in plans
  - payment
  - termination of contractor responsibility
  - guarantee against defective work
  - dispute resolution.

Note that all of these provisions relate to the relationships among the parties and the legal and management issues attendant to the construction contract but not to the technical matters related to design. These latter issues come later in the technical specifications.

Special conditions

If the general conditions could apply to any project of the type being designed and built, there will surely be some special circumstances associated with the non-technical aspects of this particular project. The *special conditions*, variously known as *special provisions*, *supplementary general conditions* or *particular conditions*, cover these project-specific matters. These sections may either add to or amend provisions in the general conditions. Examples of additions to the general conditions might include requirements for completion date, owner-provided materials and the name and address of the design professional. If the special conditions modify a section of the general conditions, the wording might be 'subparagraph X.XX of general conditions shall be modified as follows:', with a requirement that insurance coverages are different from those in the general conditions, that cleanup requirements are different, that the owner will furnish a different number of contract documents to the contractor or that the owner, rather than the contractor, will procure certain permits and consents. The New Zealand *Conditions of Contract for Building and Civil Engineering Construction* manual (Standards New Zealand Paerewa Aotearoa, 1998) provide, as its first schedule, *Special Conditions of Contract*, with each section numbered to coincide with the relevant general conditions. Among the special conditions are the following:

- name and address of principal;
- type of contract;
- whether or not this is a roading contract;
- number of sets of contract documents supplied free of charge to the contractor;
- contractor's bond requirements;
- principal's bond requirements;
- date when contractor can access the site;
- name and qualification of the engineer;
- insurance requirements;
- completion date information, include allowances for inclement weather;
- producer statement requirements;

- liquidated damages amount;
- bonus payment information;
- defects liability period;
- required guarantees;
- advance payment allowances for materials, plant and temporary works not yet on site;
- progress payment retainages;
- cost fluctuation adjustments.

This schedule acts as a convenient checklist for the design professional to be certain that all such provisions are covered; in actual use, the document can simply be photocopied and marked upon, as there are many instructions to 'delete provisions which do not apply' throughout.

Technical specifications

The *technical specifications* are the portion of the documents the layperson usually thinks of when referring to the specifications, or 'specs'. They contain the detailed technical provisions related to installation or construction of the several portions of the work and the materials incorporated therein. The order and grouping of the technical specifications vary depending on the type of work and the organisation preparing the documents, but most such materials tend to be grouped by the general order of construction. The Construction Specifications Institute (CSI) (Construction Specifications Institute, 2001) and Construction Specifications Canada (Construction Specifications Canada–Devis du Construction Canada, 2001) have developed a format for technical specifications that is especially suited to the construction of buildings, although it has been used in other types of construction, such as heavy construction. The format is based on 16 divisions, each related to a technical specialty such as concrete or electrical. The numbers assigned to each division are always used, even if a project does not use a certain division; for example, CSI Division 15 is always the mechanical specifications. Each division is subdivided into sections, which are defined for the particular project. In Figure 3.3, we show the 16 divisions, with a sampling of the kinds of section topics that might be included in a typical project.

There may be some confusion over the inclusion of general requirements (Division 1) in the technical specifications, because the totality of the documents also includes general conditions and supplementary general conditions. The distinction is that Division 1 general requirements contain *technical* requirements that apply generally to the project and thus to more than one of the divisions. All technical sections in the CSI format are divided into three parts, always in the same order, as follows: (1) general, covering scope, related work, testing and inspection, standards and certification; (2) products, including requirements for materials, equipment and fabrication, often including the names of allowable manufacturers; it is common practice to allow 'or equal' products, with the burden on the contractor to demonstrate that the product is at least equal to the named allowable products; and (3) execution, which includes explicit workmanship standards for installation, erection and construction, required finishes, special instructions, testing requirements and closeout requirements.

While the divisions listed in Figure 3.3 can be applied to any type of construction, the breakdown is most suitable for buildings. For a highway construction project, the breakdown in Figure 3.4, as suggested by Clough and Sears (1994), might be appropriate. Another approach sometimes used in transportation construction is for the agency to publish a manual of 'standard' general conditions and technical specifications that could apply to any typical highway, bridge

Division number	Division title	Sample sections
1	General requirements	Quality controls, temporary structures, startup and commissioning
2	Sitework	The site materials and methods, earth work, tunnelling
3	Concrete	Forms, reinforcement, cast in place, pre-cast, grouts
4	Masonry	Masonry units, stone refractories, corrosion-resistant masonry
5	Metals	Framing, joists, metal finishes, railroad track
6	Wood and plastics	Rough carpentry, finish carpentry, fasteners and adhesives, plastic fabrications
7	Thermal and moisture protection	Dampproofing, waterproofing, vapour and air retarders, insulation
8	Doors and windows	Metal doors and frames, entrances and storefronts, metal windows, hardware
9	Finishes	Plaster, gypsum, tile, terrazzo, flooring, wall finishes
10	Specialties	Display boards, pedestrian control devices, toilet and bath accessories
11	Equipment	Maintenance equipment, library equipment, security and vault equipment
12	Furnishings	Casework, artwork, rugs and mats, furniture, seating
13	Special construction	Air-supported structures, ice rinks, solar energy systems
14	Conveying systems	Dumbwaiters, elevators, lifts, scaffolding
15	Mechanical	Piping, plumbing fixtures, heat-generation equipment, air handling
16	Electrical	Wiring, electric power, lighting, communications, controls

Figure 3.3 Masterformat technical specification divisions for buildings, from Construction Specifications Institute and Construction Specifications Canada.

or street improvement project. Then, for a particular project, a set of 'special provisions' is issued that (1) amends and adds to the general conditions (as in the special conditions discussed earlier) and (2) provides technical construction details only for those provisions in the standard specifications that must be modified.

Division number	Title	Division number	Title
1	Obstructions	30	Painting
2	Clearing and grubbing	31	Rubbing masonry
3	Earthwork	32	Riprap
4	Erosion control and preparatory landscaping	33	Concrete slope paving and aprons
5	Subgrade	34	Concrete curbs and gutters
6	Watering	35	Portland cement concrete sidewalks
7	Finish roadway	36	Right-of-way monuments and freeway access opening markers
8	Cement-treated subgrade	37	Concrete barrier posts
9	Road-mixed cement treated base	38	Guard railing
10	Plant-mixed cement treated base	39	Pipe handrail
11	Untreated rock surfacing	40	Culvert markers, clearance markers and guideposts
12	Crusher run base	41	Fences
13	Penetration treatment	42	Reinforced concrete pipe culverts and siphons
14	Seal coats	43	Corrugated metal pipe culverts and siphons
15	Bituminous surface treatment	44	Field-assembled plate culverts
16	Armour coat	45	Non-reinforced concrete pipelines
17	Non-skid surface treatment	46	Sewer pipelines
18	Road-mixed surfacing	47	Underdrains
19	Plant-mixed surfacing	48	Spillway assemblies and down drains
20	Side forms	49	Salvaging and re-laying existing drainage facilities
21	Bituminous macadam surface	50	Reinforcement
22	Asphalt concrete pavement	51	Portland cement concrete
23	Portland cement concrete pavement	52	Asphaltic paint binder
24	Timber structures	53	Paint
25	Concrete structures	54	Asphalts
26	Steel structures	55	Liquid asphalts
27	Piling	56	Asphaltic emulsions
28	Treatment of timber and piles	57	Expansion joint filler
29	Waterproofing		

Figure 3.4 Sample technical specification outline for highway construction (from Clough and Sears, 1994).

It is common practice to refer in the technical specifications to codes and standards developed by industry associations, rather than to repeat those standards in the specifications themselves. For example, a UK technical specification for concrete testing might refer to the provisions of BS EN 12350 *Testing Fresh Concrete*, which is a European standard test method developed by the European Committee for Standardisation and published by the British Standards Institution (British Standards Institution, 2001). If this reference is part of the concrete technical specifications, the contractor will be obligated to follow its provisions for such practices as sampling, slump tests and air content testing, even though the requirements are not contained directly within the specifications.

Schedule of quantities

The practice of providing material quantities in the contract documents varies in different parts of the world. For all unit-price (measure-and-value) contracts, these quantities must be provided, because the tenders are based on a common set of quantities. In the case of lump-sum contracts, however, the quantity surveyor is sometimes engaged by the design professional to prepare a schedule of quantities used by contractors as they prepare their fixed price tenders. In this case, this schedule is an important part of the project manual containing all contract documents. The practice differs worldwide, however, with the responsibility for generating these quantities through material take-offs often falling on individual contractors and their estimating staff. In Chapter 4, we shall deal with the use of schedules of quantities, wherever prepared, in the tender preparation process.

Invitation to tender

Unless the project's contractor has already been selected, owners must publicise the availability of their projects to prospective tenderers. The procedure used on most publicly funded projects and some private ones is for the owner to issue an *invitation to tender*, also known as an *advertisement for tenders* or *notice to tenderers*, with basic information about the project. A sample invitation is shown in Figure 3.5, containing information on the name and owner of the project, the time for receipt of tenders, the location where tender documents can be obtained and their cost, the bid deposit and bonding requirements and the name of the design professional. Most invitations contain basic information about the magnitude of the project, the nature of its primary materials, and its estimated cost. The invitation is published in the public media and may be sent by mail to contractors who have expressed prior interest. The invitation in Figure 3.5 was also posted on the City of Yellowknife's website at http://city.yellowknife.nt.ca/. With this information, contractors can determine whether they are interested in pursuing further the prospect of obtaining this work.

Instructions to tenderers

Instructions to tenderers expand on the basic information contained in the invitation to tender. These instructions communicate to contractors the technicalities of the tendering process. Although they repeat some of the information contained in the invitation to tender, they also provide information on such matters as modification of tenders once submitted prior to opening,

Invitation to Tender
2001 Water and Sewer Upgrade

Project No. 01-002
Contract B

Sealed tenders marked City of Yellowknife, 2001 Water and Sewer Upgrade Contract B will be received at City Hall, City of Yellowknife, NT until 1:30:00 p.m. local time on April 4, 2001 at which time the Tenders will be opened in public.

Tender documents may be obtained on or after 1:00 p.m. March 14, 2001 at the office of the Owner at City Hall, Yellowknife, NT and at the office of the Engineer in Edmonton, Earth Tech (Canada) Inc. (Reid Crowther & Partners Limited), 17203 -103 Avenue, Edmonton, Alberta.

One set of the Contract Documents will be available for each Tenderer. A deposit of One Hundred Dollars ($100.00) in the form of cash or a cheque in favour of the City of Yellowknife is required for each set of contract Documents. Deposit will be refunded if drawings and specifications are returned to the point of origin complete, undamaged, unmarked and reusable within thirty (30) days of the Tender closing date.

Contract Documents may be examined at the offices of Reid Crowther & Partners Ltd. in Yellowknife and Earth Tech (Canada) Inc., Edmonton; at Yellowknife City Hall; at the NWT Construction Association office in Yellowknife and the Edmonton Construction Association.

Each Tender must be accompanied by a Bid Bond in the amount of Ten Percent (10%) of the Tender amount plus a Consent of Surety, or a certified cheque in the amount of (10%) of the Tender amount.

The successful Tenderer will be required to provide a Performance Bond in the amount of Fifty Percent (50%) of the contract price, and a Labour and Material Payment Bond in the amount of Fifty Percent (50%) of the contract price, or a Security Deposit in the amount of Ten Percent (10%) of the contract price.

The Owner reserves the right to reject any or all of the Tenders or to accept the Tender most favourable to the interest of the Owner.

Reid Crowther & Partners Limited
4916 - 47th Street
PO Box 1259
Yellowknife, NT
X1A 2N9

City Administrator
City of Yellowknife
PO Box 5809
Yellowknife, NT
X1A 2N4

Figure 3.5 Sample invitation to tender.

alterations to the tender, tender withdrawals, tender rejections, performance guarantees, methods for handling alternative proposals and order of precedence in the case of discrepancies between numbers and words. The American Institute of Architects (1997b) has prepared a standard document entitled *Instructions to Bidders*, often used in the tendering process for private projects. However, because these are general instructions, they usually must be modified through issuance of supplementary instructions. Some public agencies in the USA have adopted this document as well and they provide additions, deletions and amendments as part of their agency instructions common to all projects.

Tender form

A complete set of construction contract documents will usually contain a form to be used by the contractor to submit the tender. The form's details will depend upon the type of contract. For example, if the owner anticipates a lump-sum contract, the form will be rather simple, providing for the single price and a few other details. If the contract will be a unit-price/measure-and-value type, there must be places to indicate a unit price corresponding to each bid item. In our consideration of proposal preparation and submittal in Chapter 4, we shall show some samples and discuss their preparation.

Agreement

To allow all tenderers to know the form of the agreement the successful tenderer will be asked to sign, a copy of the agreement is included with the other documents in the project manual. Various organisations have developed standard agreement forms, including FIDIC (Fédération Internationale des Ingénieurs-Conseils, 2001b) and the American Institute of Architects (1997c). Typically, the agreement contains sections setting forth the name of the owner, contractor and design professional, a list of the contract documents, a description of the work encompassed within the contract, the dates of commencement and substantial completion, the contract sum, details regarding payment procedures and provisions related to termination or suspension of the contractor. We shall deal further with this form in our chapter on contractor selection.

Surety bond forms and insurance certificates

During the project mobilisation phase, the contractor may be required to acquire surety bonds as a means of assuring the owner that the contractor will perform the work and pay its obligations. We shall define and describe these bonds in Chapter 5, but here we simply indicate that the forms to be used are often included in the package of contract documents, for the contractor's use if and when the contract is awarded. Likewise, the general conditions are likely to require the contractor to carry certain types of insurance and those forms may be included in the project manual as well.

Summary of planning and design

This chapter has described that part of the project life cycle that is directed by the design professional and results in a set of project documents ready for the tender process. All non-trivial construction projects, of any size, type or complexity, follow this general step-by-step procedure, beginning with the preliminary planning and feasibility study and leading on through schematic design, design development and construction contract document preparation. The intent of the process is a directed series of discussions by the design professional and the owner about the project. Each stage pursues increasingly detailed inquiry into smaller pieces and fewer options. Early in the process, a question might be, 'How big will the facility be?' Later, one might ask, 'What kind of materials shall we use for this wall?' As the design nears completion, the owner may be confronted with 'What colour paint shall be selected?' During the early part of this phase, there will tend to be much back-and-forth, feedback type of discussion and consideration, as various options are proposed and evaluated. Somewhere during the schematic design stage – it varies with every project – there should be enough clarification to allow succeeding design work to proceed in a straight-ahead, linear fashion; the design manager does well to come to this point as early as possible, so that all parties can proceed efficiently, knowing that minimal changes will henceforth be made to the basic design plans. It is also well to remember that the several phases discussed in this chapter are part of a design–build project, or a phased-construction project, although each of the phases will tend to follow the cycle described.

Some common tasks tend to occur during each of the stages, in addition to the technical planning and design efforts. We note three such tasks briefly below.

Cost estimates

The cost estimates prepared during the planning and design phase are the responsibility of the design professional, unlike those prepared later by the contractor as part of the tendering process. As we have noted, they are a part of the schematic design and design development stages. As the design becomes more refined, estimates are expected to be more precise. Whereas the owner can only be assured that the cost estimate during preliminary planning is within, say, 30% of the eventual cost, an estimate as close as 10% may be possible after the completion of design.

Constructability analysis

In our discussion of planning and feasibility studies, we mentioned the use of *constructability analyses and studies* as a means of making the construction process more efficient. It is important that this attitude be present throughout the planning and design phase. The presence of a construction expert, perhaps by virtue of the early engagement of a construction manager, can greatly assist this process. Constructability has been defined as 'the extent to which a design of a facility provides for ease of construction yet meets the overall requirements of the facility' (Wideman, 2001). Continuous constructability studies throughout the planning and design phase can help to anticipate potential problems involving material compatibility, access issues, sequencing problems, dewatering, weather or delivery issues, unnecessary complexity, new or proprietary installation methods and long-term performance. A study by the US Federal Transit Administration (2001) of a bus maintenance facility, rail station and parking area project revealed the following actual problems that could have been identified by an early constructability analysis.

- The site geometry of the bus maintenance facility was conveyed inaccurately on the contract drawings, impacting quantities of sitework and utilities.
- Sprinkler design was not adequately presented to comply with local codes.
- Electrical requirements for owner-supplied equipment were not conveyed on the drawings.
- Lack of coordination between plumbing and architectural drawings resulted in numerous inconsistencies.
- Fire protection and fire alarm systems did not meet with local codes.
- Conflicts between light pole foundations and large diameter pipes comprising the underground detention basins under the parking lots necessitated redesign.

A term closely related to constructability analysis is *value engineering*, value analysis or value management. One definition (Fisk, 2003) is 'a systematic evaluation of a project design to obtain the most value for every dollar [or other denomination!] of cost'. Some construction contract documents provide incentives for the contractor to suggest 'value engineering changes' to the design after the construction contract has been signed, with arrangements for sharing of any cost savings. The practice of value engineering has crept backwards in the project life cycle and is now a regular part of the planning and design phase, even before the engagement of the contractor. The Public Buildings Service of the US General Services Administration generally contracts with an independent consultant to conduct two value engineering studies, the first at the completion of schematic design and the second at the end of design development (General Services Administration Architecture and Construction, 2000).

The Society of American Value Engineers defines a five-phase process for such studies incorporating information, creative, judgment, evaluation and development phases. Under this approach, a study of a US$ 75 million water pipeline project in California, USA resulted in a US$ 5.5 million saving; the saving was reported to be over 68 times the cost of the study. Among the most significant design modifications were the realignment of a roadway to minimise utility relocation and the use of rip rap in lieu of added depth at stream crossings (Johnson et al., 1998). The distinction between constructability analysis and value engineering, if any, is that while both practices focus on quality and costs, constructability includes more emphasis on safety and scheduling; also, constructability tends to involve collaboration of designers and constructors from the outset of the project life cycle (Dorsey, 1997).

Involvement of public agencies

The nature of the project determines the extent to which various public agencies will be involved. Again, there may be activity in this regard throughout the planning and design phase. Reference to Figure 3.2 indicates that, for this university classroom and laboratory building, 'agency review' is an expected part of schematic design and design development and 'agency approval' is required as part of the construction document stage. Activities might include the preparation of environmental impact statements, procedures involved in utilising wetlands, building code compliance reviews and fire safety and public health considerations. Providers of utilities such as electric power and water services may be involved if the proposed project impinges on easements or rights of way.

Discussion questions

1 Interview a design professional in your community.
   (a) In what types of projects is this person involved?
   (b) How do they define the stages of the planning and design process?
   (c) Are outside consultant specialists used? What factors are considered in decisions about the use of outside versus in-house speciality services?
   (d) What bases are used for determining the design fee?
   (e) What other aspects of this person's practice are particularly interesting?
2 Perform calculations to confirm the weighted score of 4.89 for Apex Consultants in Table 3.1. How would the ranking of the five firms change if each of the nine criteria was weighted equally?
3 You have been engaged as the project manager by the owner of a proposed 90 km natural gas pipeline that would be located in a rural and suburban area. Two early tasks will be the production of a project brief and the development of a programme statement. List several items that will be included in the brief and several questions that will be answered in the programme statement. In what ways might the public be involved in the early stages of this project and how might you, as project manager, coordinate this involvement?
4 For the gas pipeline project in Question 3, what sorts of alternatives might be studied at various stages of the planning and design phase?

5 List some constructability issues that might be considered for the gas pipeline project in Question 3.
6 Prepare a conceptual cost estimate for the construction of a three-storey concrete parking garage containing spaces for 120 vehicles, to be built in your area. You will need to obtain a unit cost per vehicle space from an appropriate source. To the cost of the structure, add the estimated cost of land acquisition, based on an assumed required land parcel size.
7 For the parking garage in Question 6, make a rough estimate of the annual cost of operating and maintaining the facility, assuming that spaces will be rented daily to vehicle owners. List the various elements of annual cost and their amounts.
8 For the parking garage in Questions 6 and 7, suppose your client is considering investing in the facility and desires a 12% annual rate of return on their investment. If the garage has a service life of 30 years and no salvage value at the end of 30 years, determine the required daily rental rate per vehicle assuming an average 75% occupancy rate. Is such a rental rate reasonable for your area? (The capital recovery factor for these conditions is (A/P, 12%, 30) = 0.12414.)
9 Answer Question 8 if the assumed occupancy rate is 90%.
10 What costs are omitted from Table 3.2 that the developer of a potential project would need to consider?
11 A website (Davis Langdon Australasia, 2002) gives international cost comparison data for the third quarter of 2000 for average standard high-rise office buildings, as follows: Auckland, 732; Hong Kong, 1430; Singapore, 950; Kuala Lumpur, 425; Manila, 600; Johannesburg, 320; San Francisco, 890; London, 2463; New York City, 970. All of these data are given in terms of US$ per square metre of floor area, for comparison purposes. Discuss reasons for these widely varying conceptual cost unit price estimates in different cities.
12 Obtain the contract documents for a recent or current construction project in your area.
   (a) List the major divisions in the technical specification section and one specific item in each.
   (b) Note three specific items in the general conditions.
   (c) Note three specific items in the special conditions.
   (d) What other sections are included in the manual?
   (e) List the several types of drawings, for example, civil, architectural.
13 Find an invitation to tender ('invitation to bid') in the classified advertising section of your local newspaper or a construction trade magazine. List the various pieces of information contained in the invitation and compare them with those discussed in the text.

References

American Institute of Architects. 1997a. *Standard Form of Agreement between Owner and Architect*. AIA Document B141.

American Institute of Architects. 1997b. *Instructions to Bidders*. AIA Document B141.

American Institute of Architects. 1997c. *Standard Form of Agreement between Owner and Contractor*. AIA Document A101.

Bachner, J.P. 1991. *Practice Management for Design Professionals*. John Wiley.

Bennett, F.L. 1996. *The Management of Engineering: Human, Quality, Organizational, Legal, and Ethical Aspects of Professional Practice*. John Wiley.

British Standards Institution. 2001. *British Standards Online*. http://www.bsi-global.com/Standards+Commercial/index.xalter.

Clough, R.H. and G.A. Sears. 1994. *Construction Contracting*, 6th edn. John Wiley.
Construction Specifications Canada–Devis du Construction Canada. 2001. http://www.csc-dcc.ca/.
Construction Specifications Institute. 2001. http://www.csinet.org.
Cornell University. 2001. *Duffield Hall Project.* http://www.duffield.cornell.edu/duff_proj.cfm. Timeline at http://www.duffield.cornell.edu/timeline.pdf.
Davis Langdon Australasia. 2002. *Cost Data: International Cost Comparison.* http://www.davislangdon.com.au.
DeWitt, S. 1999. Constructability reviews and post construction reviews. In *American Association of State Highway and Transportation Officials Subcommittee on Construction, National Survey Results*, New Orleans, Los Angeles, USA, August 1–5.
Dorsey, R.W. 1997. *Project Delivery Systems for Building Construction.* Associated General Contractors of America.
Fédération Internationale des Ingénieurs-Conseils. 2000. *The FIDIC Contracts Guide*, 1st edn.
Fédération Internationale des Ingénieurs-Conseils. 2001a. *Quality-based Selection.* http://www.fidic.org/resources/selection/qbs/.
Fédération Internationale des Ingénieurs-Conseils. 2001b. http://www.fidic.org/.
Fisk, E.R. 2003. *Construction Project Administration*, 7th edn. Prentice-Hall.
Fivecat Studio. 2001. *The Five Phases of Architecture: The Process of Design.* http://www.fivecat.com.
General Services Administration Architecture and Construction. 2000. *Value Engineering for Design and Construction.* http://gsa.gov/pbs/pc/gd_files/value.htm.
Hicks, T.G. and J.F. Mueller. 1996. *Standard Handbook of Consulting Engineering Practice.* McGraw-Hill.
Johnson, P.W., A.J. Perez, B.K. Yu and C. de Leon. 1998. Value engineering savings on pipeline project. In Castronovo, J.P. and J.A. Clark, ed. *Pipelines in the Constructed Environment. Proceedings of 1998 Pipeline Division Conference, American Society of Civil Engineers*, San Diego California, USA, August 23–27, pp. 195–202.
Means Square Foot Cost Data 2002 Book. 2002. 23rd edn. Means.
Rawlinson's Australian Construction Handbook. 2001. Rawlhouse.
Robbins, T.E. 1996. Specifications. In Merritt, F.S., M.K. Loftin and J.T. Ricketts (eds) *Standard Handbook for Civil Engineers*, 4th edn. McGraw-Hill.
Royal Incorporation of Architects in Scotland. 2001. *Clients Advice Document.* http://www.rias.org.uk/Client%20Advice%20Doc.htm.
Standards New Zealand Paerewa Aotearoa. 1998. *Conditions of Contract for Building and Civil Engineering Construction.* New Zealand Standard NZS: 3910: 1998.
State of Mississippi. 2000. *Planning and Construction Manual, Bureau of Buildings, Grounds and Real Property.* http://www.dfa.state.ms.us/building/Manual/S600A.pdf.
US Federal Transit Administration. 2001. *Lessons Learned Program. United States Department of Transportation. The Benefits of Constructability Reviews.* http://www.fta.dot.gov/library/program/ll/man/ll39.htm.
Wideman, M. 2001. *Max's Project Management Wisdom, Issues and Considerations: Constructability.* http://www.maxwideman.com/issacons/index.htm.

4

Contractor selection phase

Introduction

We now turn to the selection of the construction organisation that will assemble the project in the field. The basic steps are generally the same, whether the contract will be lump sum, unit price or cost plus. If a project manager is involved, that party will assist the owner in choosing a contractor. The construction manager, if part of the team, will have to be chosen. If the design–build approach is used, that organisation will be chosen before the design is undertaken, but some of the discussion in this chapter will apply to that selection process. After we deal with various methods the owner may use for selecting the contractor, the balance of this chapter and the others that follow will be presented from the viewpoint of the construction organisation rather than the owner or design professional.

Methods for contractor selection

The owner has some choices in its approach to selecting a contractor. Will any and all interested contractors be permitted to submit proposals? Will the list be limited to a selected group of contractors, and, if so, how will the list be identified? Will the owner negotiate with only a single contractor? Will some sort of pre-qualification or post-qualification be used? As we shall see, these questions may be related; we shall also see that certain approaches tend to be more appropriate for certain contract types.

Pre-qualification/post-qualification

To limit the number of firms allowed to submit tenders on a project, some owners require that contractors be *pre-qualified*. They must submit information about their experience, competence and financial condition, after which the owner decides whether they are qualified. The advantages are obvious; as suggested by Bockrath (1999):

> As a general proposition, it is preferable to disallow unqualified contractors to bid at all rather than to refuse to award the contract after they have gone to the trouble and expense of putting together proposals. In addition, the prequalification

procedure may prevent some bidders from being awarded a contract that, because of its scope and complexity, would likely prove disadvantageous for them, not withstanding their personal beliefs to the contrary.

But there are arguments on the other side as well; some feel that in a free business world, all interested contractors should be 'given a chance'. Also, if bonding is required, one might argue that only qualified contractors will be underwritten by surety companies and these companies protect the owner sufficiently if the contractor fails to perform or fulfil its financial obligations. In addition, pre-qualification tends to lengthen the contractor selection process by as much as a month or two and it also precludes consideration of the well-qualified contractor who appears, for some reason, at the last moment.

Three approaches can be used.

- An owner anticipating several projects over a certain period, say a state transportation department, might pre-qualify contractors for that period. In this case, for example, highway contractors wishing to be invited to bid on projects in 2005 would submit information by a specified deadline in 2004; the qualifications would be evaluated and successful contractors would be placed on 'the list' for 2005.
- Pre-qualification could be accomplished on a project-by-project basis; those wishing to submit tenders for a particular project would submit completed questionnaires and then await an evaluation that determines whether they are qualified to tender for that one project.
- Each contractor submitting a tender on a particular project could be asked to submit an individualised qualification outline with its tender. Clough and Sears (1994) call such an outline 'essentially a sales document [containing] information designed to enhance the contractor in the eyes of the owner'.

According to Russell (1996)

> . . . a properly designed pre-qualification process should:
>
> 1 Assure that the constructor and major subcontractors, vendors, and material suppliers will be competent, responsible, and experienced, with adequate resources to complete the job.
> 2 Eliminate constructors with limited financial resources, overextended commitments, and/or inadequate or overly inexperienced organisations.
> 3 Maximize competition among qualified constructors and subcontractors.

The Virginia Department of Transportation in the US state of Virginia (VDOT) includes the following statement on its website (Virginia Department of Transportation, 2001):

> VDOT's Contractor Prequalification Program ensures that all contracts for the construction, improvement and maintenance of Virginia's transportation system are awarded to the lowest responsible bidder. Virtually all construction and maintenance contracts advertised and let by the department require prospective bidders to be prequalified. In addition to prime contract work, contractor prequalification is also necessary for subcontractors unless otherwise noted in the contract specifications.

VDOT utilises a 12-page form (Virginia Department of Transportation, 2001) for pre-qualifying contractors; applicants must provide the following information:

- company's contact information
- firm's legal status and history
- ownership information
- names of individuals authorised to transact business with VDOT
- information on affiliated, financially associated and subsidiary companies
- requested classes of work (drilling and blasting, landscaping, major structures and so on)
- prior work experience
- equipment available to the applicant
- financial information
  - balance sheet
  - gross receipts summary
  - income statement
- notarised affidavit.

A similar 11-page form is used for pre-qualifying contractors for any of several projects sponsored yearly by the Louisville and Jefferson County Metropolitan Sewer District (Metropolitan Sewer District, 2001); in addition to the information outlined above, it requests safety and insurance data and information on employee training programmes.

An interesting study conducted at Hong Kong University (Palaneeswaran, 2000) ranked pre-qualification criteria for design–build projects. The results were based on a questionnaire survey of 101 respondents from 11 countries, 80% of whom were from the construction industry. The following criteria, in order, emerged as the five most important from among a total of 16 criteria: (1) past experience, (2) past performance, (3) financial strength, (4) quality and (5) organisation and management systems. The study also ranked various subcriteria within each of the 16 criteria.

Certainly this approach can be used in the private sector as well, with similar procedures and information, although the process might be somewhat less formal than in the public arena.

Post-qualification is another option. If a contractor is the apparent low tenderer for a project, it will then be asked to submit information demonstrating its qualifications. The disadvantages of such an approach include the potential for wasted effort throughout the tendering process, if the low bidder is found not to be qualified, and the prospect of favouritism in rejecting the low bidder by claiming unjustly that it is not qualified. However, owners do have the right to choose 'responsive' and 'responsible' tenders, according to all well-written contract documents, so there is always the chance for claims of unfairness when the owner decides whether a contractor's tender is 'responsible'. In most cases, an open *pre*-qualification process would appear to be the fairer method for all concerned. Certainly pre-qualification requires contractors to meet only a minimum level of qualification. In evaluating tenders, owners can then rank contractor qualifications according to specified criteria in deciding which one of the qualified contractors to select; these procedures will be considered in a later section.

Open tender

The discussion of contractor qualification in the preceding section is closely related to limitations on the pool of tenderers. If there is no pre-qualification requirement, then any interested contractors are allowed to submit tenders. An invitation to tender is issued, as discussed in Chapter 3, and tenders are received and evaluated. Decisions regarding qualification are deferred until after receipt of tenders. If the criteria for selection are based on price alone, the owner must decide whether the low bid is both 'responsive' to the tender announcement and 'responsible', meaning that the contractor is qualified to do the work. If criteria in addition to price are used to judge the tender, then measurements of qualification will be a part of this process and the end result should be an agreement with a qualified contractor.

Invited tender

The pre-qualification process described above, conducted prior to the submittal of tenders, will naturally lead to a limited number of firms considered qualified. They are the firms that will be invited to offer proposals to carry out the construction work. Whether they were qualified for all work in a certain category during the time period or were qualified for this project only, the firms on this limited list will be the only recipients of the invitation to tender. Once the tenders are received, they are evaluated in a fashion similar to those received though the open tender process.

Another version of the invited tender, sometimes used in the private sector, involves the owner, or its design professional or project manager, in inviting tenders from firms it believes to be qualified, based on their reputations, past experience and other criteria. This approach lacks the formality of qualification forms and evaluations and is probably not appropriate in the public sector, where accountability to ratepayers and taxpayers requires credibility and transparency.

Negotiation

Negotiation of construction contracts can occur at two points in the process of selecting the contractor. It might be used *in lieu of* the tendering process. If the owner has had good experience with a certain constructor, the owner or owner's representative can invite that one firm to prepare and offer a proposal, after which the parties negotiate the agreement, with rejections, counter-offers and other steps regarding price, scope of work and all other contract issues. If the negotiation is successful, an agreement for constructing the project is eventually signed; if it is not, another contractor may then be asked to offer a proposal and the process is repeated.

Negotiation may also be part of the formal tendering process. Open or invited tenders could be received and evaluated and the most highly ranked firms identified. Then the owner could negotiate with this 'short list' of, say, three firms, beginning with the top-ranked firm. Issues such as work content, schedule, price, personnel and risk sharing might be part of this negotiation. If an agreement is not reached with the top-ranked firm, negotiations proceed down the list until a satisfactory agreement emerges. This procedure is similar to that described in Chapter 3 for selecting design consultants, including initial proposals, evaluation and negotiation until a consultant agreement is reached. If the owner intends to negotiate with top-ranked tenderers, this intention must be made clear to all, via the invitation to tender.

The contractor selection method is generally dictated by the type of delivery system and the type of contract. For example, under the 'traditional' design–tender–build delivery system, with a lump-sum or unit-price contract, it is likely that tenders, either open or invited, will be called.

If low price is the only criterion, there is likely to be no negotiation following evaluation of tenders, especially in the public sector. Cost-plus contracts usually involve negotiation. One example might be a contract between the owner and construction manager signed before the completion of design; negotiations would consider the fee, definitions of reimbursable costs, schedule matters, scope of services and risk and liability issues.

Design–build contracts may involve either competitive proposals and evaluations or negotiation (Beard et al., 2001). Usually a fixed price is not specified at the time the design–build firm is chosen; if the contract for the project is to be fixed price, a common practice is to lock into that sum at the end of preliminary design, after 30–40% of the design work is complete (Halpin and Woodhead, 1998). Design competitions, as described in Chapter 3, are sometimes held as part of the design–build proposal process.

The contractor's tender decision

Now we begin to look at prospective projects from the contractor's viewpoint. Initially, we ask two questions: (1) Where can contractors find out about projects they might be interested in? and (2) What are some factors that contractors might consider in deciding whether to prepare and submit a tender?

Project information sources

For projects with open tenders, it is usually to the owner's advantage to advertise the project widely, so as to generate sufficient competition to assure tenders that are reasonably priced and from qualified, responsible contractors. For publicly funded projects, there may be legally mandated means for notifying the construction community that projects are available for tender.

In our discussion of the invitation for tender, we saw an example of such a document (Figure 3.5) that was posted on the owner's website. A great variety of electronic means are used to inform the contracting community about up-coming tenders, including the following.

- Various on-line construction publications, with sections devoted to new projects. Examples are *Construction Industry Times On-Line* (2002) in the UK and *Engineering News Record* (2002) in the USA.
- On-line publications with information confined to procurement opportunities and awards. One such publication with a worldwide coverage is *United Nations Development Business Online* (2002). This electronic document provides information on business opportunities generated through the World Bank, regional development banks and other development agencies. While engineering and contracting firms make heavy use of the publication, it also is used by consultants, manufacturers, wholesalers and exporters of many types of products. For the construction industry, it includes general procurement notices, followed by specific invitations to bid, to pre-qualify or to submit consulting proposals. It also publishes announcements of recently awarded World Bank and Asian Development Bank construction contracts.
- On-line publications whose only role is to advertise tender opportunities. An example is the *New Zealand Tenders Gazette* (2002), a sample from which is shown in Figure 4.1. To

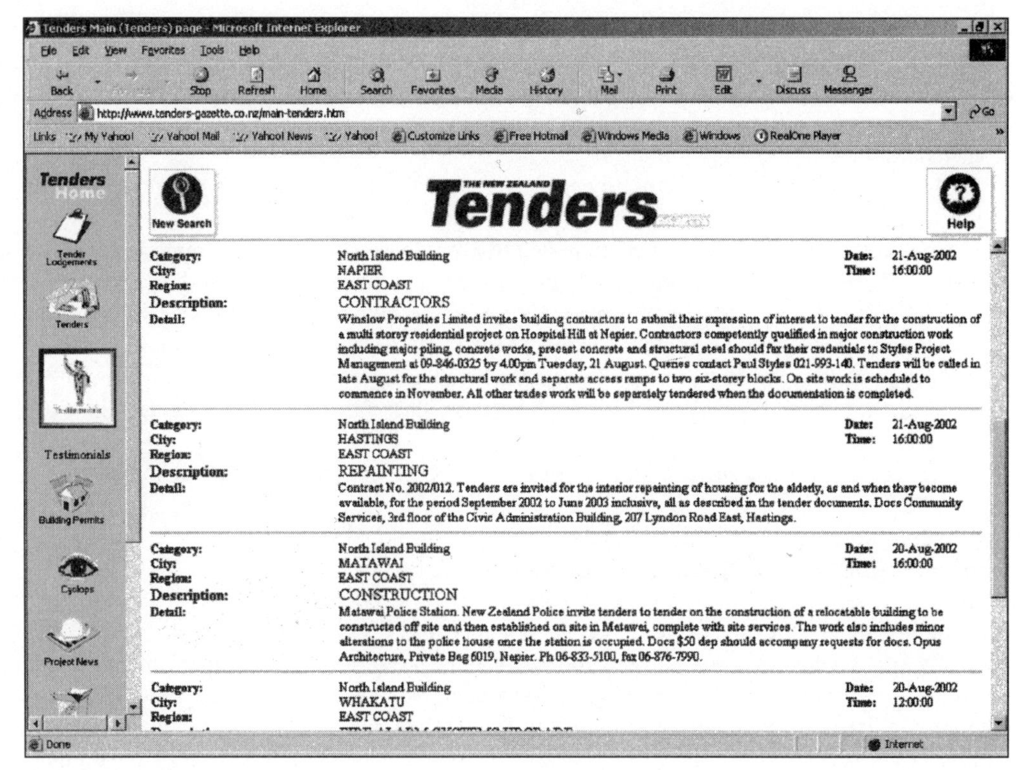

Figure 4.1 Sample portion of projects listing in *New Zealand Tenders Gazette* (2002).

use this listing, one selects the type of tender – building construction, other construction, supplies and services, motor vehicles, machinery sales, building sales and leases or 'all' and the location. Note that the listings are basically very limited invitations to tender, including contacts for further information. This website does contain some additional information, such as recent building permits and a small number of articles of interest to project personnel, but the dominant purpose is to support the tender process. Owners can lodge new tenders electronically and the site is updated daily.

- Internet plan rooms, such as one operated by the Carolinas (USA) Associated General Contractors, where contractors can 'search for up to 21 types of jobs, search specifications by key words, view plans, zoom in and out, make proprietary notes on the plans, use on-line estimating plans, and print or order plans' (Martin, 2001). This virtual information centre provides the same kind of service as that of traditional plan rooms, sometimes called plan services, where contractors visit in person for information on upcoming projects. At this writing, the cost for members of the Carolinas Associated General Contractors is US$ 2800 per year for all project listings in the states of North and South Carolina.

Some of the more traditional means for informing contractors about jobs available for tendering are listed below (based on Bennett, 1996); it will be noted that the electronic methods already discussed had their beginnings in some of these hard-copy equivalents.

- Classified advertisements in local and regional newspapers. 'Invitations to bid' or 'invitations to tender' are standard categories in most classified advertising sections.
- Notices in regional, national and international trade magazines. *Construction Industry Times* and *Engineering News Record*, noted in the previous section, also publish print editions that include information on new projects.
- Publications of contractor organisations, such as newsletters distributed by local chapters of the Associated General Contractors.
- Listings at local or regional construction plan rooms, whose electronic counterparts are noted above. These facilities are often privately owned and may be sponsored by contractor federations. They provide a central place for owners to deposit plans and other information about new construction work, thus making them easily accessible to contractors.
- Notices posted to contractors who are 'on the list' for the particular type of project, either because they have been pre-qualified or because they have expressed interest in such projects.
- In the case of some private projects, informal contacts by letter, telephone or in person.

Considerations in deciding to tender

It is not possible for the contractor to prepare a tender for every available project. Even in lean economic times, when jobs are scarce, some means must be employed by company management to decide whether to prepare a cost estimate and tender for a particular project. Some factors to consider are listed below (based on Clough and Sears (1994) and Pratt (1995)); items 1 through 12 relate to project characteristics, 13 through 16 to company status and its strategic positioning and 17 and 18 to external conditions.

1. Project type. Although some contractors claim to provide all kinds of construction services, in fact most specialise in certain types in which they have been successful. A builder of high-rise buildings is unlikely to be interested in tendering on a remote oil pipeline.
2. Project size. Project size may be measured by a rough estimate of contract value or by some physical measure such as floor area or volume of concrete. The project under consideration may be too large or too small to be of interest.
3. The estimated cost of preparing the tender. Large projects can cost a contractor several million US$ for this effort.
4. Project location. Some contractors specialise in projects in remote regions, such as rural Alaska or mining districts of South America, while others confine their work to their home cities and surrounding regions.
5. Specialised work. Even within project types, there may be specialties for which the contractor lacks expertise. A concrete building contractor might decline to tender on a project with especially complicated or innovative formwork or another might avoid a project requiring specialised heavy lifting equipment.
6. Construction contract document quality. Poorly prepared contract documents may be the precursor of a difficult and conflict-prone construction effort, with design problems to be resolved as construction proceeds. Most contractors try to avoid such jobs, but some see them as opportunities to be exploited.
7. Contract terms related to contractor responsibility and liability. All contract terms are important, but those related to the potential degree of risk to be borne by the contractor are of special interest.

8 Design professional reputation. Distinct from Item 6 above, which relates to the documents for the particular project, some designers have reputations for fairness, timeliness, thoroughness and all the other characteristics that tend to make for satisfactory jobs. Others do not.
9 Owner reputation. Similar to Item 8, an owner with a reputation for slow payments, constant haggling over minute details and general surliness will have more difficulty attracting interested contractors than an owner with a supportive organisational culture. In addition, contractors appreciate owners with at least a minor degree of technical understanding as related to the project.
10 Anticipated construction problems. Many contractors choose not to tender on projects in which soil conditions, water problems, environmental contamination, complex coordination of subcontractors and other parties, extremely tight time schedules or even severely negative public reaction may lead to costly conflicts and delays. Judgment about this concern is related to Item 6, construction document quality and Items 8 and 9, design professional and owner reputations. As in Item 6, however, some tenderers view anticipated construction problems as challenges to be met and overcome, rather than avoided, with a potential lucrative outcome.
11 Safety considerations. Projects that appear to involve abnormally high safety risks, such as those in swiftly flowing streams, high vehicular or air traffic areas or high-rise buildings or those located near operating machinery or high voltage power lines, may not attract great contractor interest, especially from those with poor safety records.
12 Completion date. This consideration relates to the coordination within the company of all its projects and the deployment of personnel, equipment and other resources. But even if this is the company's only project, an unreasonably tight deadline would decrease the project's attraction.
13 Amount of other work currently underway. The degree to which the contractor 'needs the work' is always a consideration. It is important to maintain a reasonably steady level of company activity, both from the standpoint of those performing field work and those preparing tenders. At a time of high activity, the contractor will be less interested in new work, but the real challenge is to predict the impact of this project on company activity in the months and years ahead, if the company is selected.
14 Bonding capacity. Those who provide surety bonds on behalf of contractors set limits on the total value of projects they will cover at any one time, based primarily on the contractor's financial capabilities. If the project requires the contractor to furnish surety bonds, its estimated cost might exceed the limit, if it were to be awarded to the company, thus preventing the contractor from submitting a tender.
15 Potential for becoming involved in prestigious projects that might enhance the contractor's reputation, such as the Euro Tunnel or the Trans Alaska Pipeline.
16 The opportunity that the project might provide for working into new markets, even though the financial reward for this project might not be great.
17 Probable competition. If 20 tenderers are expected, each has a 5% chance of success (other things being equal!). In such a case, the contractor may choose to spend its efforts on preparing proposals for projects with a smaller field of competition.
18 Labour conditions and supply. The contractor will want to consider whether required qualified labour is available in sufficient quantity. Also, potential difficulties in working with the local labour market and its collective bargaining agreements may weigh against submitting a tender.

Preliminary job planning

If the contractor decides, based on some or all of the considerations in the previous section, to prepare a tender for a certain project, the first step will be to obtain one or more sets of the contract documents. The invitation to tender will specify where they are available – with the owner, the design professional, a plan room, or, increasingly, over the Internet – and the cost of the documents. The major effort, if it is fixed-price or unit-price/measure-and-value contract, will be to compile an accurate estimate of the project's costs and then to convert this estimate into a priced proposal. But there are other tasks to be undertaken, as a preliminary plan is put together, including a method statement, jobsite visits, constructability studies, initial scheduling and various meetings.

After obtaining the tender documents and evaluating them carefully, once more a decision must be made regarding whether to proceed with the tender process or to suspend any further activities on the project. Certainly obtaining and reviewing the documents does not obligate the contractor to develop and submit a tender.

Method statement

An important part of the contractor's preliminary planning is the method statement, which sets forth in words and sketches the steps and special considerations that will be involved in assembling the project in the field, if the contractor is selected. Every contractor will have its unique approach to this statement. Depending on the contractor and the size and complexity of the project, the statement may consist of only a few pages or as many as several hundred. On the basis of this statement, the cost estimators will be able to provide the expected cost of each operation. In essence, the project is built 'on paper' prior to submitting the tender. If the tender is successful, the method statement becomes the basis for the detailed job planning that takes place after the award of the contract and before the start of field work. Important activities leading to the development of the statement include a constructability analysis, the development of a schedule, equipment and labour requirements analyses, environmental protection and safety planning, and transport and traffic needs forecasting.

For some contracts, the contractor does not prepare the method statement only for its use; rather the tender process may require that contractors submit their method statements as part of their proposals, to be judged as one factor among the selection criteria.

Consider the construction of a bridge, tunnel and embankment project across an open water harbour, perhaps linking two countries in northern Europe. Surely the project sponsors and their financial backers will require each interested contracting organisation to submit a method statement with its tender. Each statement would probably be as long as this book, but we can indicate some of the topics to be addressed in the section on procurement, transport and assembly methods, as follows:

- anticipated sources of supply and fabrication of major bridge components and their method of transport to the site;
- methods for dewatering and installing bridge abutment and pier foundations;
- plans for concrete supply, either batched on site or transported from off site;
- steps required to install piers and abutments;
- steps required to install all bridge superstructure components;
- sources of embankment materials and their methods of transport and placement;

- dredging procedures down to bedrock and along the navigation channels that intersect the bridge;
- temporary navigation re-routing while obstructing the navigation channels;
- efforts not to disrupt fish migrations during construction;
- plans for removal of debris generated during construction from the sea bottom;
- contingency plans in case of ship–bridge and aircraft–bridge collisions;
- considerations of wind impacts on structural components during erection.

Without this information, the contractor will be unable to assemble a meaningful cost estimate, nor can a reasonable preliminary schedule be put together.

Constructability analysis

We have already discussed these analyses in our chapter on planning and design, but now we add this note to indicate that the contractor will also be concerned about possible alternative construction methods. The method statement will contain the results of these analyses, as it will state the intended methods to be used. The basis for these statements will be some detailed analysis of options.

Consider briefly our major waterway crossing. In the case of the superstructure components, the analysis is likely to consider the extent of prefabrication. There are cost advantages to producing relatively large components in a more controlled production environment, but there are also extra costs involved in the transport and lifting of these heavier elements. The placement of under-sea concrete can be accomplished in several different ways; the contractor will want to study and evaluate these various methods. In a similar way, alternative dredging methods, including disposal of spoil material, will warrant analysis.

Jobsite visits and checklists

To carry out meaningful preliminary job planning, intimate knowledge of the construction site is required. If the undertaking is a highway project through a wilderness area, a visit on foot or by tracked vehicle will bring the contractor in close contact with site conditions. If the project is our waterway crossing, time spent on the water, on the nearby shorelines and in adjacent communities will be essential. And these visits must be far more than casual observations. They must be well planned and thorough.

Clough and Sears (1994) offer the following about the jobsite visit:

> After preliminary examination of the drawings and specifications, the construction site must be visited. Information is needed concerning a wide variety of site and local conditions. Some examples are as follows:
>
> 1 Project location
> 2 Probable weather conditions
> 3 Availability of electricity, water, telephone, and other services
> 4 Access to the site
> 5 Local ordinances and regulations
> 6 Conditions pertaining to the protection or underpinning of adjacent property

7 Storage and construction operation facilities
8 Surface topography and drainage
9 Subsurface soil, rock, and water conditions
10 Underground obstructions and services
11 Transportation and freight facilities
12 Conditions affecting the hiring, housing and feeding of workers
13 Material prices and delivery information from local material dealers
14 Rental of construction equipment
15 Local subcontractors
16 Wrecking and site clearing.

In Figure 4.2 we show an example of a two-page job site analysis checklist produced and copyrighted by the R.S. Means Company, Inc. (*Means Forms for Building Construction Professionals*, 1986). Note that it includes those suggestions in the above list plus many other items.

Preliminary schedule

In preparing its proposal, the contractor will need to develop a preliminary project schedule. This schedule is not to be confused with the detailed schedule that will be prepared after the job is awarded. This later effort, described in Chapters 5 and 6, is used both for in-depth planning of the multitude of individual project activities and for the control of the schedule as the project proceeds. Instead, the preliminary schedule breaks the project into relatively few broad activities, to provide essential information to those assembling the contractor's cost estimate.

Beyond the occasional owner-imposed requirement that such a schedule be submitted with the proposal, the contractor needs this preliminary timetable for it own reasons. First, the tender documents often specify a project completion date or total project duration and the tenderer needs a rough idea of whether this requirement can be accomplished. Also, many project overhead costs, such as supervision, the project office and site security, vary directly with project duration; thus an estimate of that duration is essential if these costs are to be assessed. Furthermore, the time during which various pieces of equipment are needed must be approximated, in order to include these costs in the proposal. In addition, there are issues regarding the sequencing of operations and their seasonal impacts. For example, will concrete placement occur during severe winter weather? If so, the costs of extra protection and lowered productivity must be incorporated. Another example might be the additional costs of overtime labour or multiple shift operations during certain portions of the schedule (Clough et al., 2000). Finally, it is likely the contractor will have to provide its own financial resources early in the project, until payments for partial completion are received from the owner. The cost of this financing, which must be part of the cost estimate, will be proportional to those early activity levels as identified on the preliminary schedule.

Figure 4.3 is a hypothetical example of a preliminary project schedule, in bar chart form, that can assist the cost estimator in confronting some of the issues described in the previous paragraph. Note that the breakdown is truly 'broad-brush', consisting of only 11 activities. Nonetheless, from it we learn several things: (1) the project's fieldwork is expected to last a total of 5 months, (2) cast-in-place concrete work will take place from late October through mid-November, necessitating cold weather protection if the project is located high in the northern or low in the southern latitudes and (3) heavy lifting equipment will be needed for structural erection between early November and mid-December and, perhaps, for mechanical system installation intermittently during December and January. In Chapter 5, we shall have further occasion to consider project scheduling, in considerably more detail.

Pre-tender meetings

Two types of meetings are common during the time the tender proposal is being prepared. One type is internal to the contractor's organisation. Attendees include the cost estimating staff, personnel expected to be in supervisory positions on the project if the contractor is selected and representatives from general company management. During a series of such meetings, topics comprise all those discussed above under preliminary job planning, plus the cost estimate as it is developed and the final version of the tender before it is submitted. In addition, in some cases, much time and effort in these meetings is devoted to quantifying the various project risks as the contractor continues to study whether to proceed with the tender.

The other type of pre-tender meeting is conducted by the owner and/or its project manager or design professional. All general contractors who have obtained contract documents, thus signifying their interest in the project, are invited; sometimes major subcontractors and material suppliers are included as well. Held part way through the tender preparation process, when contractors have had a chance to become familiar with the project, its purpose is to review the general project requirements, clarify technical and procedural matters and answer questions. The contractor is well advised to prepare for this meeting by reviewing the documents thoroughly and developing a list of questions. To treat each tenderer equally, the owner or its representative must provide clarifications in writing to all; if changes in the documents result from the meeting, they are issued as addenda to all who hold bidding documents.

The contractor may be involved at this stage in other meetings such as those with local authorities, political figures and interested joint venture partners.

Cost estimating

As suggested above, the major effort in tender preparation for a lump-sum/fixed-price or unit-price/measure-and-value contract is the development of the cost estimate. We shall describe the several elements of such a cost estimate and the process for putting an estimate together, for both of these types, using realistic though oversimplified examples. We shall also consider the use of estimating software and show an example based on a product currently available to contractors. In addition, we shall revisit the topic of value engineering, this time from the standpoint of the contractor involved with proposal preparation.

The term *estimate* is curious. It implies that the numbers are approximations, representing someone's idea of what the final project costs will be, but subject to some limited accuracy. All that is true, but whether accurate or not, the cost number(s) generated for a lump-sum or unit-price estimate become contractual obligations if the contractor's proposal is accepted. The contractor agrees to do the work, not for 'about' that amount, but for *exactly* that amount!

First, we review the various levels of detail in which estimates are prepared and the use of each.

JOB SITE
ANALYSIS (GENERAL CONTRACTOR)

SHEET NO.

PROJECT | BID DATE

LOCATION | NEAREST TOWN

ARCHITECT | ENGINEER | OWNER

Access, Highway	Surface	Capacity
Railroad Siding	Freight Station	Bus Station
Airport	Motels/Hotels	Hospital
Post Office	Communications	Police
Distance & Travel Time to Site		Dock Facilities
Water Source	Amount Available	Quality
Distance from Site	Pipe/Pump Required?	Tanks Required?
Owner	Price (MG)	Treatment Necessary?
Natural Water Availability		Amount
Power Availability	Location	Transformer
Distance	Amount Available	
Voltage Phase	Cycle	KWH or HP Rate
Temporary Roads	Lengths & Widths	
Bridges/Culverts	Number & Size	
Drainage Problems		
Clearing Problems		
Grading Problems		
Fill Availability	Distance	
Mobilization Time	Cost	
Camps or Housing	Size of Work Force	
Sewage Treatment		
Material Storage Area	Office & Shed Area	
Labor Source	Union Affiliation	
Common Labor Supply	Skilled Labor Supply	
Local Wage Rates	Fringe Benefits	
Travel Time	Per Diem	
Taxes, Sales	Facilities	Equipment
Hauling	Transportation	Property
Other		
Material Availability: Aggregates		Cement
Ready Mix Concrete		
Reinforcing Steel	Structural Steel	
Brick & Block	Lumber & Plywood	
Building Supplies	Equipment Repair & Parts	
Demolition: Type	Number	
Size	Equipment Required	
Dump Site	Distance	Dump fees
Permits		

Figure 4.2 Jobsite analysis checklist. From *Means Forms for Building Construction Professionals*, by R.S. Means, Kingston, MA, 1986. Reprinted with permission from R.S. Means.

Clearing: Area	Timber	Diameter	Species
Brush Area	Burn on Site		Disposal Area
Saleable Timber	Useable Timber		Haul
Equipment Required			
Weather: Mean Temperatures			
Highs		Lows	
Working Season Duration		Bad Weather Allowance	
Winter Construction			
Average Rainfall	Wet Season		Dry Season
Stream or Tide Conditions			
Haul Road Problems			
Long-Range Weather			
Soils: Job Borings Adequate?		Test Pits	
Additional Borings Needed	Location		Extent
Visible Rock			
U.S. Soil & Agriculture Maps			
Bureau of Mines Geological Data			
County/State Agriculture Agent			
Tests Required			
Ground Water			
Construction Plant Required			
Alternate Method			
Equipment Available			
Rental Equipment			Location
Miscellaneous: Contractor Interest			
Subcontractor Interest			
Material Fabricator Availability			
Possible Job Delays			
Political Situation			
Construction Money Availability			
Unusual Conditions			
Summary			

Figure 4.2 Continued.

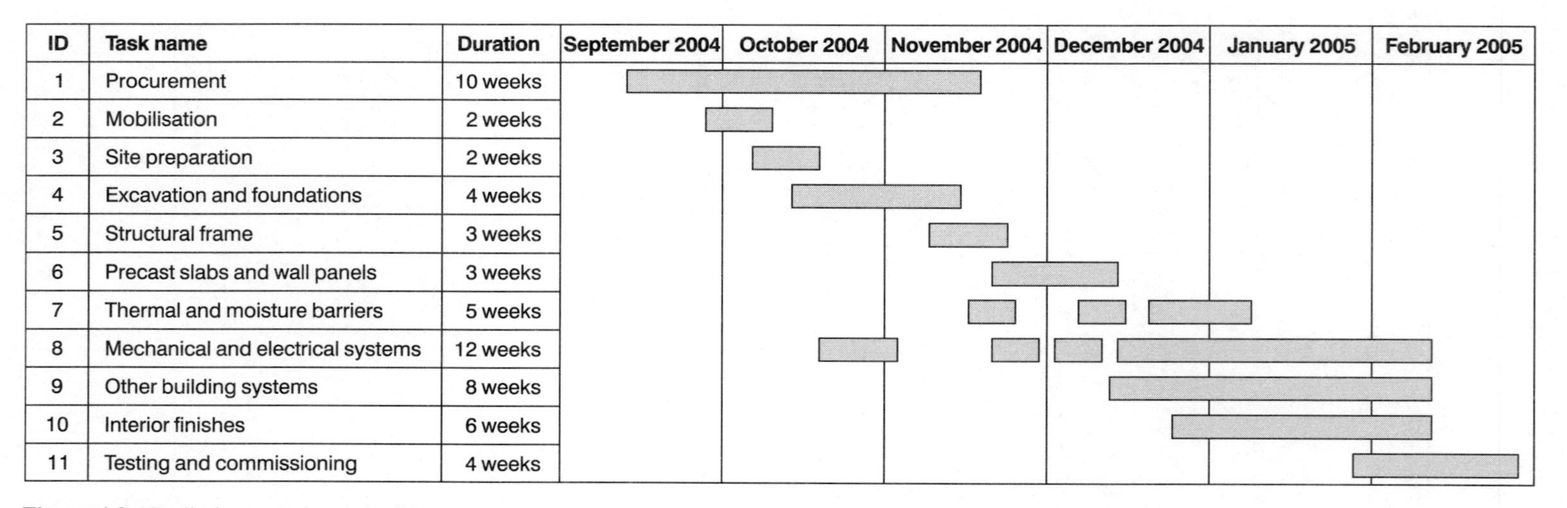

Figure 4.3 Preliminary project schedule.

Levels of detail

In Chapter 3, we noted that construction cost estimates are prepared in various levels of detail for different purposes at different points of the project life cycle and we showed, in Table 3.2, a sample of a preliminary cost estimate for a hotel project, based primarily on its floor area and number of rooms. Proceeding from least detailed and least accurate to most detailed, we describe below many of the commonly used approaches. All of them use some measure of gross unit costs from previously completed construction work, updated by the use of factors or indices that recognise cost differences due to changes over time, location variations or any peculiarities of the job being estimated.

Rough order of costs

These estimates are based on limited knowledge of the project and very little, if any, design or drawings. They are important in determining initial feasibility of the project, assisting the owner in deciding whether to proceed with additional studies and design work.

- Cost per function. This estimate is based on historic cost data per unit of use, such as cost per vehicle in a car park or garage, per patient in a hospital facility or per 1000 books in a library. For a manufacturing facility, it might be based on productive capacity, such as per tonne of ore processing capability or per annual number of vehicles assembled in an automobile factory.
- Unit area cost and unit volume cost. The estimate per unit area is found by multiplying the estimated gross floor area by the historic cost per area for the type of facility and its location. Similarly, the per-unit-volume estimate is based on the anticipated volume multiplied by a cost per unit volume. Unit-area estimates are often used in the preliminary stages of residential and other building design, while unit-volume estimates are well suited to warehouses and industrial facilities.

Preliminary assessed costs

These estimates can be carried out at various stages throughout the design process, prior to completion of the construction contract documents. They allow the owner to track the cost impacts of design development, to decide at every stage whether to proceed and to begin to arrange financing.

- Panel unit cost. This method is appropriate only for buildings. It is based on per-unit costs for the several panels in the building, such as exterior perimeter length, interior partition wall length or area, roof area and floor area.
- Parameter cost. The cost of each of several building components, or parameters, is calculated, based on their estimated unit costs multiplied, respectively, by the appropriate units, such as floor area, building perimeter and storey height. Among the common components for buildings are site work, foundations, plumbing and roofing. The Association for the Advancement of Cost Engineering has developed a parametric building cost model (Association for the Advancement of Cost Engineering International, 1996) that uses the following input parameters: floor area, floor height, number of floors, percentage of area used as office, wet laboratories and dry laboratories, percentage of area heated and cooled, substructure and superstructure strength/duty factors, exterior and interior finishes quality, mechanical and electrical services quality, and escalation, location and labour productivity factors. Figure 4.4 shows the results of the use of this Internet-based model for a sample building project.

Firm estimate of costs

These estimates are based on completed, or nearly completed, construction drawings. Thus, they become the 'final' estimates of costs, from the viewpoint of the design professional or the contractor. They allow the owner to give final approval to advancing to the construction phase.

- Partial take-off. This estimate uses quantities of major work items from the design drawings. Examples are the cubic metres of all of the structural concrete multiplied by an average cost for concrete, the weight of all of the structural steel multiplied by an average unit cost and the length of all communications cable multiplied by an average cost per linear metre. The partial take-off estimate is less detailed than that needed for the contractor's tender proposal. It could be listed under preliminary assessed costs, as it is not as 'firm' as the estimate described below. When the design professional produces an 'engineer's estimate' at the completion of design, it is likely to be based on this kind of partial take-off.
- Final cost. The 'final' cost estimate is the last one prior to construction, *not* the final accounting of actual costs after the project has been completed. It is based on a thorough review of all work quantities necessary to carry out the construction operations and produce the end result. The cost information developed in this process is the principal ingredient in the contractor's tender proposal. The next section describes the development of the final cost estimate by the contractor.

The estimating process

We begin with a statement about the limitations and diversity of approaches of cost estimating for construction projects, as carried out by contractors, taken from Clough et al. (2000):

> It must be recognized that even the final construction estimate is of limited accuracy and that it bears little resemblance to the advance determination of the production costs of mass-produced goods. By virtue of standardized conditions and close plant control, a *manufacturer* [emphasis added] can arrive at the future expense of a unit of production with considerable precision. *Construction estimating*, by comparison, is a relatively crude process. The absence of any appreciable standardization, together with a myriad of unique site and project conditions, make the advance computation of exact construction expenditures a matter more of accident than design. Nevertheless, a skilled and experienced estimator, using cost accounting information gleaned from previous construction work of a similar nature, can do a credible job of predicting construction disbursements despite the project imponderables normally involved. The character or location of a construction project sometimes presents unique problems, but some basic principles for which there is precedent almost always apply . . . There are probably as many different estimating procedures as there are estimators. In any process involving such a large number of intricate manipulations, variations naturally result. The form of the data generated, the sequential order followed, the nature of the elementary work classifications used, the mode of applying costs – all are subject to considerable diversity. Individual estimators develop and mold procedures to fit their own context and to suit their own preferences.

Parametric Building Cost Model

Case Description: Sample Research and Production Facility; Rural US

Results:

CSI Account Description	Estimated Cost	Cost per Sq.Ft.	% of Total Cost
Foundations	$7,286.00	$2.43	2.62
Substructures	$5,097.00	$1.70	1.83
Superstructures	$25,995.00	$8.66	9.33
Exterior Closures	$13,771.00	$4.59	4.94
Roofing	$11,225.00	$3.74	4.03
Interior Construction	$73,795.00	$24.60	26.49
Elevators	$8,317.00	$2.77	2.99
Mechanical	$51,523.00	$17.17	18.49
Electrical	$34,303.00	$11.43	12.31
Specialties	$4,626.00	$1.54	1.66
Architect/Design Fee	$42,674.00	$14.22	15.32
TOTAL:	**$278,612.00**	**$92.87**	**100**

Input Parameters:

Description:	Value:	Description:	Value:	Description:	Value:
Total Floor Area [s.f.]	3000	Ave. Floor Height [ft.]	10	Number of Floors	2
% as Office Space	15%	% as Wet Labs	25%	% as Dry Labs	20%
% Heated	100%	% Cooled	0%	Building Corners	6
Substructure Strength	3	Superstructure Strength	4	Exterior Finish	6
Interior Finish	6	Mechanical Services	5	Electrical Services	5
Escalation Factor	1.28	Location Factor	0.70	Productivity Factor	1.10

Note: The costs *include* all labor and material as well as an allowance for contractor's overhead and profit. The results *exclude* the following significant items: site improvements (land, landscaping, parking, utilities, etc.); furnishings and production equipment; contingency allowance. All cost estimates should include a contingency allowance to cover unidentified but expected cost occurences within the project scope. Contingency is usually estimated through the application of risk analysis techniques.

AACE International - the Association for the Advancement of Cost Engineering

209 Prairie Avenue, #100, Morgantown, WV, 26505, USA
Phone: 800.858.COST or +1.304.2968444
Fax: +1.304.2915728

E-mail: 74757.2636@compuserve.com
CompuServe: GO TCMFORUM
Internet: www.aacei.org

Figure 4.4 Parametric building cost model output (Association for the Advancement of Cost Engineering International, 1996).

Given so many different approaches, our desire here is to set forth a generalized method for cost estimating that contains at least some of the commonly used approaches. We begin with a discussion of the various elements of the tender price and then we combine them into a hypothetical estimate for a sample project. Although every tender price contains the same basic elements, terminology describing them varies with the individual construction organisation. We adopt a taxonomy suggested by Pilcher (1992), with minor modifications; many contractors use this classification. The following outline will guide our discussion:

Total tender price
- (1) Net project cost
  - (1.1) Construction cost
    - (1.11) Direct costs
      - (1.111) Labour
      - (1.112) Materials
      - (1.113) Equipment
      - (1.114) Subcontract work
    - (1.12) Site or project or job overheads or oncosts
  - (1.2) Company, general or home office overheads
- (2) Markup or margin
  - (2.1) Profit
  - (2.2) Contingency or risk

The 'estimate' usually means the estimated net project cost, consisting of the construction cost and company overheads. The contractor's cost estimating staff is responsible for compiling that net project cost. We defer until the section on 'turning the estimate into a tender' the discussion of markup and the various considerations involved in establishing that value. First, we describe the several elements of cost.

Elements of net project cost

Direct costs are those costs required to conduct operations that result directly in the installation of some component of the physical project, such as applying paint or connecting electrical cables or some other direct production work, such as soil excavation and hauling. The 'big four' among the elements of direct cost are labour, materials, equipment and subcontractor work. Sources of cost data to be used in the estimates will be the contractor's historic cost records, if any, for similar work items. A variety of published cost estimating guides and databases also provide such information, but caution must be exercised to assure the numbers are based on reasonably similar conditions.

Labour

Predicting the cost of *labour* for a construction activity is probably the most challenging part of the estimator's job. It has been said that these construction expenses are 'inherently variable and the most difficult to estimate accurately' (Clough et al., 2000). Both the productivity – the rate at which personnel complete the work activity – and the cost per hour that must be paid for that personnel effort contain many elements. Productivity varies with the size of the work crew, availability of materials and equipment, weather conditions, effectiveness of supervision, coordination with concurrent operations in the same location and the degree of caution required due to safety considerations. Labour cost per hour is influenced by union agreements and local and national legislation, which provide for such cost elements as insurance, training, holidays, sick days, travel and overtime, in addition to the regular straight time pay rate.

To estimate the labour cost for an item of work, one of three approaches can be used. All are based on the idea that the work item will consist of an estimated number of units, or quantity, and the total labour cost will be that quantity multiplied by the estimated labour cost per unit. The particular approach depends on the way the contractor keeps its historical cost records. Consider, for example, the calculation of the estimated labour cost of placing concrete in a cast-in-place foundation wall 2.5 m high, 0.3 m thick and 20 m long. The total volume of concrete in the wall will be 15 m^3. (We neglect the slight concrete volume reduction due to the reinforcing steel, as do most cost estimators.)

- If the contractor's cost records for placing concrete for this type of application are kept on the basis of labour US\$/m^3, we simply multiply the 15 m^3 by the historical unit labour cost, perhaps modified to incorporate inflationary cost changes. Thus, if the labour cost is US\$ 47.00/m^3, the labour cost estimate is simply (15 m^3)(US\$ 47.00/m^3) = US\$ 705.00.
- Perhaps a better way to keep labour effort records is by labour hours instead of labour dollars. The argument in favour of this method is that labour hours/m^3 of foundation wall concrete are not likely to change over time, whereas the cost/hour will change as labour agreements are modified. Using the same 15 m^3 example, suppose the record shows that 2 labour hours are needed to place 1 m^3 of concrete (either 2 labourers working 1 hour each, 4 labourers each working half an hour or some other combination). If the cost/labour hour is US\$ 23.50, the estimated labour cost of the work item is (15 m^3)(2 labour hours/m^3)(US\$ 23.50/labour hour) = US\$ 705.00. Note that the US\$ 47.00/m^3 in the first approach, above, is simply the product of 2 labour hours/m^3 and US\$ 23.50/labour hour.
- A third approach to this calculation is based on records of 'productivity' for the particular work item, where productivity is defined as the quantity installed per unit of input (in this case per labour hour). The productivity in our example is simply the reciprocal of the labour hours/m^3. In the second approach, above, the 2 labour hours/m^3 translate into a productivity of 0.5 m^3/labour hour. So, the calculations are (15 m^3)(US\$ 23.50/labour hour)/(0.5 m^3/labour hour) = US\$ 705.00.

The direct cost of labour also includes *indirect labour costs* (a rather confusing terminology!). These costs are those expenses paid by the employer that are in addition to the basic hourly rates, including employer contributions to various welfare and social security funds, payroll-based insurance expenses, pension plans, paid vacations and apprenticeship programmes. Such expenses can be substantial, often exceeding 50% of the basic hourly rate. Some contractors use a single labour rate, for each labour classification, that includes both basic hourly wage and indirect labour. Others add indirect labour separately. Whichever approach is used, it is essential to include these costs in the cost estimate! We assume, in our 15 m^3 example, that the US\$ 47.00/hour includes indirect labour costs as well as the basic hourly wage.

Materials

The term 'materials' includes all physical objects that become part of the finished structure. Thus materials include not only such common and traditional items as soil, concrete, steel, timber, paving, air ducts and electrical cable, but also various pieces of permanent equipment such as conveyors, medical diagnostic devices, bank vaults and kitchen stoves and refrigerators. We emphasise this point because another element of direct cost, *equipment* (sometimes called *plant*) is used to refer only to construction equipment deployed to install various elements of the project but not to become part of the completed structure.

In the process of assembling its cost estimate, the contractor will receive priced proposals from material suppliers for furnishing many of the materials required in the project. The contractor may supply some of the materials from its own inventory; examples might include a gravel source or a stock of timber purchased in large quantities for use on several projects. Whatever the potential source, the cost will be reduced to a cost per unit to conform to the estimate format, so that the cost for a work item can be calculated as the product of its estimated quantity and its cost per unit. The material cost for our $15\,m^3$ concrete example is straightforward; we expect concrete suppliers to propose furnishing concrete meeting this specification on a cost/m^3 basis. If the proposed price of timber is given on a volume or area basis (cost per board-foot is common in the USA) and if the quantity of a certain timber item is given as a length measure, some conversion of the price will be necessary. It is important that all material costs have a common basis, such as delivered and off loaded at the jobsite, with no sales tax included. As we shall see in our examples, sales tax is usually added as a single item toward the end of the estimating process.

Equipment

Construction *equipment*, or *plant*, is used to move, raise, fasten, excavate and compact the various materials installed on the project. A substantial portion of the cost of highway and heavy construction projects is for equipment, including bulldozers, scrapers, trucks, cranes and compactors. For building projects, the proportionate cost of equipment is less. Equipment may be owned by the contractor or it may be rented or leased for the project. Contractors rent equipment on a short-term basis, often month to month. Leasing is for longer terms of a year or more, with a common provision that the contractor has the option of purchasing the equipment at the end of the lease period.

An often-used process for estimating the cost of equipment is similar to that for labour, wherein a work item quantity is multiplied by the estimated equipment cost per unit. If the contractor will use its own equipment, the contractor may have sufficiently accurate unit-cost data in its historical cost database for some of the items, but for others it will be necessary to calculate the cost. This procedure will involve recognising the capital recovery costs, including both acquisition and interest, and the operating costs, including fuel and oil, maintenance, insurance, storage, taxes and licensing. Various contractors differ in their handling of the costs of equipment operators; some include these costs in the equipment category, whereas others recognise equipment operator expense as labour. Whatever the method used and whether the equipment is owned, leased or rented, the calculated cost will usually be on a daily or hourly basis. It will then have to be converted to a cost per unit for the particular work item, based on an analysis of the equipment's production rate on this item, plus some measure of non-productive time for which the contractor must pay.

Instead of including the cost of construction equipment in each bid item, the contractor may calculate and add a single amount toward the end of the process, almost as an 'overhead' item. If a fleet of equipment is expected to be utilised throughout most of the project, this cost, based on hourly, daily or monthly lease or rental rates and the total cost of any owned equipment, will appear as one item on the cost estimate summary.

Subcontract work

The final category of direct cost includes work that will be performed by *subcontractors*. Typically these are specialty contractors that will sign a contract with the general contractor for work in such areas as painting, plumbing, electrical work, concrete finishing and traffic control, if their proposals are accepted and if the general contractor is chosen. Generally, the contractor wants to engage subcontractors whose prices are lowest, but there are other considerations. If four subcontract proposals are received for electrical work, the contractor will want to inquire about the qualifications and reputations of each, in a similar way that the owner is interested in the contractor's qualifications. Further, the instructions to tenderers may provide that the owner has the right to accept or deny the contractor's choice of a subcontractor.

Beyond all that, the contractor must be very diligent to ascertain exactly what each subcontractor has included in its proposal. It is unlikely the electrical subcontractor will make an offer to 'provide all materials and perform all work contained in Division 16' (electrical work in Construction Specifications Institute (CSI) specification format). In nearly every case, subcontractors will provide qualifications or exceptions in their offers. One may exclude all testing and any concrete bases for electrical equipment. Another may attach a condition that the general contractor will unload any delivered materials, while another may exclude (1) the provision of a piece of gear it knows will be difficult to acquire and (2) all testing. Considerable analysis may be required to determine the overall cost impact of these qualifications and thus to select the subcontractor whose price will be included in the overall cost estimate.

Subcontract offers may be provided to the general contractor either on a lump-sum or unit-price basis. If a lump sum, as would be likely with electrical work, they are simply added to the other items. If they are given on a unit-price basis, the product of that proposed price times the estimated quantity becomes the estimated cost of the item; examples might be rock excavation on a per-cubic-metre basis or concrete slab finishing based on the area in square metres.

Provisional and prime cost allowances

If the precise nature of a portion of the work is not defined at the time of tendering, the tender documents may include a requirement that the tenderers include in their total prices a certain amount as a *provisional sum*, as recognition that the successful contractor will be required to perform this work. All tenderers include this same amount, which is the owner's best estimate of its cost, with the understanding that the parties will change the contract amount later if the actual cost is different from the provisional sum. *Prime cost sums* are similar, but they are confined to materials to be provided by the contractor or a subcontractor but whose exact nature or quantity cannot be defined at the time of tendering.

Site overheads

Construction cost includes not only the direct costs defined above but also various costs that occur on the construction site but are not directly associated with the various work items. They are referred to as *site overheads*, *project overheads*, *job overheads* or *oncosts*; some contractors use the term *indirect costs*, but the use of this term together with *indirect labour* costs described earlier can lead to considerable confusion. Among the types of costs included here are job supervision, site office facilities, other temporary buildings, temporary utilities, permits and fees, load tests, project scheduling and surveys. Clough and Sears (1994) list 48 typical items of site overheads, ranging, alphabetically, from badges and barricades to winter operation, worker housing and worker transportation, while the Kiewit Construction organisation includes about 75 items in its checklist (Milne, 1989).

Depending on the job, such costs may range between 5 and 15% of direct costs (Clough et al., 2000) and some contractors include a simple percentage of direct costs as their estimate for site overheads. However, a much more reliable method is to carry out an item-by-item enumeration of site overheads for the job under consideration. The importance of the preliminary schedule is evident in the preparation of such a compilation, because most of these costs are related to the durations of various portions of the schedule. In estimating site overhead costs for the project whose preliminary schedule is shown in Figure 4.3, for example, we would need, perhaps, 5.5 months of project manager time, 10 weeks for the procurement manager, 4+ months of shelter and rest facilities for construction workers and cold weather protection for the time period from the expected onset of winter until mid-December, when the building should be enclosed, as well as a multitude of other site overhead efforts. Our examples later in the chapter show the inclusion of site overhead costs in the overall cost estimate.

Company overheads

The final element constituting the net project cost includes a proportion of the cost of operating the company's home office. The terms *company overheads*, *general overheads* and *home office overheads* refer to such home office costs as officer salaries, payroll processing, information technology, advertising, legal costs, office utilities, travel, dues, donations and all other costs involved in operating that facility. They are costs incurred in supporting the company's overall construction programme that cannot be charged to any particular project. The costs of tender preparation are usually considered part of company overhead, because those efforts support the company's overall efforts to acquire project work, not just the projects it is successful in tendering. Depending on the number of projects underway in a year, the company's total home office expenses are likely to be between 2 and 8% of the company's annual business volume (Clough et al., 2000). If the total home office expenses for 2005 are estimated at US$ 720 000 and the estimated value of all project work to be completed in that year is US$ 18 000 000, a reasonable amount to add to each estimate for company overheads is (US$ 720 000)/(US$ 18 000 000) = 0.04, or 4% of the estimated construction cost.

Steps in the process

The process of compiling an estimate of net project cost, though time consuming and subject to a variety of unknowns, is relatively simple and straightforward. First, a quantity survey is performed. As indicated earlier, for some projects, the bill of quantities is compiled by a quantity surveyor engaged by the owner or design professional. For others, the first step by the contractor's cost estimator, after reviewing the drawings and other contract documents, is to perform this survey, or *take-off*. This step requires measuring and counting, assuring that all items are accounted for and none is duplicated. It requires organisational skills and the ability to produce a document that others can interpret easily. Most contracting organisations have standardised forms and procedures for performing quantity surveys. Even if the owner has provided a bill of quantities, it is good practice for the contractor's personnel to check at least some of those numbers by performing their own quantity take-off.

Once the quantities are available, the estimator begins applying direct-cost data to each item. As explained in our discussion of direct costs, such data can come directly from the contractor's historic cost database of unit costs, perhaps updated to recognise changes in such costs with

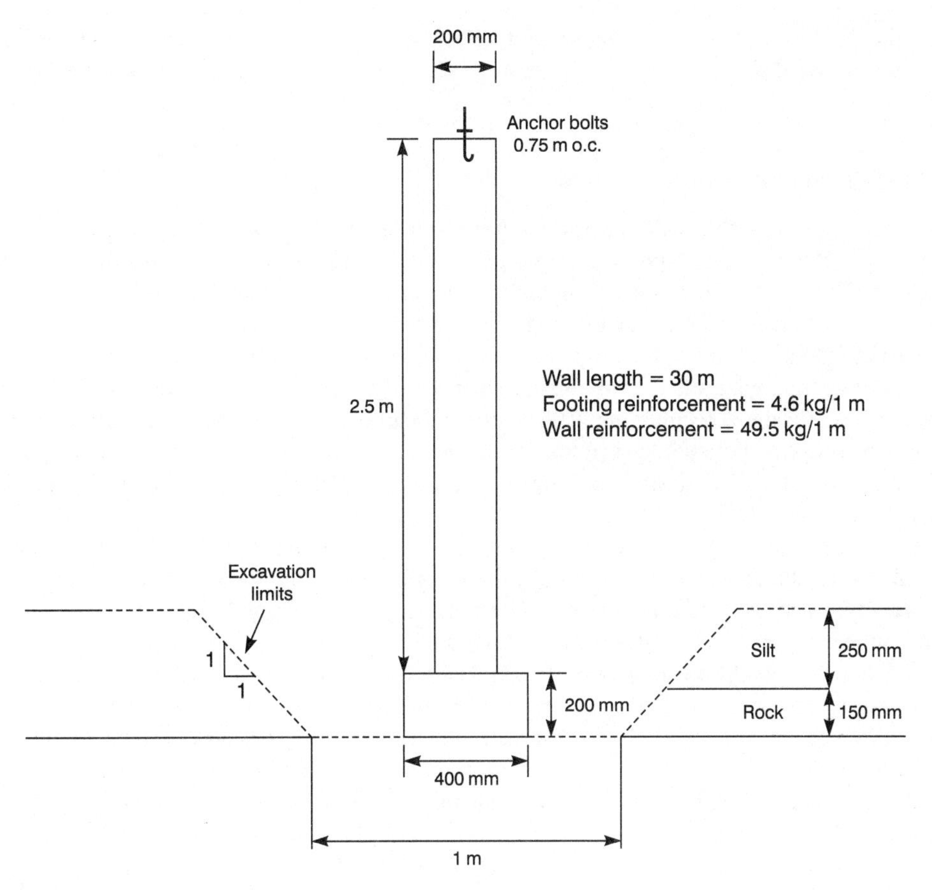

Figure 4.5 Cross-section of concrete foundation wall.

time. If a certain item is unique to the contractor's experience, an analysis will be required to estimate the unit costs, assuming a particular method and productivity rate. Material costs may come from supplier proposals, equipment costs may be based on proposals from lessors and cost estimates for subcontract work will be provided by subcontractors interested in the job. Cost data can also be found in published construction cost databases, either in hard copy or electronic form. Examples include *Means Square Foot Cost Data 2002 Book* (2002), National Construction Estimator (2001) and *Rawlinson's Australian Construction Handbook* (2001). In addition, various pieces of cost-estimating software include cost databases. In using these data, one must be certain of the conditions upon which the numbers are based. For example, does the cost per cubic metre for soil excavation include the labour cost for the operator? What kind of soil was excavated? From what depth? With what kind of machine?

A handy means of organising the direct-cost estimate and one that tends to avoid overlooking items, is on the basis of the outline of the technical specifications. Thus, all of Division 1 (in CSI format) would be listed first, followed by Division 2 and so forth.

To the direct costs, the estimator adds site overheads and general overheads, to complete the net project cost. We shall consider the establishment of markup, or margin, in a later section. Two other cost elements must be mentioned, bonds and taxes. They are both costs, but they are not added until after markup, because their values may be based on the total after inclusion of markup. As we develop the two example cost estimates, we shall observe all of these various elements of cost.

A lump-sum/fixed-price example

Consider a 'project' that will construct a cast-in-place concrete foundation wall, as shown in Figure 4.5. We desire to compile a single price for submittal to the owner for this lump-sum contract. The work will consist of mobilisation; clearing and grubbing an area of 400 m^2; excavating through silt and rock to a depth of 400 mm as shown and removing the excavated material from the site; placing footing forms, footing reinforcement and concrete for the 400 mm wide footing; removing the footing forms; placing wall forms, wall reinforcement, anchor bolts and concrete for the 200 mm wide wall; removing wall forms; backfilling on both sides of the new wall with new gravel to the original ground level; and cleanup and move-out. Note that this is a rather 'minimalist' design, lacking in specification details about concrete, reinforcing, anchor bolts and backfill.

First, we prepare a bill of quantities, with the result shown in the quantity and unit columns of Table 4.1. The unit measure represents the units by which the various items are measured. Mobilisation and cleanup and move-out are both measured as simply one job, with a lump-sum amount for each. The square metre (m^2) and cubic metre (m^3) units should require no explanation, as should the tonne (= 1000 kg) units for steel reinforcement. Anchor bolts are measured by the number of items or 'each'. The unit *smca* stands for *square metre of contact area*, the area of contact between formwork and concrete. In the case of both footing and wall, this area is the area of the vertical formwork at the sides and ends of each. As a sample of the calculations required to determine quantities, consider the excavation quantities, as follows.

Assuming the end slopes and clearances are the same as the side slopes,
rock excavation: average width = 1.15 m; average length = 30.75 m; thickness = 0.15 m
volume = $(1.15\,\text{m})(30.75\,\text{m})(0.15\,\text{m}) = 5.3\,\text{m}^3$
silt excavation: average width = 1.55 m; average length = 31.15 m; thickness = 0.25 m
volume = $(1.55\,\text{m})(31.15\,\text{m})(0.25\,\text{m}) = 12.1\,\text{m}^3$.
As a check, find the total excavation volume:
$(1.4\,\text{m})(31\,\text{m})(0.4\,\text{m}) = 17.4\,\text{m}^3 = 5.3\,\text{m}^3 + 12.1\,\text{m}^3$ (OK).

Verification of the other quantities is left to the reader. We then apply unit costs from our company database or other sources, or we develop unit costs based on a study of the operation and assumptions about productivity and costs or we apply costs from material and subcontract proposals. The result is the subtotal of US$ 18 534, as shown in Table 4.1. The two overheads are added next, site overheads of US$ 2250, the result of an analysis of supervisory, site office, utility and other expenses not shown here and company overheads stipulated by the company as 6.5% of the second subtotal or US$ 1351. The result is a net project cost of US$ 22 135. A markup of 10.5% of the net project cost (US$ 2324), sales tax on materials (US$ 183) and bond costs of 1.5% (US$ 370) complete the total tender price of US$ 25 013.

Some comments on the presentation in Table 4.1 are in order. The items labelled 'mobilisation' and 'cleanup and move-out' could be considered as applying to the entire project and thus not individual work items; in that case, they would be included as part of site overheads and not shown as direct costs. Note that we intend to subcontract the excavation and backfilling

Table 4.1 Cost estimate for concrete wall project

Item number	Item	Quantity	Unit	Labour		Material		Equipment		Subcontract		Total
				Unit cost (US$)	Extension (US$)	Unit cost (US$)	Extension (US$)	Unit cost (US$)	Extension (US$)	Unit cost (US$)	Extension (US$)	
1	Mobilisation	1	job	550.00	550			120.00	120			670.00
2	Clearing and grubbing	400	m^2	0.33	132			0.21	84			216.00
3	Silt excavation	12.1	m^3							42.00	508	508.20
4	Rock excavation	5.3	m^3							133.50	708	707.55
5	Backfill	13.8	m^3							23.70	327	327.06
6	Footing forms – place and remove	12.2	smca	31.32	382	16.45	201					582.79
7	Footing reinforcement	0.138	tonne	507.10	70	617.30	85					155.17
8	Footing concrete	2.4	m^3	62.50	150	110.80	266	19.75	47			463.32
9	Wall forms – place and remove	151	smca	49.20	7429	17.70	2673					10 101.90
10	Wall reinforcement	1.335	tonne	507.10	677	617.30	824					1501.07
11	Anchor bolts	40	ea	4.40	176	4.02	161					336.80
12	Wall concrete	15	m^3	26.40	396	126.70	1901	15.20	228			2 524.50
13	Cleanup and move-out	1	job	320.00	320			120.00	120			440.00
	Direct totals				10 282		6110		599		1543	
									Subtotal – direct costs			18 534
14									Site overheads			2 250
									Subtotal – construction cost			20 784
15									Company overheads @ 6.5%			1 351
									Subtotal – net project cost			22 135
16									Markup @ 10.5%			2 324
									Subtotal			24 460
17									Sales tax, 3% on materials			183
									Subtotal			24 643
18									Bonds @ 1.5%			370
									Total tender price			25 013

Table 4.2 Quantities for embankment and roadway estimate

Item	Description	Quantity	Unit
201(3A)	Clearing and grubbing	0.65	hectare
203(3)	Unclassified excavation	13 577	m^3
203(5)	Borrow	23 625	m^3
301(2)	Aggregate base course	3915	m^3
301(4)	Aggregate surface course	2460	m^3
401(4)	Asphalt concrete – type II; class A	2942	tonne
603(1–300)	300 mm corrugated steel pipe	80	lm
603(1–500)	500 mm corrugated steel pipe	252	lm
615(1)	Standard signs	14	m^2
630(1)	Geotextile, separation	26 100	m^2
670(3)	Traffic marking, wide stripe	4.5	km

work and pay for that work on a price-per-cubic-metre basis. The order in which some of the 'add-ons' are added is chosen by the tenderer as a matter of policy; in our example, markup is not calculated on sales tax, but sales tax could be added earlier. Also, this compilation assumes bonding costs are based on the entire amount, including markup and sales tax. The question of whether sales tax and bond costs should be classified as direct costs, overhead or markup is probably trivial. Sales tax on materials only is probably a direct cost, but if the jurisdiction extracts sales tax on the entire tender price, the tax is more on the order of an overhead cost. The order of presentation of the add-ons in Table 4.1 is typical of many such estimates.

A unit-price/measure-and-value example

Our second example of a cost estimate is for the construction of an embankment and roadway. Once again, the example is oversimplified but still includes the basic concepts. In this example, the result of our effort will be a series of unit prices for each of the specified bid items, as required by the tender instructions. These unit prices, multiplied by their respective quantities, will determine the total tender price, which will be compared with those from other tenderers. Our approach will be to calculate the total tender price using methods similar to those employed for the previous example, to allocate that total to each of the various bid items and to divide by the respective estimated quantities to determine the unit prices.

The tender documents would include a schedule of quantities such as those shown in Table 4.2. It is important to understand that the items listed here are the only items for which the owner will pay. Thus all project expenses not included under the listed items must somehow be channelled into the unit prices for the listed items. If the contractor's proposal is accepted by the owner, its payments will be based on the actual quantities installed multiplied by their respective unit prices, as explained in Chapter 2. The estimated quantities are used only to determine a total tender price and thus compare the tender prices from each tenderer.

Table 4.3 shows the cost estimate for the project whose quantities are listed in Table 4.2. First, we develop the direct costs for each of the 11 pay items. Note that the contractor intends to perform clearing and grubbing, earthwork, culvert piping and geotextile installation with its own forces and subcontract the paving, signage and traffic marking.

Table 4.3 Embankment and roadway estimate

Item	Description	Estimated quantity	Unit	Labour cost (US$)	Material cost (US$)	Subcontract cost (US$)	Direct cost (US$)	Tender total (US$)	Unit price (US$)
201(3A)	Clearing and grubbing	0.65	hectare	2 763			2 763	5 232	8049.21
203(3)	Unclassified excavation	13 577	m^3	77 406			77 406	146 575	10.80
203(5)	Borrow	23 625	m^3	76 545	100 406		176 951	335 072	14.18
301(2)	Aggregate base course	3915	m^3	40 364	161 298		201 662	381 865	97.54
301(4)	Aggregate surface course	2460	m^3	25 953	112 299		138 252	261 792	106.42
401(4)	Asphalt concrete – type II; Class A	2942	tonne			201 380	201 380	381 331	129.62
603(1–300)	300 mm corrugated steel pipe	80	lm	653	3 250		3 903	7 391	92.38
603(1–500)	500 mm corrugated steel pipe	252	lm	2 278	19 903		22 181	42 002	166.67
615(1)	Standard signs	14	m^2			7 462	7 462	14 130	1009.28
630(1)	Geotextile, separation	26 100	m^2	24 273	44 370		68 643	129 982	4.98
670(3)	Traffic marking wide stripe	4.5	km			3 365	3 365	6 372	1415.98
			Totals	250 235	441 526	212 207	903 968	1 711 743	
				Equipment			314 064		
				Mobilisation			17 500		
				Layout			7 600		
				Traffic control			36 000		
	Factor = US$1 711 743/903 968 = 1.893588			Demobilisation			12 000		
				Subtotal – direct costs			1 291 132		
				Site overhead			135 600		
				Subtotal – construction cost			1 426 732		
				General overhead @ 5.8%			82 750		
				Subtotal – net project cost			1 509 482		
				Profit and contingency @ 9.6%			144 910		
				Subtotal			1 654 393		
				Bonds			15 600		
				Subtotal			1 669 993		
				Sales tax @ 2.5%			41 750		
				Total tender price			1 711 743		

For each element of direct cost, we compile a cost estimate. The contractor's cost records may contain sufficient unit cost information to allow simply multiplying that unit cost by the estimated quantity. Clearing and grubbing might be an example of this approach. In other cases, the item's cost may be developed as a combination of several different components, each with its unit costs. For example, the material cost for corrugated steel pipe might be a combination of the costs of pipe, connectors, bedding material and hardware, even though the pay item is based solely on length of pipe. Another approach is to conduct an analysis of the work's details, setting forth, in the case of labour, the estimated time required, based on crew size and productivity and the cost per hour or day. The labour involved in base course installation might include studies of loading, hauling, dumping, spreading, watering, compacting and final grading.

Whatever method is used, we arrive at a direct cost for each element of each item. In the case of aggregate base course, for example, the estimated direct costs are US$ 201 662, based on US$ 40 364 for labour and US$ 161 298 for material. In this example, we have elected to show equipment costs as a single item, rather than as an element of each item. We noted this approach in our previous discussion of equipment costs. The rationale in this case might be that we expect our equipment fleet to be assigned to the project for its duration or specified periods within its duration and the project will be responsible for the fleet's costs continuously during the time the equipment is on the job. We thus calculate these costs based on their cost per time (say, per month) and the duration each is expected to be assigned to the project.

In our example, the total of the direct costs (excluding equipment) for all pay items is US$ 903 968. However, there are other costs that relate directly to construction operations. Whether they are classified as direct or overhead costs may be arguable, but in any case, they must be included in our cost estimate! For this project, we expect to have mobilisation, surveying and layout, traffic control and demobilisation expenses in the amounts shown. Because no pay items include these expenses, we will allocate them to the items in a process to be explained shortly. The subtotal for all direct costs is US$ 1 291 132. To this subtotal we add costs for site overhead, general overhead, profit and contingency, bonds and sales tax, in a manner similar (though not in the same order) to that explained for our lump-sum estimate.

The 'bottom line' total tender price is US$ 1 711 743. If this were a lump-sum tender, we would submit this single number as our proposed price. However, in the case of a unit-price (measure-and-value) tender, we also furnish unit prices for each item listed in the schedule of quantities. The final steps in our calculations result in unit prices for each of the 11 items which, when multiplied by their respective estimated quantities, give a sum equal, or nearly equal, to US$ 1 711 743. One way to approach the task is to allocate the sum of all the amounts below the US$ 903 968 of direct costs in Table 4.3 to each of the 11 items in proportion to that item's listed direct cost. The amount to be added is the difference between the total tender price of US$ 1 711 743 and the US$ 903 968 of direct costs, or US$ 807 775. To accomplish this calculation, we find the ratio of US$ 1 711 743 to US$ 903 968, or 1.893588 and multiply that ratio by the direct cost for each item to arrive at the 'tender total' for the item. In the case of unclassified excavation, the calculation is (US$ 77 406)(1.893588) = US$ 146 575.

As shown in Table 4.3, the total of these 11 'tender totals' is our total tender price of US$ 1 711 743, as expected. Finally, we divide each tender total by its respective estimated quantity to find its unit price. Because the unit price has been derived in this fashion, the total price in our tender, which shows unit prices multiplied by their estimated quantities, will give the desired total tender price.

Table 4.4 summarises the results of the various calculations in Table 4.3; the pricing section of the contractor's tender would consist essentially of the information in Table 4.4. Note that

Table 4.4 Unit price schedule for embankment and roadway project

Item	Description	Estimated Quantity	Unit	Unit price (US$)	Estimated amount (US$)
201(3A)	Clearing and grubbing	0.65	hectare	8049.21	5 231.99
203(3)	Unclassified excavation	13 577	m^3	10.80	146 631.60
203(5)	Borrow	23 625	m^3	14.18	335 002.50
301(2)	Aggregate base course	3 915	m^3	97.54	381 869.10
301(4)	Aggregate surface course	2 460	m^3	106.42	261 793.20
401(4)	Asphalt concrete – type II; class A	2 942	tonne	129.62	381 342.04
603(1–300)	300 mm corrugated steel pipe	80	lm	92.38	7 390.40
603(1–500)	500 mm corrugated steel pipe	252	lm	166.67	42 000.84
615(1)	Standard signs	14	m^2	1009.28	14 129.92
630(1)	Geotextile, separation	26 100	m^2	4.98	129 978.00
670(3)	Traffic marking, wide stripe	4.5	km	1415.98	6 371.91
			Final tender price		1 711 741.50

rounding of the unit prices to the nearest US$ 0.01, as would be required in a tender offer, results in some minor differences between the 11 'tender total' numbers in Table 4.3 and the 'estimated amounts' in Table 4.4, although the final tender price, at US$ 1 711 741.50, is remarkably close to our target of US$ 1 711 743.

The calculations explained above are an example of 'balanced' unit-price tendering, in which the allocations of the various overheads and other non-direct charges are allocated proportionately to each tender item's direct cost. There are a variety of other ways we could split the total tender price of US$ 1 711 743 among the 11 items and still arrive at the desired total. For example, we could reduce the amount for asphalt concrete by US$ 10 000 and add an equal amount to the clearing and grubbing item. The resulting unit prices would be US$ 23 434 per hectare and US$ 126.22 per tonne. This practice of 'unbalanced' tendering might be used by a contractor interested in increasing its cash flow early in the project by receiving higher revenue on clearing and grubbing; note that the total revenue is unchanged, assuming the actual quantities are the same as the estimated quantities. A contractor might also employ unbalanced tendering if it believed the actual quantities for some items were going to be different than estimated. (This is a good reason to perform independent calculations to check the quantities furnished in the schedule of quantities.) Items whose quantities were expected to be higher than estimated would be assigned higher than proportional unit prices, while those anticipated to be lower than estimated would have their unit prices reduced. An exercise at the end of this chapter demonstrates this effect. Caution is advised in using unbalanced tendering for this latter purpose; if the actual quantities are not as anticipated, the results may be unfavourable.

Cost-estimating software

The process described above for compiling construction cost estimates might be described as simple, routine, time consuming, even boring, but also essential, with requirements for accuracy, organisation, no omissions and no duplications. The spreadsheets used to prepare Tables 4.1

through 4.4 are examples of the use of information technology to manage effectively and efficiently the large amounts of cost-estimate data in a simple, logical manner and many contractors prepare their estimates using this well-suited application. Beyond the use of spreadsheets, several commercial software packages provide convenient ways of data management by combining spreadsheet concepts with the use of cost databases and other features.

The basic structure of a cost-estimating program includes a spreadsheet-like table, with one line for each item, a means for calculating item quantities, databases of cost information, routines for determining and adding various add-ons, capability for all required cost calculations and means to interface with software that will be used to account for project costs as the project proceeds (Bennett, 2000). A sampling of available packages includes the Global Estimating® program developed and supported by BuildSoft Pty. Ltd (2002), the Precision Estimating Collection® (Timberline Software Corporation, 2002), which integrates the process from conceptual estimate to final bill of materials and also interfaces with its Gold Collection for project cost accounting and Everest®, an estimating and cost planning program available from Construction Software Services Partnership Pty. Ltd (2002). Several software developers, including BuildSoft, have taken the process a step further with on-line estimating systems, by which the user can upload schedules of quantities to a website, where subcontractors and material suppliers can attach their proposed prices electronically.

Another package is the WinEst® 6.01 program (WinEstimator, Inc., 2002). We have utilised this package to produce an estimate for our concrete wall example, whose spreadsheet version was presented in Table 4.1. Details for each item are shown in Table 4.5, including take-off quantity and the unit cost for each element of cost, plus the several items of overhead and markup as discussed previously. Table 4.6 contains the same information in summary form; it is organised by major CSI section and includes the total, rather than the unit, cost for each major section.

Value engineering

In our consideration of the role of the project planner and designer, we have discussed the importance of constructability analysis and value engineering. Formal value engineering is an essential activity throughout the construction process as well. In this process, the contractor is invited to suggest changes to the design that will result in cost savings; if a proposal is accepted, the contractor and owner share the savings in a manner provided in the contract documents. A common contract provision (Office of Federal Procurement Policy, 2002) reads as follows:

> The Contractor is encouraged to develop, prepare, and submit value engineering change proposals (VECP's) voluntarily. The Contractor shall share in any net acquisition savings realized from accepted VECP's, in accordance with the incentive sharing rates in paragraph (f) of this clause.

Note that if the contractor saves money by implementing a different construction technique than originally envisioned, with no change in design, the cost savings accrue solely and fully to the contractor. While such analyses by the contractor are usually presumed to take place after the contract is signed (the 'Contractor' in the above quotation means the contractor that has already been selected and is under contract), value engineering proposals made during the tendering process may provide benefits to both the owner and contractor. Thus, the cost estimator may be

Table 4.5 Concrete wall estimate, showing unit prices for each item, from WinEst® 6.01 program

CSI	Item description	Takeoff qty	Unit	Labour $/unit	Mat $/unit	Subs $/unit	Equip $/unit	Other $/unit	User $/unit	Total $/unit	Grand total
1505	Mobilisation	1	job	550.00			120.00			670.00	670.00
1710	Clean up and move out	1	job	320.00			120.00			440.00	440.00
2110	Clearing and grubbing	400	m^2	0.33			0.21			0.54	216.00
2220	Silt excavation – subcontractor	12	m^3			42.00				42.00	508.20
2220	Rock excavation – subcontractor	5	m^3			133.50				133.50	707.65
2224	Backfill – subcontractor	14	m^3			23.70				23.70	327.06
3115	Continuous footing edge forms – place and remove	12	smca	31.32	16.45					47.77	582.79
3135	Wallo framework – place and remove	151	smca	49.20	17.70					66.90	10 101.90
3211	Footing reinforcement	0	tonne	507.10	617.30					1124.40	155.17
3213	Wall reinforcement	1	tonne	507.10	617.30					1124.40	1 501.07
3310	Footing concrete	2	m^3	62.50	110.80		19.75			193.05	463.32
3318	Wall concrete	15	m^3	26.40	126.70		15.20			168.30	2 524.50
5050	Anchor bolts	40	each	4.40	4.02					8.42	336.80
					Grand total						18 534.37
					Net costs subtotal						18 534.37
	12.14%				Site overheads						2 250.00
					Subtotal						20 784.37
	6.50%				Company overheads						1 350.98
					Subtotal – net project cost						22 135.35
	11.18%				Markup @ 10.5% of net project						2 324.40
					Cost						
					Subtotal						24 459.75
					Bonds @ 1.5%						369.65
					Total estimate						24 829.40
	0.75%				Sales tax, 3% on materials						183.40
					Total estimate with taxes						25 012.80

Courtesy of WinEstimator, Inc. (2002).

Table 4.6 Concrete wall estimate, showing summary by major Construction Specifications Institute section, from WinEst 6.01 program

CSI	Major section	Labour total	Mat total	Subs total	Equip total	Other total	Grand total
1500	Constr. facilities and temp constr.	550.00			120.00		670.00
1700	Contract closeout	320.00			120.00		440.00
2100	Site preparation	132.00			84.00		216.00
2200	Earthwork			1542.81			1 542.81
3100	Concrete formwork	7 811.30	2873.39				10 684.69
3200	Concrete reinforcement	746.96	909.28				1 656.24
3300	Concrete material and placing	546.00	2166.42		275.40		2 987.82
5050	Metal fastening	176.00	160.80				336.80
	Grand total	**10 282.26**	**6109.89**	**1542.81**	**599.40**		**18 534.37**
				Net costs subtotal			18 534.37
	12.14%			Site overheads			2 250.00
				Subtotal			20 784.37
	6.50%			Company overheads			1 350.98
				Subtotal – net project cost			22 135.35
	11.18%			Markup @ 10.5% of net project			2 324.40
				Cost			
				Subtotal			24 459.75
				Bonds @ 1.5%			369.65
				Total estimate			24 829.40
	0.75%			Sales tax, 3% on materials			183.40
				Total estimate with taxes			25 012.80

Courtesy of WinEstimator, Inc. (2002).

involved in pricing various alternative designs and developing proposals for their implementation, in addition to performing the more standard estimating work. The resulting tender contains a base price related to the original design, plus proposed prices associated with the proposed design alternatives.

Proposal preparation, submittal and opening

The final steps in the contractor selection phase involve decisions about the size of the markup to be added to the net project cost, the submittal and opening of all tenders, the selection of the successful contractor and, at last, the *notice to proceed*, which directs the contractor to begin work.

Turning the estimate into a tender

Pilcher (1992) defines *tendering* as

> the process whereby a contractor, given the net cost, converts this to the sum that will actually be submitted to the client, together with any qualifications that are seen to be required. At this stage the principal discussions are concerned with the profit and the risk, together known as the *margin* or the *markup*.

Although we have already illustrated the addition of markup, or 'profit and contingency', to net project cost in the determination of total tender price, it is important that we consider some of the issues involved in determining the markup amount. Also, the separation of the rather mechanical process of determining the net project cost from the judgment-based setting of markup is an important part of the thought process. Furthermore, while the estimating staff will perform most or all of the assembly of net project cost, the setting of markup is the responsibility of upper management. If the net project cost estimate is a relatively accurate prediction of what the project will cost, the decision about markup will 'make or break' the project. For those reasons, we set this section apart from the 'cost estimating' section in this presentation.

Hendrickson and Au (1989), in their discussion of the principles of competitive bidding, state that most contractors 'exercise a high degree of subjective judgment' in the setting of markup. The process is far from exact. In a competitive tendering situation, two opposing objectives are at work: (1) the desire to be selected as the winning tenderer and (2) the desire to make a decent profit from the project. The lower the markup, the higher will be the probability of the contractor having a sufficiently low price to be selected. On the other hand, the higher the markup, the greater will be the profit if the contractor is selected, other things being equal. It is upper management's responsibility to set an 'optimum' markup that balances these two objectives for any particular tender.

The factors that may be involved in setting the markup amount have some resemblance to the factors the contractor considers in making the decision whether to spend the effort to prepare the tender, discussed earlier in this chapter. The factors are related to the contractor's potential risk if the project is won, the degree of need and desire to win the project and the expected competition. Among all the factors a contractor might take into account are the following.

- The owner and design professional and the likelihood they will cause difficulties for the contractor.

- Stipulations in the contract documents for delays in payments or retention of moneys owed to the contractor.
- Disclaimer clauses that place on the contractor most or all of the risk for unknown physical conditions at the site, especially underground conditions.
- Clauses making the contractor responsible for any delay in the project, even if not caused by the contractor.
- Other clauses providing for procedures the contractor may believe to be unreasonable, for such matters as change orders (variations), contract claims and the rendering of binding decisions in case of disputes.
- The extent to which the contractor may be liable for any worker safety-and-health problems or labour law violations.
- The project's location, size and complexity.
- The amount of work to be done by the contractor's own forces in comparison to work to be done by subcontractors. In general, contractors believe that more risk is associated with doing the work yourself; some apply a higher markup percentage to that work than to subcontracted work.
- How 'hungry' the contractor is, based on the number of projects the contractor already has under contract and the potential for other new projects. This degree of desire to win the project can be a major influence on markup.
- The expected competition, including the number of tenderers and the characteristics of each. In general, the competition will be more intense as the number of tenderers increases and the contractors will need a smaller markup to have a high chance of success. Also, if some of the other contractors have reputations for offering low-priced proposals, this fact may influence our contractor's markup decision. Of course these other tenderers are considering the same factors when pricing their proposals, including their current and future workloads, so attempts to analyse their potential competitiveness will be inexact at best.

Most of these factors are quite intangible, which is why the evaluation of these risks and the decision on markup is the responsibility of those personnel with considerable experience and judgment skills. In our two examples earlier in this chapter, we used markups of 9.6% and 10.5%, before adding taxes and bond costs. In good times, these may be reasonable percentages, but it is well known that some contractors add only a few per cent for profit and contingency when the competition is intense and they are highly desirous of obtaining the work.

One other comment is in order with regard to markup. Sometimes this term is defined to include general overhead, as well as profit and contingency. In that case, of course, a higher percentage would be used. Clough et al. (2000) suggest that markup may vary from 5% to more than 20%, especially if it includes general overhead.

Submittal and opening process

Recall the first paragraph of the invitation to tender in Figure 3.5:

> Sealed tenders marked City of Yellowknife, 2001 Water and Sewer Upgrade Contract B will be received at City Hall, City of Yellowknife, NT until 1:30:00 p.m. local time on April 4, 2001 at which time the Tenders will be opened in public.

All of the contractor's efforts described so far in this chapter have been targeted towards the tender-opening occasion! The sentence quoted above contains three important pieces of information: (1) the nature of the envelope that contains the tender, (2) the time and place of the opening and (3) the fact that there will be a public opening (and presumably reading) of the offers. The contractor completes the required tender proposal form and places it in a sealed envelope, addressed as stipulated and labelled as a proposal for the specified project. If the form of the tender proposal has been included with the construction contract documents, the contractor is obligated to use it for its submission. The total submission will include that form, a certified cheque or tender bond as security to assure the owner that the contractor will sign the contract without delay if the contractor is selected and, if required, information about the contractor and its plans for the project that will assist the owner in evaluating the non-price aspects of the proposal. In Figure 4.6 we show a typical proposal form for a very simple lump-sum project; note that the form requires the tenderer to provide the total amount, as well as the amount for each of the two items, in both words and figures.

The contractor is responsible for delivering the envelope on time and at the specified place for the tender opening. Common practice is to deliver the tender in person shortly before the required time, although the document may also be dispatched by post or messenger service. If the tender is late, even by a few minutes, many jurisdictions in the public sector disallow its being considered. Whether or not the instructions to tenderers contain words like ‘late submissions will not be considered’, the contractor is well advised to be early with tender submissions. The case study on ‘Is that a late tender?’ at the end of this chapter is an interesting recap of two diverging rulings in the case of tenders submitted within one minute after the stated deadline.

Public opening of tenders is a requirement in nearly all public contract policy and it is common on private projects as well. The meeting is usually attended by general contractors who have submitted tenders, subcontractors, material suppliers, the design professional, the owner and others who may be interested in the project. The process involves the owner, or its representative, opening each envelope in turn and reading pertinent information, including the total offer price and, in the case of a unit price contract, the amount of each item. A checklist for the opening of tenders, as used by the Republic of Slovenia (Republika Slovenija, 2002), is shown in Figure 4.7. Tabulation forms are often provided to attendees, so they can track the numbers as they are announced. At many tender openings, the cost estimate prepared by the design professional is made available for comparison with contractors’ prices.

There can be only one winner, so the tender-opening occasion is usually accompanied by many ‘sweaty hands’ on the part of those who have spent much time and effort in hopes of being selected. If the contractor is to be selected using other criteria in addition to price, nothing can be said at the tender opening about a winning contractor. Even if price is the sole criterion, the owner still may not comment on the prices and will certainly go no further than announcing an *apparent low tenderer*, because it will want to review all submitted materials before making an award.

Selecting the successful contractor

For some construction projects, selection of the contractor is based on the lowest tender price, provided the contractor is qualified. For others, the criteria are wider than price alone. On public work in the USA, for example, despite contract provisions that include words such as ‘the owner reserves the right to reject any and all bids, and to award the contract to other than the lowest bidder’, it is unusual for the owner to award to other than the low bidder, if that low bidder is qualified. Many legal battles have resulted when contractors with lowest proposed prices have been rejected; the key in these proceedings is the court’s interpretation and application of the term ‘qualified’. Privately funded projects give the owner more flexibility in evaluating

NENANA PHASE II WATER & SEWER PROJECT
INSTALLATION OF UNDERGROUND STEEL PIPE CASINGS **BID FORM**
MUNICIPALITY OF NENANA

BID FORM

MUNICIPALITY OF NENANA, ALASKA

The undersigned Bidder hereby certifies and represents that it has examined and thoroughly understands the Request For Bids, Instructions To Bidders, Bid Form, Drawings, General Requirements, Contract, and any subsequently issued Addenda (hereafter collectively referred to as the "bid documents"), including the following (If no addenda have been received state "none").

Addenda No. Date

In strict accordance with the bid documents, the five underground steel casings shall be provided and installed on or before May 31, 2002, for the following amount.

Item	Description	Bid Amount (Words)	Bid Amt. ($)
1	Mobilization & Demobilization		
2	Provide & Install Steel Casings		

TOTAL BID AMOUNT:____________________ ($________)
WORDS FIGURE

Pursuant to and in compliance with the bid documents, Bidder hereby proposes to furnish all labor, equipment and materials specified by Nenana in accordance with the terms and conditions of the bid documents for the basic bid amount stated above. The undersigned further certifies that he or she is lawfully authorized to submit this bid on behalf of the Bidder, and that Bidder will obtain the required insurance and otherwise comply with all of the terms and conditions of the Contract included in the bid documents.

Bidder's Name:____________________

Signature of Bidder's Representative: ____________________

Printed Name of Bidder's Representative: ____________________

Title of Bidder's Representative: ____________________

Telephone: ____________________ Fax: ____________________

Address: ____________________

BID FORM - PAGE 1

Figure 4.6 Proposal form for lump-sum project at Nenana Alaska. Courtesy of Design Alaska Inc.

WORKS TENDER OPENING CHECKLIST

PUBLICATION REF:____

Step	✔
Preparatory session	
1. Chairman and Secretary check documentation for proper approval of all Evaluation Committee members and any observers	
2. Chairman describes the scope of the proposed contract, identifies the organisations responsible for preparing the tender dossier, and summarises the essential features of the tender procedure to date, including the evaluation grid published as part of the tender dossier	
Tender opening session	
1. All tender envelopes are handed over to the Chairman	
2. All tender envelopes must be numbered according to the order in which they have been received	
3. Chairman verifies that all tender envelopes which have been received are available at the tender opening session	
4. Chairman & Secretary verify that all tender envelopes were sealed and in good condition	
5. Chairman & Secretary open the tender envelopes in order of receipt. They mark the tender envelope number on the front page of each document	
6. For each tender envelope, the Chairman & Secretary announce and check that the summary of tenders received correctly records: ❑ The registration number on the envelope ❑ The name of the tenderer ❑ The date (and time, for those received on the last date for submission of tenders) of receipt ❑ The condition of the outer envelope ❑ Whether or not the tenderer has duly completed a tender submission form ❑ The total financial offer and any discounts applicable (exact working as in the tender submission form) ❑ Overall decision regarding suitability of tenders for further evaluation	
7. Declarations of impartiality and confidentiality are signed by all members of the Evaluation Committee and any observers	
8. Chairman signs the Summary of tenders received	
9. For each tender considered suitable for further evaluation (as determined on the Summary of tender received), the Chairman initials the first page of every copy of the tender and all pages of the financial offer	
10. The Tender opening report is signed by all the members of the Evaluation Committee	

Figure 4.7 Works tender opening checklist (from Republika Slovenija, 2002).

proposals. In many other English-speaking countries, contract documents for public projects require the use of several criteria, in addition to price, in the evaluation process.

Criteria

First, we consider the use of *price alone* as the selection criteria. If those contractors submitting tenders have already been pre-qualified, the owner may simply compare their prices and select that contractor with the lowest price. Even in that case, the invitation for tenders may have requested that proposals include prices for alternatives, in addition to the basic work. If the available funds make it possible to include some of those alternatives, in addition to the basic work, the owner may analyse combinations of several alternatives to decide which proposal can provide the greatest benefits at the lowest price consistent with available funds. If no pre-qualification process has been used to identify qualified contractors, the owner might still award to the low tenderer, but it is more likely that a post-qualification process would be used to ascertain the lowest qualified tenderer.

Often, the contract is awarded to the *lowest qualified tenderer.* Sometimes tendering documents will specify that the contract will be awarded to the 'lowest, responsive, responsible tenderer'. These three adjectives have special meaning. *Lowest* means the lowest price, either for the basic work or for the basic work plus some or all of the alternatives. *Responsive* means that the tender submittal is completely regular, with all signatures in place, all blanks filled in, all attachments such as the tender bond included and all addenda acknowledged and that it was on time. *Responsible* means qualified in the sense we have already used the term.

The Texas Department of Transportation (2002), in its standard contract language for the award and execution of unit price contracts, states the following:

> 30–01 CONSIDERATION OF PROPOSALS. After the proposals are publicly opened and read, they will be compared on the basis of the summation of the products obtained by multiplying the estimated quantities shown in the proposal by the unit bid prices. If a bidder's proposal contains a discrepancy between unit bid prices written in words and unit bid prices written in numbers, the unit price written in words shall govern.
>
> Until the award of a contract is made, the owner reserves the right to reject a bidder's proposal for any of the following reasons:
>
> (a) If the proposal is irregular as specified in the subsection titled IRREGULAR PROPOSALS of Section 20.
> (b) If the bidder is disqualified for any of the reasons specified in the subsection titled DISQUALIFICATION OF BIDDERS of Section 20.
>
> In addition, until the award of a contract is made, the owner reserves the right to reject any or all proposals; waive technicalities, if such waiver is in the best interest of the owner and is in conformance with applicable state and local laws or regulations pertaining to the letting of construction contracts; advertise for new proposals; or proceed with the work otherwise. All such actions shall promote the owner's best interests.
>
> 30–02 AWARD OF CONTRACT. The award of a contract, if it is to be awarded, shall be made within 30 calendar days of the date specified for publicly opening proposals, unless otherwise specified herein.

Award of the contract shall be made by the owner to the lowest, qualified bidder whose proposal conforms to the cited requirements of the owner.

A proposal is considered irregular, according to Section 20,

(a) If the proposal is on a form other than that furnished by the owner, or if the owner's form is altered, or if any part of the proposal form is detached.
(b) If there are unauthorized additions, conditional or alternate pay items, or irregularities of any kind which make the proposal incomplete, indefinite, or otherwise ambiguous.
(c) If the proposal does not contain a unit price for each pay item listed in the proposal, except in the case of authorized alternate pay items, for which the bidder is not required to furnish a unit price.
(d) If the proposal contains unit prices that are obviously unbalanced.
(e) If the proposal is not accompanied by the proposal guaranty *[tender bond or other security]* specified by the owner.

A tenderer may be disqualified, according to Section 20, if (1) it submits more than one proposal from the same partnership, firm or corporation under the same or different name, (2) there is evidence of collusion among bidders or (3) it is considered to be in 'default' for any reason specified in a separate subsection.

The foregoing quotation is a good illustration of many of the issues we have covered in this section. We should also note that the Texas documents require that 'each bidder shall furnish the owner satisfactory evidence of his/her competency to perform the proposed work' at the time the proposal is submitted. Note, finally, that Section 30–01 above allows the owner to 'waive technicalities'; this provision makes it possible for the owner to ignore minor irregularities in tender submittals, if it is in the owner's best interest to do so and if the result is legal.

The third and final approach to evaluating contractors' tenders is on a *best-value* basis, wherein factors other than low price are considered. In this case, a series of criteria are set forth, to which tenderers respond, in a manner similar to the method we described for the selection of design consultants. Those responsible for the evaluation assign a mark to each item in each response and compile the overall scores. After review, the project is awarded to the contractor with the most favourable score.

In one project in New Zealand, the construction of renovations and additions to the Queen Elizabeth II Pool complex in Christchurch, the following criteria and corresponding weightings were used to select the contractor: key personnel (40%), relevant experience (20%), quality record (20%), proposed methodology (15%) and quality assurance systems (5%). It may be surprising that price is excluded from this list! In other cases, contractors' proposed prices are included in the consideration, with a weighting percentage representative of the relative importance the owner attaches to that criterion. In any case, the owner will inform tenderers of the criteria to be used in the evaluation, but practices vary as to whether contractors are informed of the weights to be assigned to each criterion.

Table 4.7 contains a scoring matrix for contractor selection on another New Zealand project. Note at least three matters of interest, in addition to the list of criteria. First, the initial scoring was performed without reference to the tendered price. Second, two of the five tenders were eliminated after the initial scoring. Third, the tendered price was weighted at 15% of the total in the final scoring.

Table 4.7 Contractor selection criteria matrix

CONTRACT A – Carson Group

Contractor selection criteria	Weighting %	Contractor #1				Contractor #2				Contractor #3				#4		#5	
		Initial		Final		Initial		Final		Initial		Final		Initial		Initial	
		Score	Wt	Score	Wt	Score	Wt	Score	Wt	Score	Wt	Score	Wt	Score	Wt	Score	Wt
Company Health project construction experience in the last 5 years.	15	0	0	0	0	100	15	100	15	70	10.5	70	10.5	20	3	0	0
Experience capability of Tenderer's proposed senior site management on multi-million dollar projects (team list, roles and responsibilities, CVs, percentage of time on site) for each person.	15	70	10.5	70	10.5	100	15	100	15	70	10.5	70	10.5	50	7.5	50	7.5
Company's present workload and capability to support this contract (and possibly other contracts as part of the project).	10	70	7	90	9	100	10	100	10	100	10	100	10	65	6.5	70	7
References – Tenderers shall provide the names and contact details of 5 clients and 3 consultant project managers who have worked recently with the proposed team (Referees will be asked to comment on performance, attitude, negotiation stance, time and quality).	10		0	50	5		0	80	8		0	65	6.5		0		0

Table 4.7 Continued

CONTRACT A – Carson Group

Contractor selection criteria	Weighting %	Contractor #1				Contractor #2				Contractor #3				#4		#5	
		Initial		Final		Initial		Final		Initial		Final		Initial		Initial	
		Score	Wt	Score	Wt	Score	Wt	Score	Wt	Score	Wt	Score	Wt	Score	Wt	Score	Wt
Quality Control Systems.	5	80	4	80	4	90	4.5	90	4.5	90	4.5	90	4.5	40	2	40	2
Company and senior site management experience in working in a large functioning facility.	10	65	6.5	65	6.5	90	9	90	9	80	8	70	7	45	4.5	50	5
Experience in Partnering.	5	80	4	80	4	50	2.5	65	3.25	55	2.75	80	4	12	0.6	20	1
Contract resources/equipment available financial strength, bond provision capability, balance sheet and capital structure.	15	60	9	60	9	90	13.5	90	13.5	95	14.25	95	14.25	55	8.25	70	10.5
Tendered Price	15		0	85	12.8		0	95	14.3		0	90	13.5		0		0
	100		41		60.8		69.5		92.5		60.5		80.75		32.35		33

Courtesy of Carson Group, Christchurch, New Zealand.

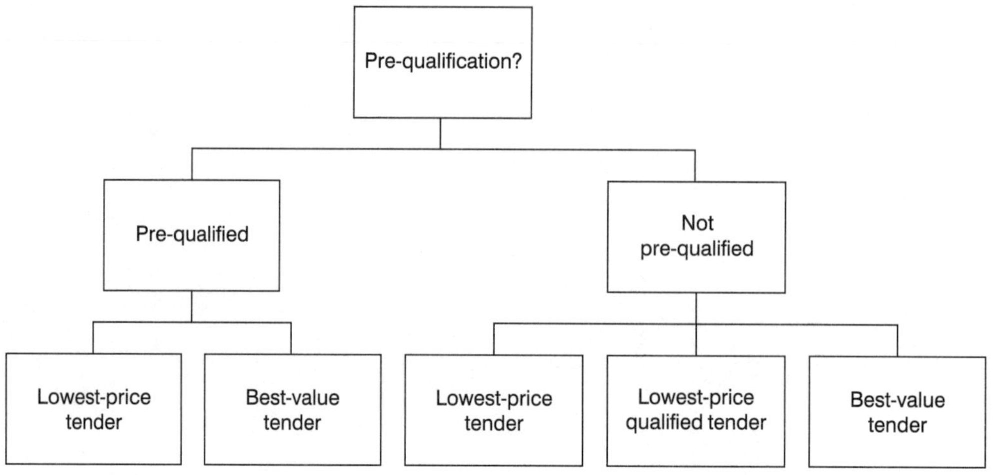

Figure 4.8 Contractor selection criteria options.

Figure 4.8 sets forth a summary of the three criteria options discussed above, both with and without contractor pre-qualification.

Qualifications

In preparing and submitting their tenders, contractors may find that they would like to change some of the requirements or conditions specified for the project and thus 'qualify' their proposals. Like so much of our discussion regarding the tendering process, practices differ in different parts of the world and between public- and private-sector projects. In some settings, whether public or private, contractors regularly attach such *qualifications*, or *tags*, to their proposals. One example is a proposal to furnish a lighting system not allowed in the specifications because it would be manufactured by a company not listed in the documents; the contractor might qualify its tender by saying that the use of a locally manufactured system would lower the tender price by a certain amount. Another example is a tender in which a contractor stated that that its price did not include the removal of obstructions such as wood and concrete in the excavation for a sewer line. When such tags are permitted to accompany tenders, the owner must evaluate their impact on the project cost in its analysis of tender prices.

In places where qualifications are not allowed, the owner would probably reject a tender containing such tags by saying it is not 'responsive' to the invitation for tenders. This culture prevails in most public-sector work in the USA, although such owners sometimes invite tenderers to propose alternative methods and materials for any portion of the work they choose. In general, it can be said that private-sector work provides much more flexibility and that the owner has more freedom to accept proposal qualifications.

Notice to proceed and contract agreement

After the tenders have been opened, the owner begins a period of evaluation. So as not to hold contractors 'in limbo' for an excessive time, most tender conditions provide for an acceptance period, often 60 or 90 calendar days long. During this time, submitted tenders may not be withdrawn. The owner must send a notice to the successful tenderer during this period; if it does not, the tenderers are released from their proposals. The *notice to proceed* is the document that notifies the contractor of the acceptance of its proposal and directs the contractor to commence work, often within a specified time such as 10 calendar days. The notice to proceed also 'starts the clock' by establishing the reference date from which the project duration is measured; often the contract will stipulate that work will be completed a stated number of calendar days after the contractor receives its notice to proceed.

It is good practice for the owner to notify the unsuccessful tenderers as well. Some public agency regulations require that unsuccessful tenderers receive notice of the name and contract amount of the selected contractor and some regulations even require that all tenderers be notified of the tags that have been accepted.

The notice to proceed implies that the site is free of encumbrances and thus available for the contractor's use. If this is not the case, because some unresolved property ownership or other matters are outstanding, the owner may issue a *letter of intent*, stating that it intends to contract with the contractor as soon as the unresolved issues are settled.

The final step in the contractor selection phase is the issuance of the formal *contract agreement*. In a single document, it binds the owner and the contractor and includes, by reference, the drawings, general and special conditions and technical specifications, as we explained in Chapter 3. The contractor may celebrate briefly upon receiving its copy of the signed contract agreement, but soon thereafter the real work begins in the next phase of the project life cycle, as the contractor begins to mobilise for the beginning of field operations.

Discussion questions

1. Make a brief study of methods for placing concrete for sub-sea bridge pier foundations. Describe each method including required equipment, list advantages and limitations of each and tell how a contractor might decide, during its constructability analysis, which method will be used.
2. You need to develop a cost per cubic metre, including operator, for a truck operation, to be used in preparing a tender and, if the tender is successful, in the subsequent monitoring and control of costs. The truck will be leased for US$ 450 per week, without operator. It is expected to be operated for 30 hours per week and to haul 50 cubic metres per hour. The operator earns US$ 18 per hour during the 30 hours the truck is operating and the contractor must also pay an additional 30% for indirect labour costs. Assume that the lease rate includes all operating costs except operator. Find the cost per cubic metre.
3. An owner and its design professional will pre-qualify general contractors for a remote marine docking facility in Northern Canada. What factors should be included in the prequalification criteria?
4. For the project in Question 3, identify five especially important factors to be considered by a prospective contractor as it investigates whether to submit a tender.
5. Prepare an outline for the method statement a prospective contractor in Question 3 might prepare prior to developing its detailed cost estimate.
6. Pick one aspect of the marine docking facility project in Question 3, such as piling installation or mining and placing gravel over frozen ground for a roadway. Identify several of the possible methods that might be studied in the constructability analysis.
7. Perform calculations to verify quantities in Table 4.1 for Item 5 – backfill, Item 9 – wall forms – place and remove, Item 10 – wall reinforcement and Item 12 – wall concrete.

8 Find an on-line source that advertises construction tender opportunities, such as the *New Zealand Tenders Gazette*. List the kinds of information that are commonly included in each listing.
9 Suppose a cost estimate is needed for a 20 m × 35 m single-storey warehouse with a flat roof and two 20 lm interior partitions. Exterior and interior walls are 6 m high. Explain how one would prepare a panel unit cost estimate. Calculate the appropriate quantities. Set up a table, with proper column headings, that could be used to determine the estimate if the unit costs were available.
10 For what purpose might the estimate in Question 9 be prepared?
11 Explain the differences between the cost estimate in Question 9 and a final cost estimate for this same project. List 20 items whose quantities might be needed in developing a final cost estimate for this project. (You will need to make some assumptions about the kinds of materials that have been specified for this building!)
12 Based on any assumptions you wish to make, prepare a bar chart for a preliminary project schedule for the warehouse project in Question 9. Include about 15 tasks in your schedule.
13 If a contractor decides to prepare a tender for the warehouse project in Question 9, what will be the primary considerations in deciding how large a figure to add for markup?
14 Consider the unit price schedule in Table 4.4. If we consider this schedule to represent a 'balanced' tender, an 'unbalanced' tender might use the following unit prices for unclassified excavation and asphalt concrete respectively: US$ 13.06 and US$ 119.19, with no changes to any other unit prices.
(a) Verify that the final tender price remains the same.
(b) Give two reasons the tenderer might decide to unbalance the tender in this way.
15 Suppose the actual quantities for the project in Table 4.4 are as shown below.

Item	Description	Actual quantity	Unit
201(3A)	Clearing and grubbing	0.65	hectare
203(3)	Unclassified excavation	15 120	m^3
203(5)	Borrow	23 800	m^3
301(2)	Aggregate base course	3 915	m^3
301(4)	Aggregate surface course	2 460	m^3
401(4)	Asphalt concrete – Type II; Class A	2 753	tonne
603(1–300)	300 mm corrugated steel pipe	80	lm
603(1–500)	500 mm corrugated steel pipe	232	lm
615(1)	Standard signs	14	m^2
630(1)	Geotextile, separation	25 950	m^2
670(3)	Traffic marking, wide stripe	4.5	km

Find the total amount the contractor will be paid for this project, using (a) the original balanced unit prices, and (b) the unbalanced unit prices given in Question 14, assuming no renegotiation of unit prices.
16 Describe an alternative way to incorporate equipment costs into the embankment and roadway estimate in Table 4.3. Which way do you prefer? Why?

References

Association for the Advancement of Cost Engineering International. 1996. *Parametric Cost Estimating Model for Buildings*. http://www.aacei.org/technical/costmodels/BuildingModel.html.

Beard, J.L., M.C. Loulakis Sr and E.C. Wundram. 2001. *Design–Build: Planning through Development*. McGraw-Hill Companies.

Bennett, F.L. 1996. *The Management of Engineering: Human, Quality, Organizational, Legal, and Ethical Aspects of Professional Practice*. John Wiley.

Bennett, F.L. 2000. Information technology applications in the management of construction: an overview. *Asian Journal of Civil Engineering (Building and Housing)*, **1**, 27–43.

Bockrath, J.T. 1999. *Contracts and the Legal Environment for Engineers and Architects*, 6th edn. McGraw-Hill.

BuildSoft Pty. Ltd. 2002. *Global Estimating*. http://www.buildsoft.com.au.

Clough, R.H. and G.A. Sears. 1994. *Construction Contracting*, 6th edn. John Wiley.

Clough, R.H., G.A. Sears and S.K. Sears. 2000. *Construction Project Management*, 4th edn. John Wiley.

Construction Industry Times On-Line. 2002. http://www.constructiontimes.co.uk/.

Construction Software Services Partnership Pty. Ltd. 2002. *Everest Estimating and Cost Planning System*. http://www.cssp.com.au/.

Engineering News Record. 2002. *Project Bids and Legal Notices*. http://www.enr.com/propf.asp.

Halpin, D.W. and R.W. Woodhead. 1998. *Construction Management*, 2nd edn. John Wiley.

Hendrickson, C. and T. Au. 1989. *Project Management for Construction: Fundamental Concepts for Owners, Engineers, Architects, and Builders*. Prentice-Hall.

Martin, S. 2001. Plan room may be relic as online service arrives. *Business Journal of Charlotte*, November 2. See also http://charlotte.bcentral.com/charlotte/stories/2001/11/05/focus3.html.

Means Forms for Building Construction Professionals. 1986. Means.

Means Square Foot Cost Data 2002 Book. 2002. 23rd edn. Means.

Milne, C.R. 1989. *Construction Cost Estimating and Bid Preparation Reference Book*. University of Alaska, Fairbanks Department of Engineering and Science Management.

Metropolitan Sewer District. 2001. *Louisville/Jefferson County (Kentucky USA) Metropolitan Sewer District*. http://www.msdlouky.org/.

National Construction Estimator. 2001. *On-line Construction Costs for Contractors*. http://www.get-a-quote.net/.

New Zealand Tenders Gazette. 2002. http://www.tenders-gazette.co.nz/.

Office of Federal Procurement Policy. 2002. *52.248–1 Value Engineering. ARNet: Acquisition Reform Network*. http://www.arnet.gov/far/current/html/52_248_253.html.

Palaneeswaran, E. 2000. Contractor selection systems for design–build projects. PhD Thesis, University of Hong Kong.

Pilcher, R. 1992. *Principles of Construction Management*, 3rd edn. McGraw-Hill.

Pratt, D.J. 1995. *Fundamentals of Construction Estimating*. Delmar.

Rawlinson's Australian Construction Handbook. 2001. Rawlhouse.

Republika Slovenija. 2002. *Tender Documents*. http://www.sigov.si/arr/6viri/1v-2.html.

Russell, J.S. 1996. *Constructor Prequalification*. American Society of Civil Engineers.

Texas Department of Transportation. 2002. *Aviation Information General Conditions*. http://www.dot.state.tx.us/.

Timberline Software Corporation. 2002. *Precision Estimating Collection.* http://www.timberline.com/products.htm.
United Nations Development Business Online. 2002. http://www.devbusiness.com/.
Virginia Department of Transportation. 2001. *Prequalification, DBE Certification and Bidding Process.* http://virginiadot.org/business/const/prequal.asp/.
WinEstimator, Inc. 2002. *WinEst 6.01 Users Reference.* See also *WinEst Pro Plus.* http://www.winest.com.

Case study: Is that a late tender?

The following paper by Evan Stregger, PQS, CArb, was first published in the December 1999 issue of Construction Economist, The Journal of the Canadian Institute of Quantity Surveyors *(Stregger, 1999) and is used with the kind permission of Mr Stregger. The paper reaches no clear-cut conclusion but, instead, demonstrates the differences of opinion in the legal community regarding the point in time at which a tender becomes 'late'. It also makes the important point that the exact wording of the tender documents will often determine the legal outcome in such a case. Mr Stregger practises in dispute resolution and legal support and as an expert witness in central and western Canada. Articles on similar topics can be found on-line at www.costex.ca.*

Controversy sometimes arises over the interpretation of the closing time for a tender. The question will arise: *'Is that a late tender?'* The answer turns upon the wording of the tender documents; depending upon how the documents are worded, the two key interpretations can be:

1 It must be received not later than the stated closing time; or,
2 It may be received at the stated closing time plus 59 seconds.

Two recent court rulings, one in British Columbia and the other in Ontario, have addressed this issue.

1 In *Smith Bros and Wilson v. British Columbia Hydro and Power Authority and Kingston Construction Ltd.* in the Supreme Court of British Columbia by Justice D.W. Shaw, time according to the Advertisement and the Instructions to Tenderers was set out as follows:

   1. The Advertisement to tender used the expression:
      'B.C. Hydro will receive tenders until 11:00 a.m. local time . . .'
   2. The Instructions to Tenderers said: 'Closing Time: Tenderers shall deliver their Tenders . . . not later than 11:00 a.m. local time . . . (The "closing time"), and Tenders which are delivered after closing time will not be considered.'

   As Justice Shaw read both the Advertisement and the Instructions, nothing implies that '11:00 a.m. local time' means the time according to B.C. Hydro's clock. There is nothing to suggest that if B.C. Hydro's clock is inaccurate that it will nonetheless prevail over accurate time. He did however note that *'There is some evidence of custom that generally the clock used by the party receiving tenders will govern. But none of that evidence goes so far as to establish that on a close disputed call, an inaccurate clock will prevail over accurate time. While a provision to cover that situation could be included in the Advertisement and the Invitation to Tender, none was used in the present case.'*

He also found that *'When the Smith Bros. tender was delivered, it was almost immediately stamped by the Widmer clock. At that time, both the clock and the stamp read 11:01 a.m. The conclusion I draw, based upon Mr. Lee's report, is that the tender was received after 11:00 a.m. and before 11:01 a.m. actual time.'*

Kingston Construction (the second bidder who was awarded the contract) submitted that whatever may have been B.C. Hydro's policy, it cannot change the clear provisions of the Advertisement and the Instructions to Tenderers. The words in each document are *'until 11:00 a.m.'* and *'not later than 11:00 a.m.'* respectively.

Justice Shaw concluded, *'In my opinion, one cannot read into the quoted words that the time for delivery of tenders will extend past 11:00 a.m. until almost 11:01 a.m.'*

2 In *Bradscot (MCL) Ltd. v. Hamilton-Wentworth Catholic School Board* heard by Justice Somers of the Ontario Court of Justice (General Division) it was determined that the tender submitted thirty seconds after the time of closing was not late. The tender documents stated *'Friday May 8, 1998 at 1:00 p.m.'* as the deadline. The Owner's Instructions to Tenderers made it emphatically clear that bids not received by the time stated *'WILL NOT be accepted by the owner'*. The official clock was a digital one, but it showed the hours and minutes only and did not record the seconds. According to the watch of the representative from Bradscot (MCL) Ltd., the second bidder, the time that the tender of the low bid was submitted was 30 seconds past 1:00 p.m. The President of the Ontario General Contractors' Association stated in a letter: *'In our opinion any tender received after the instant of 1:00 . . . is late . . . One thing for sure contractors understand the tender that is even one second late, is late, and should not be considered.'* The Board's Architect offered the opinion that where the bid closing time is stated to be 1:00 p.m., any bid received at 1:00 p.m. was delivered on time and for a delivery to be late the clock would have to register 1:01 p.m.

Justice Somers reviewed the decision in *Smith Bros and Wilson v. British Columbia Hydro and Power Authority* but did not find it particularly helpful as the clear provisions and particular words in that matter were 'not later than 11:00 a.m.' Justice Somers noted that the relevant bid deadline in the matter was set as 'Friday, May 8, 1998 *at* 1:00 p.m.'

Justice Somers then concluded: *'In my opinion when it is stated that some deed be done "at 2:00 p.m." the time is for that minute and the act is not overdue until the minute hand has moved off the 12 hand to the :01 position.'*

Clearly the answer is in the wording of the tender documents. *'Before and not later than'* clearly has a different legal meaning than *'at'*. Bidders, in my experience, are always waiting for one last, lower price or final adjustment. Submitting your tender prior to the stated closing time is the only certain way to avoid this problem.

Reference

Stregger, E. 1999. Is that a late tender? *Construction Economist, The Journal of the Canadian Institute of Quantity Surveyors*, **9** (4), December, p. 9.

5

Project mobilisation phase

Introduction

This chapter deals with many of the activities that take place between the award of the construction contract and the beginning of construction work in the field. Some of this work may have begun prior to, and in anticipation of, the award, and much of it will continue into the days and weeks during which fieldwork is beginning. But these are the actions that prepare for and set the stage for what the layperson considers 'construction', the things that people and machines do to assemble the project in the field. We consider in cursory fashion such contractual issues as permits, bonding and insurance and then we deal with the preparation of detailed project schedules, the conversion of the cost estimate into a project budget for use in controlling project costs, the organisation of the work site, acquisition of materials and the engagement of subcontractors and such staffing issues as the project management structure, collective bargaining agreements and non-union contracting. At the end of the chapter we consider briefly some special considerations for mobilising projects in remote regions.

Legal and contractual issues

The contractor may be required to make a number of arrangements for various kinds of 'paperwork' as part of the pre-construction mobilisation. Because these requirements vary from project to project and especially from country to country, we deal with them in a general way.

Permits, consents and licences

Local, regional and sometimes national authorities may require the issuance of *permits* for various aspects of the project. Approval of applications for these permits, called *consents* in some countries, must be secured prior to specified stages of the project. Whether the contractor is directly involved depends on regulations and the owner's wishes. In the case of a New Zealand building consent (or *approval to construct*), the following statement is helpful for us: 'Whilst the Building Act places full responsibility on building owners, it is normal procedure for

Figure 5.1 Contractor's view of resource consent delays (*Contractor Magazine*, 2001). Reproduced by permission of *Contractor Magazine* and with thanks to cartoonist Derek Wigzell.

the person carrying out the work to carry that responsibility as the "Owners Agent"' (*New Zealand Building Business: Your Building Guide, South Island Edition*, no date).

If not already obtained by the design professional, the contractor will be required to obtain a building permit; often this permit is issued by a state public safety official such as a Fire Marshall. Work authorised by such a permit might include the building structure, plumbing and drainage work, siteworks, demolition and relocation of existing buildings, as is the case in New Zealand, or separate plumbing and electrical permits may be required. Such permits or consents are formal authorisation that the work can proceed; they contain appropriate conditions and confirm the requirements for public authority inspections as the work is carried out. The terms and conditions of a permit are binding upon the contractor; in the case of conflict between such a permit and the project specifications, normally the permit will take precedence (Fisk, 2003).

In the USA and elsewhere, there is currently much interest in wetlands; normally the owner or the design professional will obtain any required wetlands permits after demonstrating that no damage will occur to such ground or that any damage caused to a wetland area will be balanced by the addition of an equal area to protected, undamaged wetlands status. If the wetlands permit has not already been obtained, the contractor will arrange for it. Other environment-related permits that may be required include resource consents, as specified under New Zealand's *Resource Management Act*, for such issues as noise, odour, building location, visual effects and vegetation clearance; storm water runoff permits affecting the active construction site, for projects over a certain size; and air quality permits if activities such as asphalt production or dust-producing work are anticipated. The cobwebs, birds nest and beard in the cartoon in Figure 5.1 is one contractor association's way of expressing unease at perceived delays in the permit approval process by a regional council (*Contractor Magazine*, 2001).

Although most licences are not issued for the project itself, the contractor may become involved in licensing. For example, a contractor's contracting licence and its business licence, if any, are usually issued for a time period such as a year and they must be current at the time of tendering and throughout the job. Also, the contractor will be expected to assure that its subcontractors have current licences; these might include both business and specialty contractors' licences, depending upon the jurisdiction. If the contracting organisation is a joint venture, the usual requirement is that the joint venture must have a separate licence, even if the individual joint venturers are already licensed.

Bonding

In Chapter 4, we included the cost of bonds in our construction cost estimates. Now we describe in more detail the nature and purposes of these documents. A *surety bond* is a legal instrument issued by a third party to guarantee the obligations of a first party to a second party. This *suretyship* can be thought of as a guarantee of performance. In the realm of construction, the first party, or the one that takes on the obligation, is the contractor, or the *principal*. The second party, the owner with whom the contractor enters, or anticipates entering, into a contract is the *obligee*. The third party, the bonding company that guarantees the obligation, is the *surety*. A surety bond differs from an insurance policy in that an insurance underwriter expects some percentage of its coverages to experience losses, whereas, in surety, no losses are contemplated; the premium paid in surety is a service fee for the use of the surety's credit. If the principal (contractor) fails to carry out its obligations, the surety makes good the loss to the owner, but the surety then has recourse to the contractor for reimbursement if possible. For a good review of suretyship and its usage in construction, see the website of the Surety Association of Canada (2002).

Of the three types of bonds the contractor may be required to obtain, the *tender bond* (or *bid bond*, or *proposal bond*) is issued during the contractor selection phase, whereas the other two, the *performance bond* and the *labour and material payment bond*, are obtained during project mobilisation. The tender bond provides the owner with bid security. The purpose is to protect the owner against loss if the selected contractor does not begin the project as directed. A typical section of instructions to bidders related to bid security is the following (Clough and Sears, 1994):

> Proposal and Performance Guarantees. A certified check, cashier's check, or bid bond for an amount equal to at least five per cent (5%) of the total amount bid shall accompany each proposal as evidence in good faith and as a guarantee that if awarded the contract, the bidder will execute the contract and give bond as required. The successful bidder's check or bid bond will be retained until he has entered into a satisfactory contract and furnished required contract bonds.

Thus we see that (1) the bid (tender) bond is but one way of providing bid security, the other, less common way being a certified or a cashier's cheque and (2) the bid bond guarantees two things if the bidder is selected: the contractor will sign the contract and will provide the other required bonds. Although the 5% of the total tender specified in the above example is a typical amount, it can be as high as 20% in some contract documents, including some government work in the USA. Most bid bonds are of the 'difference-in-price' type, by which the owner is protected against the cost of having to contract with the second-to-lowest bidder, up to the value of the bond, if the lowest bidder fails to enter into the contract and furnish the required bonds.

As an example, consider the following scenario. A tender requires a 5% tender bond. The lowest tenderer proposes a price of US$ 1 500 000 and next lowest tenderer proposes US$ 1 560 000. The lowest tenderer defaults and the owner is forced to contract with the second lowest tenderer for an extra US$ 1 560 000 – US$ 1 500 000 = US$ 60 000. The lowest tenderer has furnished a tender bond for 0.05(US$ 1 500 000) = US$ 75 000. The surety will reimburse the owner for the extra US$ 60 000, because it is less than the US$ 75 000 value of the bond. If the second lowest tender had been US$ 1 580 000 and thus the owner had stood to suffer an extra US$ 80 000 cost in contracting with that tenderer, the reimbursement would have been only US$ 75 000, the value of the tender bond.

It is not commonly understood that after the surety has reimbursed the owner for moneys due under the bond agreement, the contractor is liable to the surety for damages the surety has paid to the owner. If the contractor is in poor financial condition, however, there is some risk that the surety will not be able to recover from the contractor. Thus, the surety is not likely to issue a tender bond to the contractor unless it believes that the contractor has sufficient assets to cover any payments to the owner resulting from failure to enter into the contract.

We show in Figure 5.2 a sample proposal bond (Washington State Department of Transportation, 2001). Of special interest is the rather 'backwards' wording in the 'NOW, THEREFORE' paragraph. Most bond forms use this type of wording, which says that the bond obligation remains in force *unless* the contractor does what is expected, which is to enter into the contract and furnish other bonds.

Sometimes general contractors require that subcontractors furnish tender bonds, as a check on their financial stability. If a surety is willing to provide such a bond, the general contractor has some assurance that the subcontractor will enter into an agreement if the owner awards the contract to the general contractor.

Now we discuss the bonds that may be required during the project mobilisation phase. The nature of the relationships is the same as has been explained for tender bonds, involving principal, obligee and surety. A *performance bond* acts to protect the owner from non-performance by the contractor. Upon signing the contract, the owner is entitled to receive what it contracted for: a completed project in substantial accordance with the contract documents. If the contractor (the *principal*, in the parlance used above) defaults by not delivering in accordance with the contract, the surety is responsible to the owner (*obligee*) to complete the project or have it completed, at the price agreed upon between the owner and the defaulting contractor, provided that the owner has met its obligations under the contract. Normally a performance bond extends through any warranty period required by the construction contract, such as 1 or 2 years. A performance bond has a face value which sets the upper limit of the surety's obligation in case of the contractor's default; a 100% performance bond, for example, provides protection up to the total contract price. Figure 5.3 contains the form of performance bond required for construction projects at a US university (Arizona State University, 2002); like the sample proposal bond in Figure 5.2, its 'NOW, THEREFORE' paragraph provides that the surety's obligation 'remains in full force and effect' unless the contractor fulfils all of its obligations under the contract.

A *labour and material payment bond* also provides protection to the owner, but in a different way. Such a bond protects the owner against claims from subcontractors, material suppliers, workers and others that the contractor has not paid them moneys due. If the contractor fails to pay outstanding charges incurred in connection with the project, the surety will pay those debts. Thus, those owed the payments are protected by the assurance that they will be paid and the owner is protected by the assurance that it will not have to pay such claims. Suppose a subcontractor fails to pay one of *its* material suppliers; will the payment bond provide protection

Proposal Bond

KNOW ALL MEN BY THESE PRESENTS, That we, ______________________________

______________________ of ____

as principal, and the ______________________________________ a corporation duly organized under

the laws of the state of ____________________, and authorized to do business in the State of Washington,

as surety, are held and firmly bound unto the ______________________ in the full and penal sum of five (5) percent of the total amount of the bid proposal of said principal for the work hereinafter described, for the payment of which, well and truly to be made, we bind our heirs, executors, administrators and assigns, and successors and assigns, firmly by these presents.

The condition of this bond is such, that whereas the principal herein is herewith submitting his or its sealed proposal for the following highway construction, to wit:

____

____

____

said bid and proposal, by reference thereto, being made a part hereof.

NOW, THEREFORE, If the said proposal bid by said principal be accepted, and the contract be awarded to said principal, and if said principal shall duly make and enter into and execute said contract and shall furnish

bond as required by the ______________________ within a period of twenty (20) days from and after said award, exclusive of the day of such award, then this obligation shall be null and void, otherwise it shall remain and be in full force and effect.

IN TESTIMONY WHEREOF, The principal and surety have caused these presents to be signed

and sealed this ______ day of______________, ______.

____

____
(Principal)

____
(Surety)

____
(Attorney-in-fact)

DOT Form 272-001
Revised 08/01 for Local Agency Use

Figure 5.2 State of Washington highway construction proposal bond.

for that supplier? The general rule is that 'payment bonds exclude from their coverage parties who are remote from the general contractor' (Clough and Sears, 1994). The issue in any particular case, of course, is the degree of remoteness. When a payment bond is issued, it must

ARIZONA BOARD OF REGENTS
PERFORMANCE BOND
PURSUANT TO BOARD OF REGENTS POLICY 3-804D
(Penalty of this bond must be 100% of the contract amount)

KNOW ALL MEN BY THESE PRESENTS:

THAT, ______________________________, (hereinafter called Principal), as Principal, and ______________________________, a corporation organized and existing under the laws of the State of ____________, with its principal office in the city of __________, (hereinafter called the Surety), as Surety, are held and firmly bound unto the Arizona Board of Regents, (hereinafter called the Obligee) in the amount of ______ ____________________($) for the payment whereof, the said Principal and Surety bind themselves, and their heirs, administrators, executors, successors and assigns, jointly and severally, firmly by these presents.

WHEREAS, the Principal has entered into a certain written contract with the Obligee, dated __________ ______**2000** to construct and complete a certain work described as ______________________________ ___________, **ASU PROJECT, #**______________which contract is hereby referred to and made a part hereof as fully and to the same extent as if copied at length herein.

NOW, THEREFORE, THE CONDITION OF THIS OBLIGATION IS SUCH, that if the said Principal shall faithfully perform and fulfill all the undertakings, covenants, terms, conditions and agreements of said contract during the original term of said contract and any extension thereof, with or without notice to the Surety and during the life of any guarantee required under the contract, and shall also perform and fulfill all the undertakings, covenants, terms, conditions, and agreements of any and all duly authorized modifications of said contract that may hereafter be made, notice of which modifications to the Surety being hereby waived; then the obligation shall be void, otherwise to remain in full force and effect.

PROVIDED, HOWEVER, that this bond is executed pursuant to the provisions of Arizona Board of Regents Policy Section 3-804D, and all liabilities on this bond shall be determined in accordance with the provisions of this section, to the extent as if copied at length herein.

The prevailing party in a suit on this bond, including any appeal thereof, shall recover as a part of his judgment such reasonable attorneys' fees as may be fixed by a judge of the Court.

Witness our hands this _____ day of _______________, 20__________.

____
Principal Seal

By______________________________

____
Surety Seal

By______________________________

Figure 5.3 Arizona State University performance bond.

be clear which parties are covered by it. Whereas a performance bond is usually written for 100% of the contract amount, payment bonds typically cover smaller amounts, such as 50%. Figure 5.4 shows a sample payment bond for the same US university (Arizona State University, 2002); note that this bond provides protection to 'all persons supplying labor or materials to him [the contractor] or his subcontractors'. Note also that this bond must be written for 100% of the contract amount.

Just as with the tender bond discussed earlier, if the surety is required to pay on a performance or payment bond, it can then seek reimbursement from the contractor. The risk to the surety is that if the contractor has not been able to complete its performance or pay its obligations, there is some likelihood that it may have no financial assets available to reimburse the surety.

ARIZONA BOARD OF REGENTS
PAYMENT BOND
PURSUANT TO BOARD OF REGENTS POLICY 3-804D
(Penalty of this bond must be 100% of the contract amount)

KNOW ALL MEN BY THESE PRESENTS:

THAT, ______________________________, a corporation organized and existing under the laws of the State of ____________________, with its principal office in the City of , (hereinafter called the Surety), as Surety, are held and firmly bound unto the Arizona Board of Regents (hereinafter called the Obligee) in the amount of ______________________________($) the payment whereof, the said Principal and Surety bind themselves, and their heirs, administrators, executors, successors and assigns, jointly and severally, firmly by these presents.

WHEREAS, the Principal has entered into a certain written contract with the Obligee, dated __________ **2000** to construct and complete a certain work______________________________, **ASU PROJECT,**
#__________ which contract is hereby referred to and made a part hereof as fully and to the same extent as if copied at length herein.

NOW, THEREFORE, THE CONDITION OF THIS OBLIGATION IS SUCH, that if the said Principal shall promptly pay all monies due to all persons supplying labor or materials to him or his subcontractors in the prosecution of the work provided for in said contract, then this obligation shall be void, otherwise to remain in full force and effect.

PROVIDED, HOWEVER, that this bond is executed pursuant to the provisions of Arizona Board of Regents Policy Section 3-804D, and all liabilities on this bond shall be determined in accordance with the provisions of the section, to the same extent as if copied at length herein.

The prevailing party in a suit on this bond, including any appeal thereof, shall recover as a part of his judgment such reasonable attorneys' fees as may be fixed by a judge of the Court.

Witness our hands this ________ day of ______________, 20___.

____
Principal Seal

By______________________________

____
Surety Seal

By ______________________________

Figure 5.4 Arizona State University payment bond.

Tender bonds are furnished with the tender. Performance and payment bonds are furnished to the owner during project mobilisation, within a specified time limit after contract signing (or perhaps *at* the time of signing). One construction contract (Municipality of Nenana, 2002), in its performance and payment bond section, states 'the contractor shall provide the actual bonds to the City within ten (10) days of the date of award of the contract and may not begin work on the project until the city has accepted the bonds'. The cost of the bonds we have discussed can vary from less than 1 to as much as 4% of the project budget (Johnston and Mansfield, 2001). The usual practice among sureties is to provide tender bonds at no charge and then to charge the successful tenderer for the performance and payment bonds. Bond costs for relatively small projects are higher as percentages of the project budget than are those for larger projects. Costs also depend on the contractor's financial strength and its experience and the charges also vary among different sureties.

Insurance

Although the available *insurance* coverages and the regulations governing insurance vary among different countries, the contractor, in mobilising for the project, will always have to arrange for appropriate insurance. This section gives a brief review of the types of insurance utilised in the construction industry.

An insurance policy is an agreement under which the insurer agrees to assume financial responsibility for a loss or liability covered by the policy. The insurance company has the duty (1) to defend the contractor if a claim covered by the policy is brought against the contractor and (2) to protect the contractor against loss covered by the policy. As consideration for the insurance company's promise to provide such protection, the contractor pays a fee, called a *premium*. The contractor's loss record is an important determinant of the premium rates. This record is reflected in an *experience rating*. Standard, or 'manual', rates for given classes of risks are adjusted up or down based on the experience rating, so that the contractor with a low loss record benefits from savings in the costs of these experience-rated coverages.

The contract with the owner will specify certain types of required insurance coverage to be purchased and maintained by the contractor. Beyond that, some insurance may be a legal requirement even if not stipulated in the contract. Furthermore, the contractor may choose to carry other insurance not required by law or by contract. Depending on the contract, the owner may be responsible for obtaining some kinds of insurance, so the wise contractor will check to make sure it is not purchasing duplicate insurance unnecessarily.

It is helpful to divide the types of construction insurance into property insurance, liability insurance, employee insurance and other types. Details may be found in Clough and Sears (1994) and Fisk (2003). *Property insurance* is of two basic types, covering the contract works underway during construction and the contractor's own plant and other property. Builder's risk insurance provides coverage on the contract works. Such insurance may be of the all-risk type, under which protection is afforded against all risks of physical loss or damage to the works or associated materials, due to any external cause, unless an exception is stated. Or it may be a standard, or named-peril, builder's-risk policy, covering only stated perils such as fire and lightning, vandalism and windstorm. Other contract works insurance can cover damage due to earthquake, damage caused by the operation of boilers during construction and damage to bridges; installation floater policies provide coverage for permanently installed equipment from the time it is shipped to the project until it is installed and tested.

Insurance against the contractor's own plant and other property will probably not be required by the contract, but the contractor may choose to purchase this coverage. Among the types available are coverage for various buildings such as sheds, warehouses and offices, equipment (sometimes only on the site and sometimes regardless of location), materials and supplies carried on the contractor's trucks and valuable papers such as drawings, maps, books and contracts. Crime insurance is a special category that protects the contractor against loss of office equipment, money and other valuables due to theft, burglary, destruction and other crimes.

A common form of *liability insurance* is comprehensive general liability insurance, which protects the contractor, in a single policy, against liability to the public. It covers injuries to people not employed on the project and damage to the property of others arising from operations on the project. Various other kinds of coverage may be included, such as liability to the public due to actions of subcontractors, completed-operations coverage for work handed over to the owner, professional liability insurance for design and other professional services rendered by the contractor and protection for injury and property damage caused by the contractor's vehicles. If not included in the comprehensive policy, these provisions may be purchased separately. The law may require that the contractor carry workers' compensation insurance, a form of liability insurance that provides benefits to employees injured on the job or to their heirs in cases of fatal accidents.

Among common forms of *employee insurance*, some of which may be required as part of a collective bargaining contract, are employee benefit insurance, covering medical and hospital costs beyond those provided by workers' compensation, social security (in the USA) or similar government programmes that provide retirement, disability and survivor benefits to insured workers and unemployment insurance, normally a government-operated programme offering benefits during periods of unemployment between jobs. In our *other insurance* category, we note business-interruption insurance, covering losses due to times of business inactivity, key-person life insurance, used to reimburse the company for financial loss due to the death of a major officer or other designated essential employee and corporate continuity insurance, providing cash to purchase a deceased stockholder's corporate stock as a means of preventing that stock from being purchased by a party unfriendly or otherwise undesirable to the business.

As indicated earlier, some insurance coverage is required by law or by contract. Early in the mobilisation phase, the contractor must furnish evidence to the owner that it meets the contractual requirements, just as it must for bonding requirements. Beyond the required insurance, the contractor has considerable choice, as described above, about whether to protect against its various risks by acquiring other insurance.

Partnering

This section may be misplaced, because *partnering* is not really a 'legal' issue. Participation is voluntary and any partnering agreement or charter is not legally binding. But if the owner has specified the use of partnering and if the project is of the design–tender–build type, the mobilisation phase will be the time when the contractor becomes involved. If the contractor is named earlier in the process, such as in a design–construct contract, the basics of the partnering agreement will probably have been worked out prior to the completion of design.

In Chapter 3, we introduced the notion of partnering, in which the various parties involved in a construction project 'pledge to cooperate as fully as possible' beyond that cooperation anticipated by any contractual obligation. Fisk (2003) states it well:

> While the contract establishes the legal relationships, the partnering process is designed to establish working relationships among the parties through a mutually-developed, formal strategy of commitment and communication. It attempts to create an environment where trust and teamwork prevent disputes, foster a cooperative bond to everyone's benefit, and facilitate the completion of a successful project.

Partnering works best if all members of the project team, including owner, design professional, contractor, subcontractors and material suppliers, commit to the concept. Trust is a key element and all stakeholders must respect each others' views and interests. The process involves establishing mutually agreed-upon goals as early as possible, the signing of a formal charter, continuous evaluation of the process and timely response to questions and potential conflicts. Rewards and celebrations of success are appropriate elements as well (Dorsey, 1997).

Essential to the process is a partnering workshop, held as soon as key parties are identified, led by an independent facilitator and attended by all parties. At this workshop, goals and

Partnering Charter

We, ..., are committed to Partnering through the construction, administration and completion of this Project on time and within budget. We agree to make our best efforts to achieve the goals listed below, and believe that these goals reflect our intentions and commitment to the performance of this project as a team.

GOALS

1. Accident-free job site.
2. Resolve all safety issues immediately.
3. No more than two percent cost growth.
4. Complete all contract phases ahead of schedule, including punch-list.
5. By .... (date), define in writing, roles and communication lines for the partnering arrangement.
6. Execute necessary contract changes without delaying the project.
7. Foster a positive job environment.
8. No repeats on notices of deficiencies.
9. Participation of all appropriate team members in the quality control program.
10. Avoid litigation by:
    - Addressing issues and working them out as a team before they become differences.
    - Resolving differences through negotiation.
    - If all other methods fail, obtain a disinterested third party arbitrator's opinion.
11. Build a project of which we can all be proud.
12. Submittal and evaluation of all submittals to avoid delaying project progress.
13. Empower joint problem resolution at the lowest possible level.
14. Foster new ways of doing business.

Figure 5.5 Typical partnering charter (Jamieson, 2001).

expectations are expressed and clarified, potential problems are identified, communication and issue resolution processes are established, reporting, evaluation and recognition methods are agreed upon and a partnering charter is developed and signed. An example of a typical partnering charter, suggested by Jamieson (2001), is shown in Figure 5.5.

While partnering is not a panacea for eliminating all problems that arise on a construction project, it has been utilised successfully on many projects in both the private and public arena. Its principles of communication, commitment and conflict resolution help to foster a win–win approach to problem solving for all members of the project team.

Programming, planning and scheduling

In Chapter 4, we discussed the importance of a preliminary project schedule in conjunction with tender preparation and we showed, in Figure 4.3, a sample bar chart depicting such a schedule. Now, with the contract in hand and the project mobilisation phase underway, the contractor will undertake to develop a more detailed plan and schedule for the project's time dimensions. The

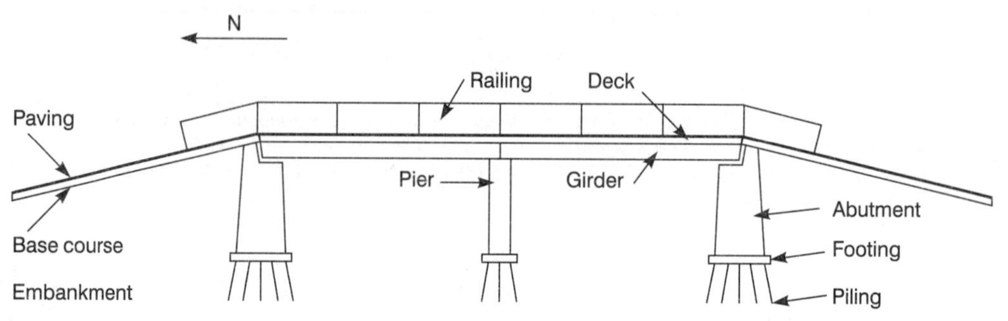

Figure 5.6 Bridge project schematic design, cross-sectional view.

result will be a document that will assist with deploying personnel and equipment, procuring materials and supplies and planning cash flows. Equally as important, it will provide the basis for monitoring and controlling project progress as the project proceeds. The terms *programme* and *schedule* tend to be used synonymously, depending on where in the world one is operating, as do the terms *programming* and *scheduling*. The overall document that depicts the project's time dimensions will be called either the *master programme* (Pilcher, 1992) or the *master schedule*.

Work breakdown structure

To begin the preparation of a detailed project schedule, the contractor will prepare a *work breakdown structure* (WBS). Its purpose is to define the various work tasks that must be completed. The WBS has been variously defined as 'a product-oriented "family tree" of work effort that provides a level-by-level subdivision of the work to be performed in a contract' (Babcock and Morse, 2001) and 'a "picture" of the project subdivided into hierarchical units of work' (Bennett, 1996). At the highest level of the hierarchy, we show the project itself. At the next level, the project's major subdivisions are shown. Each major subdivision is further subdivided into additional levels as required to identify in sufficient detail all of the work involved in the project. By defining the various work packages, the WBS allows the construction manager to (1) plan and schedule the work and thus describe the project's total programme, (2) estimate costs and budgets, (3) authorise parcels of work and assign responsibilities and (4) track schedule, cost and quality performance as the project proceeds (Bennett, 1996).

For any given project, numerous hierarchy designs could be used (Hendrickson and Au, 1989). Major subdivisions might be based on components of the project, such as abutments, roadway and deck, or by types of work as represented by the various crafts (concrete, electrical and the like), or by spatial location on the project site. To capture all project work elements, it may be necessary to employ a combination of approaches. Furthermore, there will be a question related to the amount of detail to be shown in the WBS; too much detail becomes unwieldy and overly costly, but too little detail does not permit accurate time and resource estimation or practical assignment of tasks. The manager must rely on judgment and experience when deciding on the proper amount of work breakdown schedule detail.

Suppose we are mobilising to build the bridge whose schematic design is shown in Figure 5.6. A work breakdown structure such as that in Figure 5.7 could be developed for this bridge construction project. Note that the highest level in the hierarchy is simply 'bridge project', while the second-level subdivisions are based on the various components of the work, as well as both project mobilisation and termination. This WBS also illustrates that not all branches in the hierarchy have equal numbers of levels; the 'finish work' element has only one level below it, while the 'project mobilisation' element leads to three additional levels. Colour coding is a handy way to identify the various levels. It should be emphasised that, for practicality of illustration, we have shown much less detail on some of the branches than would normally be given for a 'real-life' project. Construction of the abutment structure would be subdivided into additional detail contained in, perhaps, two more levels. Even with these limitations, we can use this WBS to demonstrate its use in the preparation of a project schedule.

With sufficient time, effort and attention to detail, the project planner can use the WBS to identify the entirety of the work involved in the project. The lowest level elements in each branch then become the work packages that are the basis for the programme, for the assignment of tasks and for the monitoring of time and cost. The 45 elements at the lowest levels of the various branches of Figure 5.7 will become the tasks that define the project schedule to be developed later in this section.

In applying the WBS methodology to projects whose scope and details evolve during the project's early phases, the planner is able to expand branches as more definition emerges. A good example might be the design–build project, in which the planning and design effort is normally well defined at the project's beginning. Because the purpose of the planning and design phase is to define and specify the construction work, the WBS will initially show little detail for the construction phases, with more elements and levels added as information about those phases is developed. As we shall continue to emphasise, project programming is a dynamic process.

Bar charts

Traditionally, *bar charts* have been the most popular methods for presenting construction project schedules, especially for small projects. Even today, many contract documents require the contractor to furnish the owner with a bar chart indicating how the various activities will be carried out so as to complete the project by the target completion date. We have already seen a bar chart schedule for a building project in Figure 4.3, suggested as a useful part of preliminary project planning during the tender preparation stage. Simple yet effective, it plots proposed progress on a horizontal time scale for each of the elements. The term *Gantt chart* is also used, named after Henry L. Gantt, an American engineer and social scientist who began using bar charts as a production control tool in 1917.

If a bar chart is used as the basis of the project schedule without first performing a network schedule analysis of the type to be described in the next section, the planner simply places the element's bar on the chart in the proper position relative to those for the other elements. The process can be done by hand or by any number of computer-based drafting methods. As we explain in the next section, a bar chart is also an effective means of displaying the results of a computerised network scheduling analysis. In addition, this graphical technique allows one to display actual progress beside the bar showing an element's planned progress, during the process of monitoring and controlling project progress.

Bennett (1996) lists the following advantages of the bar charting approach to project scheduling:

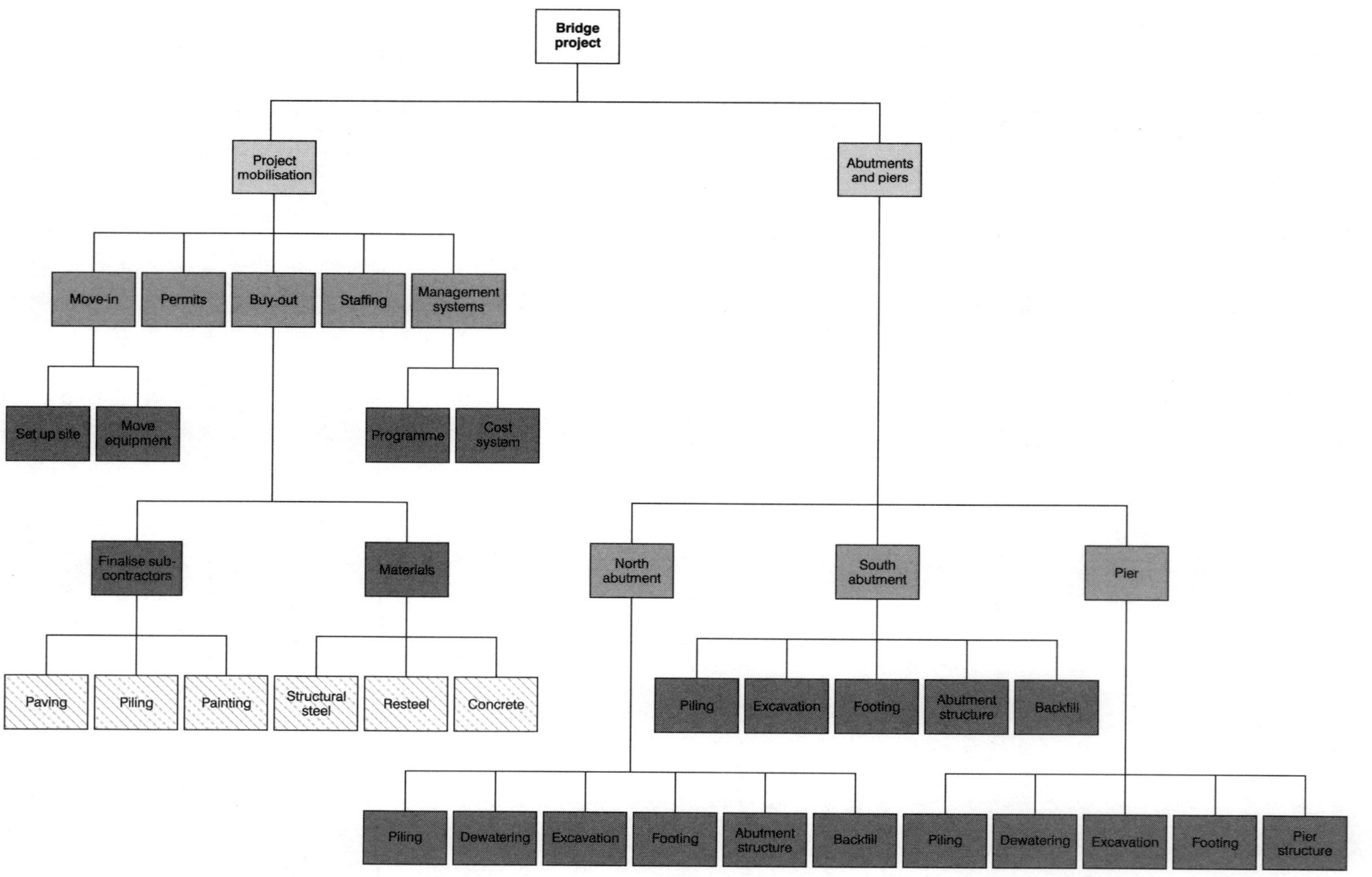

Figure 5.7 Work breakdown structure for bridge project.

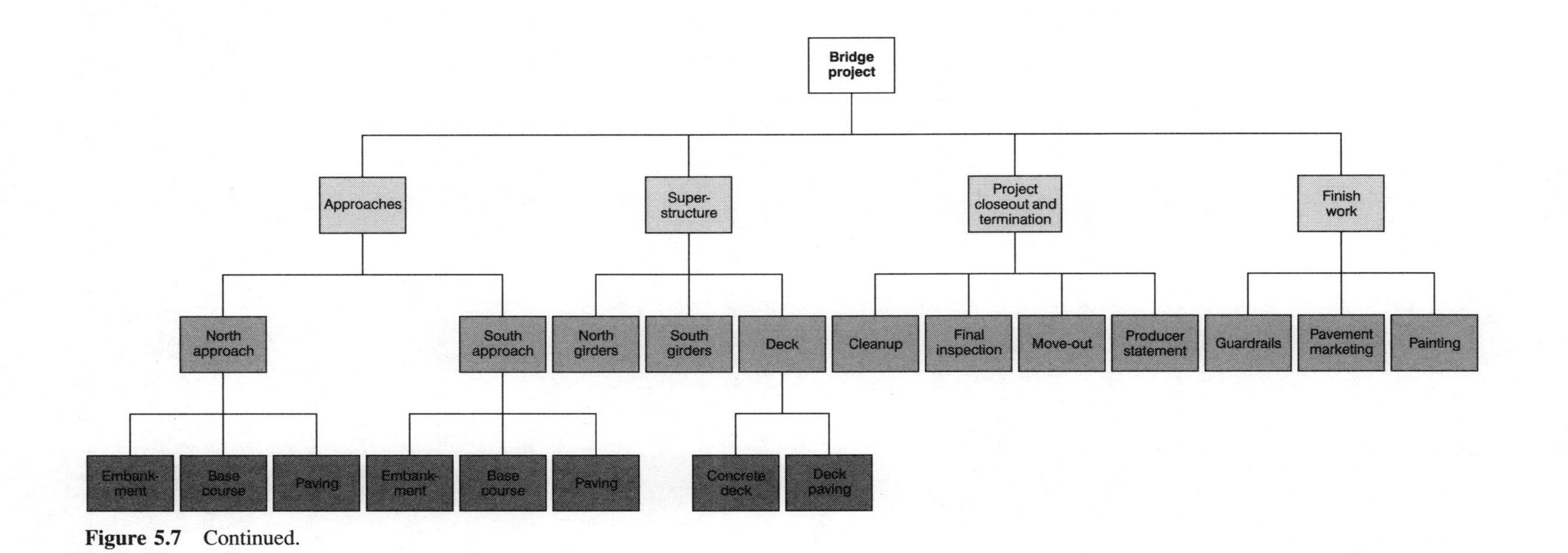

Figure 5.7 Continued.

1 simple to construct and easy to comprehend
2 convenient organisation by work breakdown structure elements
3 with computer capability, easy to update and show project progress
4 relatively inexpensive to use
5 summaries at any desired WBS level.

However, there are also disadvantages in the use of bar charts without also performing a network analysis (Bennett, 1996).

1 There is no indication of interrelationships among activities. [Some bar charting software includes lines or arrows between related activities, but they tend to be difficult to interpret.]
2 In developing the bar chart, the analyst tends to fit the tasks into the available timeframe rather than analyzing closely how much time will actually be required.
3 The chart shows expected time requirements but not level of effort requirements. If a task requires five person weeks of effort spread over eight weeks, we learn only about the eight weeks from the bar chart.
4 Bar charts are impractical for complex projects with many tasks.
5 Without a computer-driven plot generator, updating of the schedule is time consuming.

Network schedules

Since the late 1950s, a network-based approach to developing schedules has seen increasing use for all kinds of projects, including construction projects. Project activities are related to each other on a network, showing the order in which they are intended to be performed. The duration of each activity is estimated and calculations determine the timeframe within which each activity can be performed and the time at which the project can be completed. In addition, the calculations identify those activities that are most critical in establishing that completion time. Two methods have been used to develop project networks. One method uses arrows to represent the activities, with their endpoints depicting their relationships to other activities. The other method uses boxes or similar nodes to signify activities and arrows or lines to connect the nodes in the order in which the activities are planned to be carried out. Most project programmers currently use the latter *activity-on-node* method and we shall confine our discussion to this approach. In this section we describe the basics of activity-on-node scheduling and we then apply the method to the bridge construction project whose work breakdown structure is shown in Figure 5.7.

General concepts: a modest example

Consider a ‘project’ to construct a large footing that will support a bridge abutment or a piece of industrial machinery. Figure 5.8 shows the activities that might be required to carry out this project. The project begins with excavation, ordering and delivering reinforcing steel and setting up a pre-fabrication shop on the site. It proceeds through placement of forms, pre-fabrication of reinforcing steel, setting of steel in the forms, placement of concrete and removal of the pre-fabrication shop. After the concrete is placed and the shop is removed, an inspection takes place and the project is considered complete. (As usual, to stress the fundamentals, we have oversimplified the real project, omitting such tasks as curing of the concrete, removal of forms and installing inserts into the concrete footing.) In drawing the network, we consider three questions for each activity:

1 What activity or activities, if any, must take place directly prior to this activity?
2 What activity or activities, if any, must take place directly following this activity?
3 What activity or activities, if any, can take place concurrently with this activity?

We observe several characteristics of an activity-on-node network in Figure 5.8. Each activity is represented by a node, a rectangle in this case. Arrows connect the nodes to show the order in which the work is to flow. Generally the sequencing is from left to right; in a complicated network, such sequencing might be violated, but the direction of the arrow always clarifies the work flow direction. Nodes with no arrows entering from the left represent activities that can start at the beginning of the project; they have no predecessors. In Figure 5.8, 'excavate', 'order and deliver reinforcing steel' and 'set up pre-fabrication shop' can begin as soon as the project begins. Likewise, nodes with no arrows leaving from the right depict the final activities in the project; they have no successors. In our example, the final activity is 'inspection'. We also note that some work items can be done concurrently with others; for example, 'remove pre-fabrication shop' can be done while the 'set steel' and 'place concrete' activities are underway. Finally, we observe that Figure 5.8 makes no mention of the durations of the activities. At this stage, we are concerned only with the identity of the activities and their order. The position of a node in the diagram is not related to the time at which the activity takes place; the diagram is not drawn to any sort of time scale. While the planning function has many aspects, certainly one of the most important is this activity work flow definition. Much healthy discussion and debate should occur among members of the project team as they consider various alternative sequences and arrive at an agreement, albeit tentative and subject to modification, as to how the work flow will proceed. Once this diagram is complete, we proceed from this part of *planning* to the *scheduling* of the activities.

Figure 5.9 differs from Figure 5.8 in two respects. First, we have added an estimated activity duration to each node. In making these estimates, we will have to review the amount of work required by the activity, the anticipated crew size and productivity and equipment productivity. The cost estimate will be a helpful information source, as will records of actual time spent on similar tasks on previous projects.

The second addition in Figure 5.9 is the early start and early finish time for each activity. These times result from some basic calculations that will lead to an estimate of the overall project duration. An activity's *early start* time is the earliest it can begin, based on the early completions of all preceding activities; its *early finish* time is its early start time plus its duration. For activities that can begin at the start of the project (those with no predecessors), we assign an early start time of 0. For all other activities, the early start time is the latest early finish time of its immediate predecessor activities. It is the *latest* early finish time because *all* predecessor activities must be completed before the activity under consideration can be started. Thus, for any activity,

ES = early start time = 0 for all 'begin' activities; otherwise = max (early finishes of all immediate predecessors);
EF = early finish time = ES + duration.

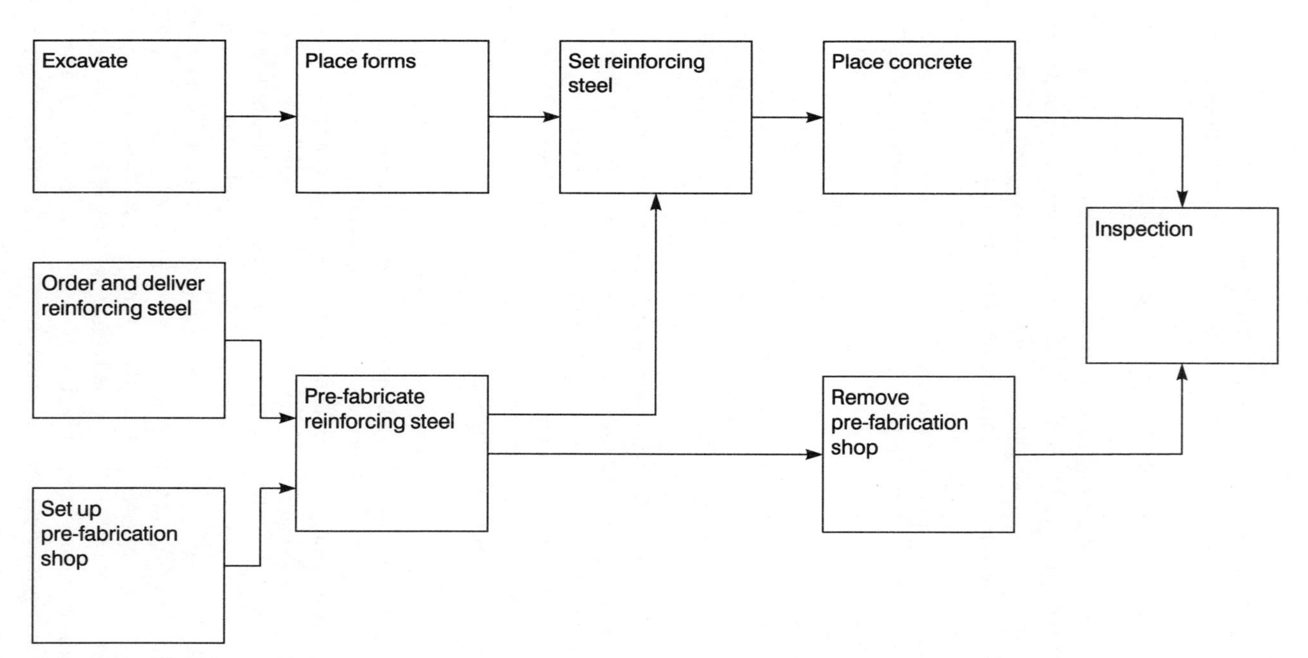

Figure 5.8 Footing schedule network.

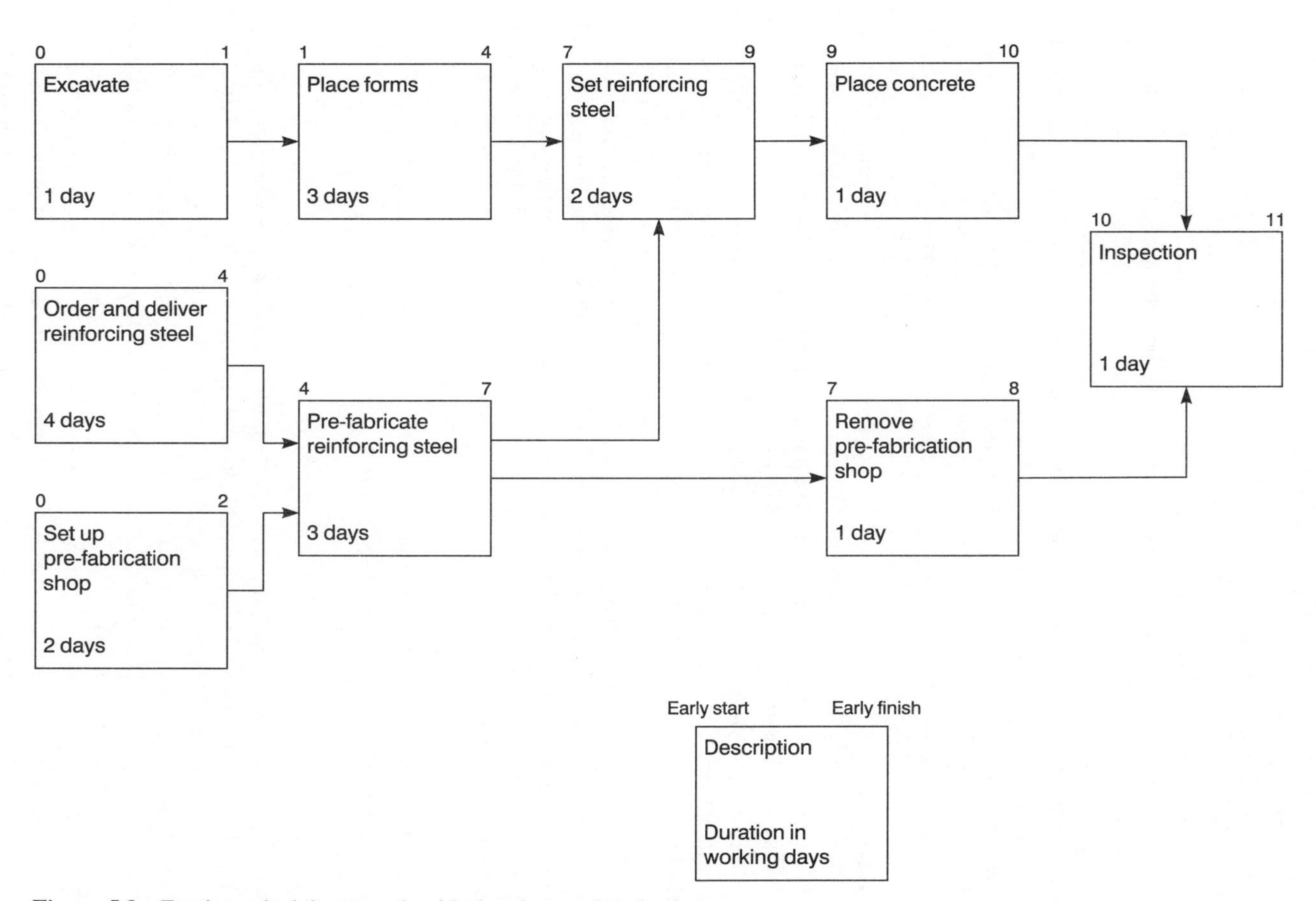

Figure 5.9 Footing schedule network with durations and early times.

In Figure 5.9, the early start time for 'excavate' is 0, because it can begin at the beginning of the project. Its early finish time is 0 + 1 = 1. For 'place forms', the early start is 1, because it has only one predecessor, whose early finish is 1. For 'pre-fabricate reinforcing steel', the early start is ES = max (4, 2) = 4. Its early finish is 4 + 3 = 7. Note that Figure 5.9 shows all early starts and early finishes just above the nodes' upper left and upper right hand corners, respectively. We proceed through the network from beginning to end, in a process sometimes called the *forward pass*, until we reach the end of the project. The result is an early finish time of 11 for 'inspection' and we conclude that the project can be completed in 11 working days if the indicated sequence is followed and the estimated activity durations are correct. If the network included more than one finish activity (that is, if another activity or activities could proceed concurrently with 'inspection'), the project duration would be defined as the maximum of the early finishes for all finish activities.

Note carefully that an activity's early start and finish times are the earliest the activity *can* start and finish, based on prior activities. Not all activities *must* start and finish at those times in order to complete the project in 11 days. The next set of calculations determines the *latest* an activity can occur without overrunning the target project completion time. First, we must designate a finish time for the project. The usual practice is to set the project duration, as determined by the early start and finish calculations (11 in the case of our sample project), as the target. We shall adopt this practice for our sample, although we could have used some other target, such as a contractually obligated completion time.

We perform a set of *backward pass* calculations to determine a latest start and latest finish time for each activity, beginning at the end of the project. Figure 5.10 includes the results of these calculations, with late starts and late finishes shown under the lower left and right hand corners, respectively, of each node. An activity's *late finish* time is the latest the activity must finish in order not to overrun the target completion time. It equals the target project completion time if it is a finish activity; otherwise, the late finish is the earliest of the late starts for all of the activity's immediate successors. It is the *earliest* late start time because this activity must be completed prior to the start of *all* successor activities. The *late start* time is the late finish time minus the activity's duration. Thus,

LF = late finish time = target completion time for all 'finish' activities; otherwise = min (late starts of all immediate successors);
LS = late start time = LF – duration.

In Figure 5.10, we have assigned 11 as the late finish for 'inspection'. Its late start is thus 11 – 1 = 10. 'Remove pre-fabrication shop' has only one immediate successor, so its late finish is 10, the late start for 'inspection', and its late start is 10 – 1 = 9. Because 'pre-fabricate reinforcing steel' has two immediate successors, its late finish is determined as EF = min (7, 9) = 7 and its late start = LS = 7 – 3 = 4. The process continues backwards to the beginning until late times have been calculated for all activities.

We observe in Figure 5.10 that the early start and late start times for some activities are identical, whereas for others, the late start time is greater than the early start time. The difference between the late start and early start is a measure of the activity's criticality or of how 'tight' is it, schedule wise. We call this difference the activity's *slack*, or *float*. For any activity, a formal definition is SL = slack = LS – ES. It can also be calculated as SL = LF – EF or as SL = LF – ES – duration. In practical terms, slack is the amount of time an activity can slip beyond its early times without causing an extension of the total project duration.

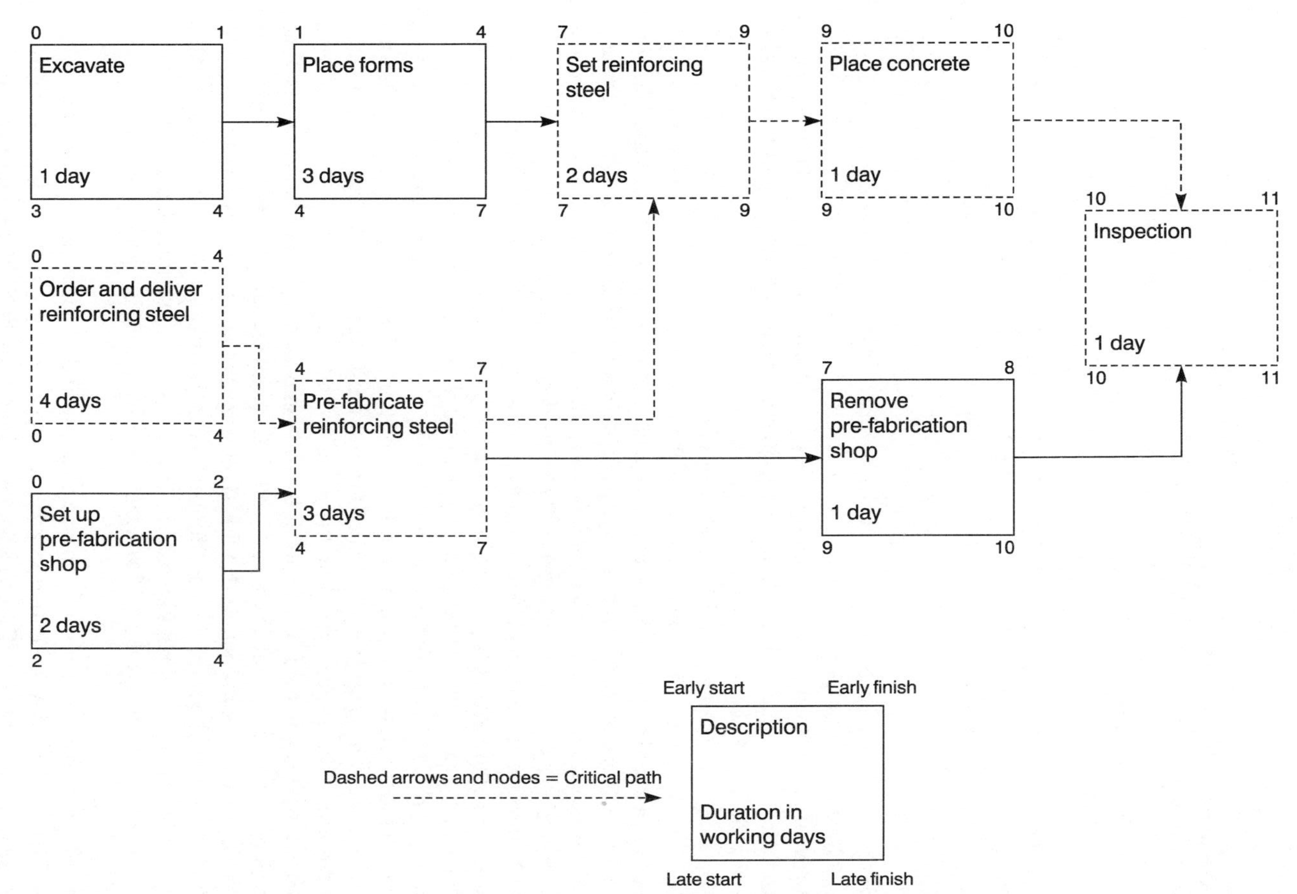

Figure 5.10 Footing schedule network with durations, early times, late times and critical path.

Table 5.1 Activity table for footing construction project

Description	Duration (days)	Early start	Early finish	Late start	Late finish	Slack	Critical?
Excavate	1	0	1	3	4	3	
Order and deliver reinforcing steel	4	0	4	0	4	0	✓
Set up pre-fabrication shop	2	0	2	2	4	2	
Place forms	3	1	4	4	7	3	
Pre-fabricate reinforcing steel	3	4	7	4	7	0	✓
Set reinforcing steel	2	7	9	7	9	0	✓
Place concrete	1	9	10	9	10	0	✓
Remove pre-fabrication shop	1	7	8	9	10	2	
Inspection	1	10	11	10	11	0	✓

If an activity has slack equal to 0, it is said to be *critical*. Critical activities cannot be allowed to slip beyond their early times with causing an overrun of the project's target completion time, because as their slacks equal 0, their late start and finish times are equal to their early start and finish times, respectively. In Figure 5.10, the critical activities have been identified; note that they form a continuous path from the beginning to the end of the project. This path formed by 'order and deliver reinforcing steel', 'pre-fabricate reinforcing steel', 'set reinforcing steel', 'place concrete' and 'inspection' is called the *critical path*. These are the activities whose time dimensions must be supervised most closely; any slippage along this path without a concomitant shortening of some other activity on the same path will cause a slippage of the total project duration. Furthermore, if it is desired to shorten the total project duration, one must focus on one or more of the critical activities; shortening only a non-critical activity, by working overtime, assigning more workers or otherwise improving its completion rate, will not result in any reduction in total project duration. In our example, we could save a day in the project duration by arranging to procure the reinforcing steel in three days instead of four, assuming no other changes were made.

Figure 5.10 provides a helpful visual picture of the results of our schedule analysis. Another way to present the results is in tabular form, as shown in Table 5.1. Here we list each activity, with its description, duration, early and late start and finish and slack. A final column indicates whether the activity is critical, based on the value of its slack.

Slack is defined for individual activities, but it is important to understand that slack is not additive along a path. In Figure 5.10 and Table 5.1, 'excavate' and 'place forms' each have 3 days of slack. We must not conclude that this short subpath has a total of 6 days of slack. Recall that slack is the amount an activity can slip beyond its *early* times without extending the total project duration. If 'excavate' slips by 3 days, 'place forms' will be forced to take place between its late start and finish times and any further slippage will cause the project to overrun its target completion time.

A bar chart provides another useful means of displaying the outcome of a network schedule analysis. Figure 5.11 shows the five critical activities in black and the four activities with positive values of slack in grey. The dashed lines extending to the right from the non-critical bars are an effective means of displaying each activity's slack. The right-hand end of the dashed line is located at the activity's late finish time and the difference between the late finish and early finish, shown by the dashed line, is the slack.

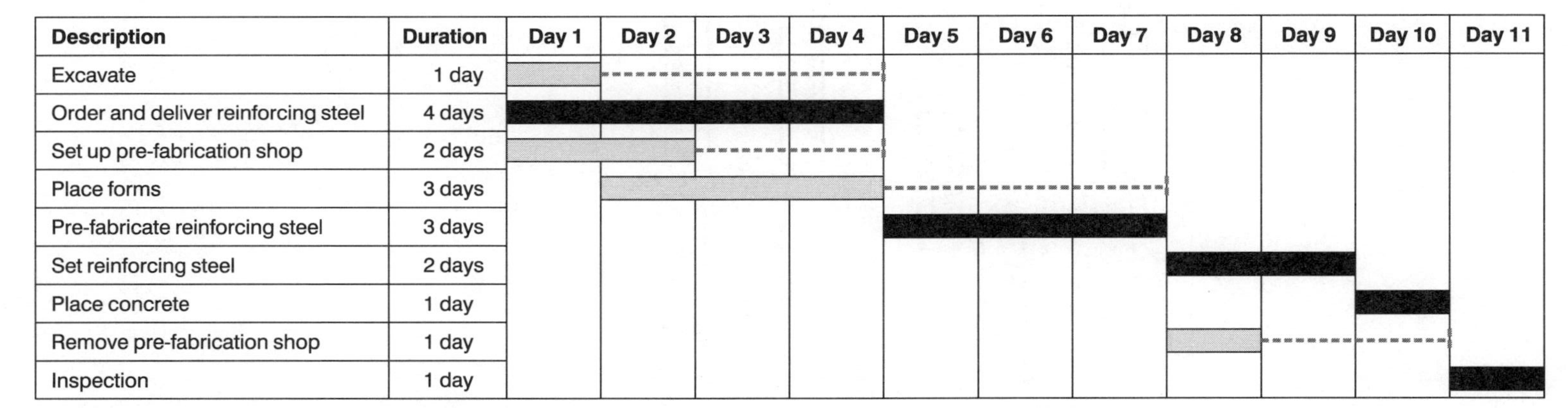

Figure 5.11 Bar chart with results of footing schedule analysis.

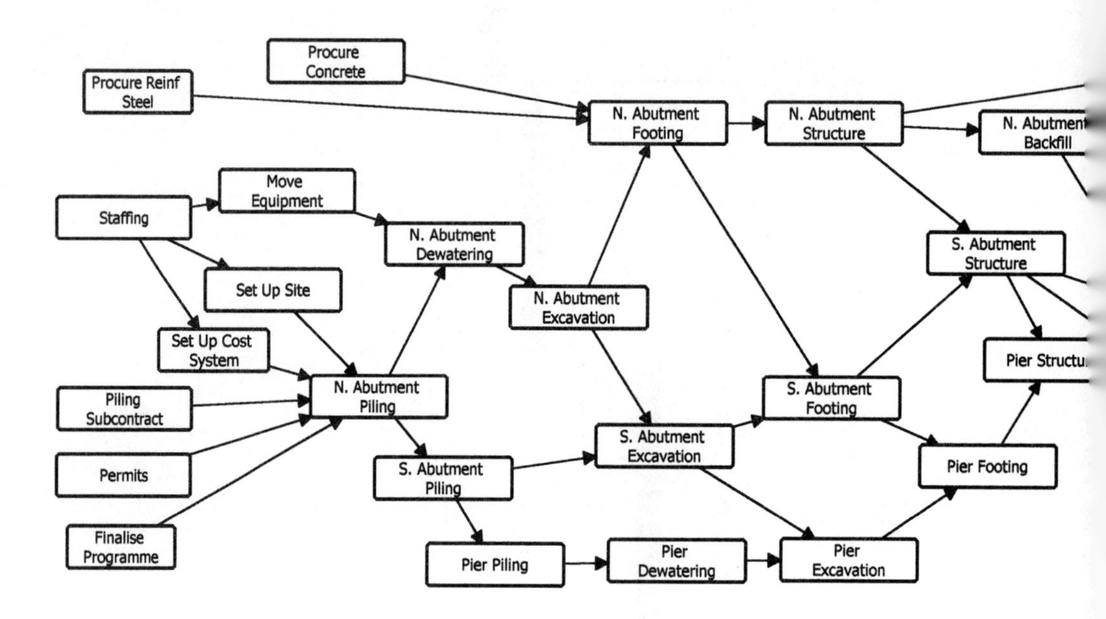

Figure 5.12 Bridge project schedule network.

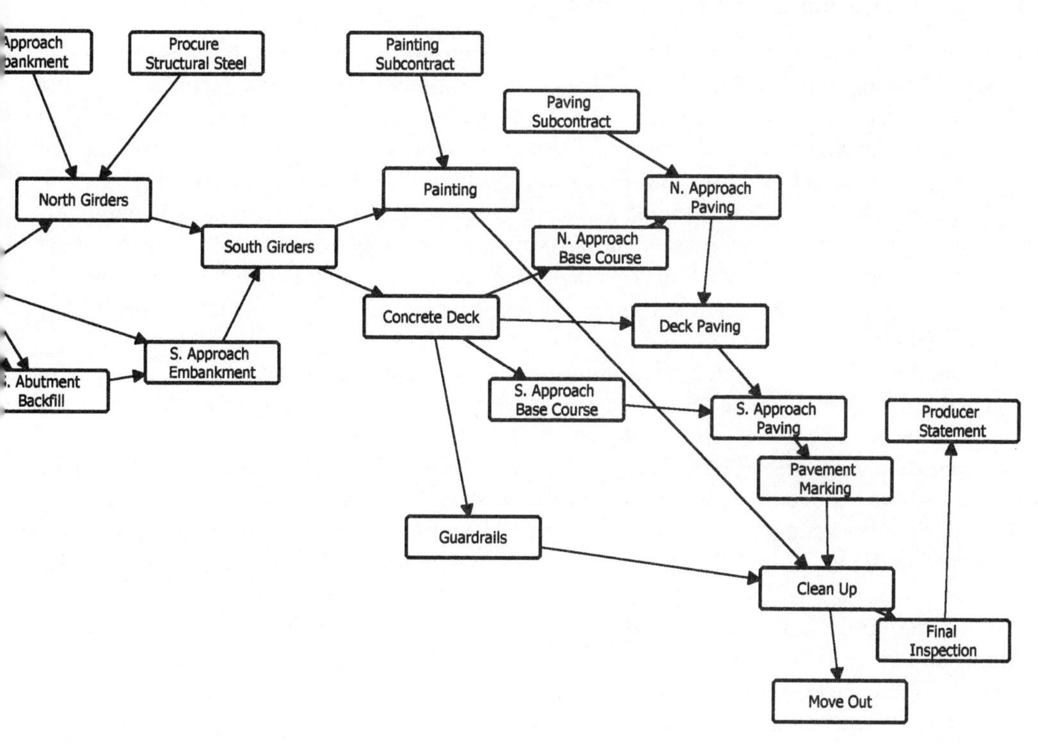

Figure 5.12 Continued.

Sometimes there is confusion surrounding the numbering system used to determine start and finish times; the bar chart may be a helpful means of clarifying the numbers. When we say the project begins at 'time 0', we mean the beginning of Day 1. On an early start–early finish basis, 'set up pre-fabrication shop', for example, begins at 'time 0', lasts 2 days and ends at 'time 2'. We could also say that it begins at the beginning of Day 1 and ends at the end of Day 2, which coincides with the beginning of Day 3. The calculated start and finish times shown in Figure 5.10 and Table 5.1 are the ends of the indicated days. 'Inspection' must start at the *end* of Day 10 (which is the *beginning* of Day 11) and must finish at the *end* of Day 11.

Computer applications: a larger example

In Figure 5.7, we showed a work breakdown structure for a hypothetical bridge construction project. Although this WBS lacks all of the details that would normally be included in such a structure for a project of this magnitude, it is sufficiently detailed to apply the principles outlined above to the development of its schedule. Figure 5.7 shows 45 elements at the lowest levels of their respective branches; these elements form the basis of our schedule network. First, we assemble an activity-on-node network, as shown in Figure 5.12. The process of assembling this plan involves as many members of the project team as possible; the result is an excellent communication tool for explaining the intended sequence of activities. There is considerable value in such an exercise, even if it is not followed by the numerical scheduling analysis we discuss below.

Figure 5.12 deserves several observations. First, this plan is only one of several orders in which the 45 activities could be carried out. It represents the best efforts of the project team, based on the information available at the time it is developed. The plan shows, for example, that the foundation work will proceed from the north abutment to the south abutment and then to the pier. It is subject to modification as the planning process and the execution of the plan proceed. Second, the plan assumes that nine activities can begin concurrently at the beginning of the project. The drafter of the diagram has taken some liberties in placing some of these nodes ('painting subcontract', for example) toward the right of the diagram rather than at its extreme left, but the principle that a node without a predecessor node is a 'begin' node still applies. Third, in a few cases, we have violated the 'left-to-right' flow of work, in the interest of a reasonably compact diagram. Finally, the project will be complete when two activities, 'move-out' and 'producer statement', have been completed.

The next step is to assign durations to each activity and perform basic scheduling calculations. We show the results of these calculations in several ways, in output from Microsoft Project 2000®. First, Figure 5.13 shows a bar chart containing our 45 activities. Each activity is shown at its earliest times, beginning at its early start time and ending at its early finish time. Critical activities are shown in black (shown in red in Microsoft Project) and arrows show the various precedence relationships. We have assumed that the project will begin on 29 March 2004 and that work will be accomplished 5 days per week. Based on the estimated activity durations and the sequencing in Figure 5.12, the project will be completed on 26 November 2004.

Microsoft Project 2000® allows the assignment of *lag* times to the connectors between activities. A lag time is a non-work duration that must occur between activities. In our project, the curing of concrete must take place after it is placed and before the structure is loaded. Although not shown on the bar charts and network diagrams, we have assigned lag times after several concrete placement activities. In particular, after the 'pier structure' is completed, there is a 20-working day (= 28 calendar-day) lag before the 'north girders' can be placed. Figure 5.13

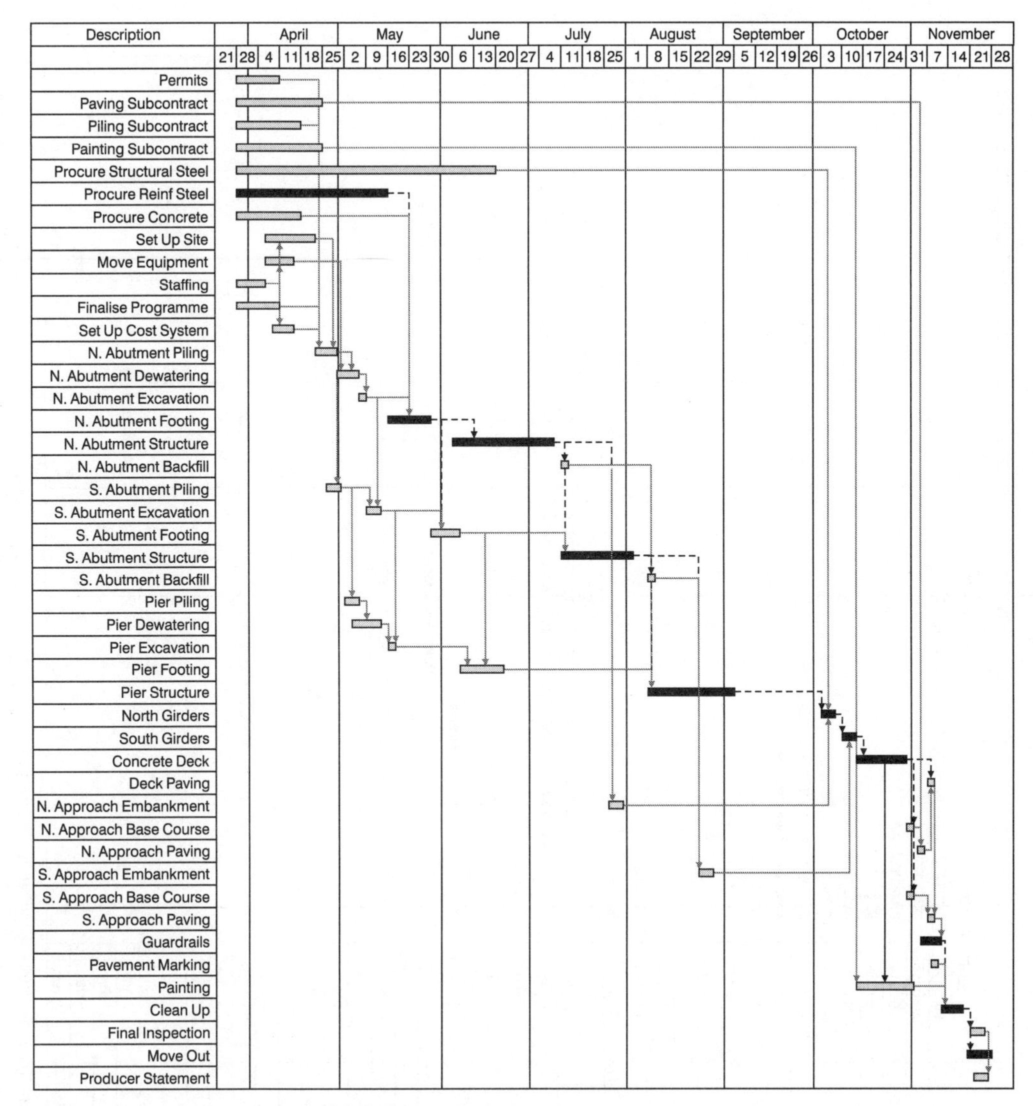

Figure 5.13 Sample bridge project bar chart.

shows this 4-week gap in the schedule clearly and indicates that no work will be occurring on the project while the pier concrete cures. Perhaps some revision of the programme could lead to a more efficient sequencing of activities during this time period!

One interesting feature of our analysis software is the capability to include all elements of the work breakdown structure, in a hierarchical fashion as shown in Figure 5.14. In fact, Figure 5.13 is a subset of the activities in Figure 5.14. The 45 lowest level elements that are the work packages or activities in our network are shown in regular text and their bars on the chart are black or shaded depending on whether they are critical (shown in colour in Microsoft Project), identical to those in Figure 5.13. All other elements are summaries, exactly as depicted in our WBS; they appear in bold text and as hatched/bars on the chart. In total, there are 62 elements, not including the one element at the top that represents the total project, of which 17 are summary elements. Other

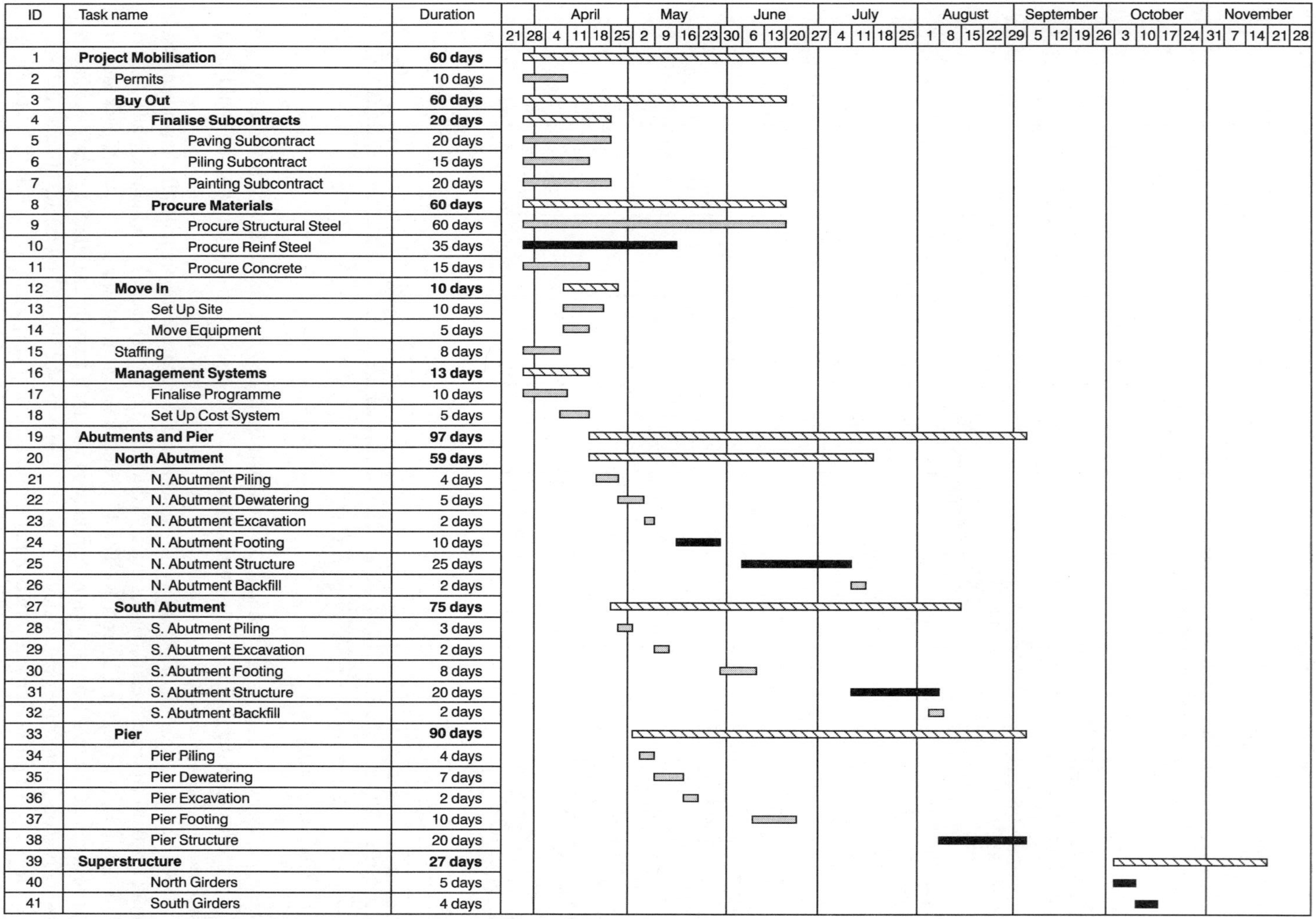

ID	Task name	Duration
1	**Project Mobilisation**	**60 days**
2	Permits	10 days
3	**Buy Out**	**60 days**
4	**Finalise Subcontracts**	**20 days**
5	Paving Subcontract	20 days
6	Piling Subcontract	15 days
7	Painting Subcontract	20 days
8	**Procure Materials**	**60 days**
9	Procure Structural Steel	60 days
10	Procure Reinf Steel	35 days
11	Procure Concrete	15 days
12	**Move In**	**10 days**
13	Set Up Site	10 days
14	Move Equipment	5 days
15	Staffing	8 days
16	**Management Systems**	**13 days**
17	Finalise Programme	10 days
18	Set Up Cost System	5 days
19	**Abutments and Pier**	**97 days**
20	**North Abutment**	**59 days**
21	N. Abutment Piling	4 days
22	N. Abutment Dewatering	5 days
23	N. Abutment Excavation	2 days
24	N. Abutment Footing	10 days
25	N. Abutment Structure	25 days
26	N. Abutment Backfill	2 days
27	**South Abutment**	**75 days**
28	S. Abutment Piling	3 days
29	S. Abutment Excavation	2 days
30	S. Abutment Footing	8 days
31	S. Abutment Structure	20 days
32	S. Abutment Backfill	2 days
33	**Pier**	**90 days**
34	Pier Piling	4 days
35	Pier Dewatering	7 days
36	Pier Excavation	2 days
37	Pier Footing	10 days
38	Pier Structure	20 days
39	**Superstructure**	**27 days**
40	North Girders	5 days
41	South Girders	4 days

Timeline: April | May | June | July | August | September | October | November

21 | 28 | 4 | 11 | 18 | 25 | 2 | 9 | 16 | 23 | 30 | 6 | 13 | 20 | 27 | 4 | 11 | 18 | 25 | 1 | 8 | 15 | 22 | 29 | 5 | 12 | 19 | 26 | 3 | 10 | 17 | 24 | 31 | 7 | 14 | 21 | 28

Figure 5.14 Sample bridge project bar chart, including summary activities.

ID	Task name	Duration
42	**Deck**	**18 days**
43	Concrete Deck	15 days
44	Deck Paving	1 day
45	**Approaches**	**78 days**
46	**North Approach**	**76 days**
47	N. Approach Embankment	3 days
48	N. Approach Base Course	1 day
49	N. Approach Paving	1 day
50	**South Approach**	**58 days**
51	S. Approach Embankment	3 days
52	S. Approach Base Course	1 day
53	S. Approach Paving	1 day
54	**Finish Work**	**21 days**
55	Guardrails	6 days
56	Pavement Marking	1 day
57	Painting	10 days
58	**Project Closeout & Termination**	**10 days**
59	Clean Up	5 days
60	Final Inspection	2 days
61	Move Out	5 days
62	Producer Statement	2 days

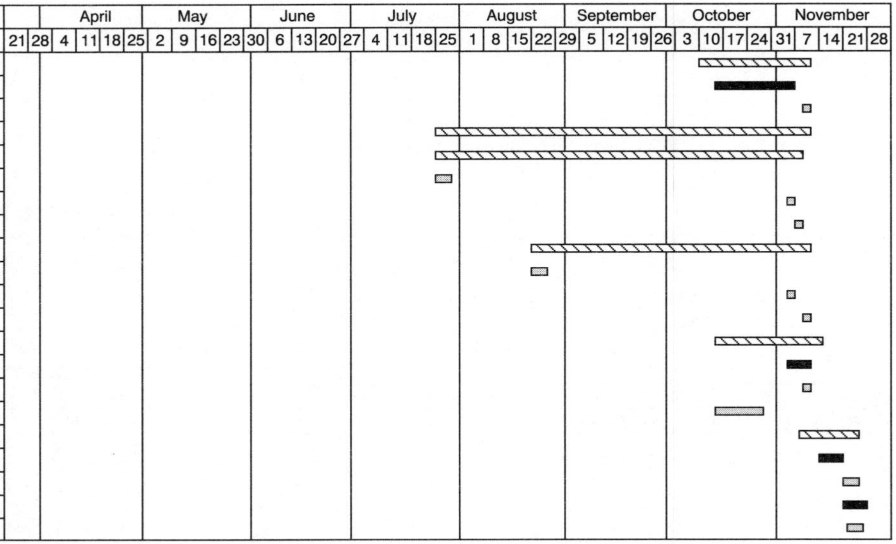

Project: Bridge Project
Date: Wed 2/6/02

Critical | Non-critical | Summary

Figure 5.14 Continued.

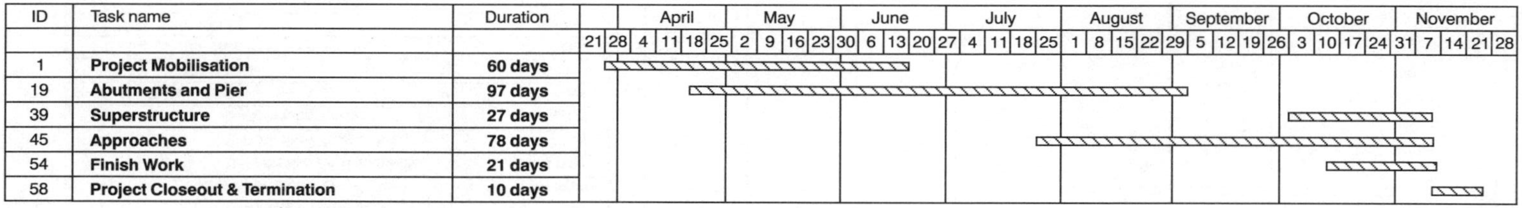

ID
Task name
Duration
April
May
June
July
August
September
October
November
21 28 4 11 18 25 2 9 16 23 30 6 13 20 27 4 11 18 25 1 8 15 22 29 5 12 19 26 3 10 17 24 31 7 14 21 28
1
Project Mobilisation
60 days
19
Abutments and Pier
97 days
39
Superstructure
27 days
45
Approaches
78 days
54
Finish Work
21 days
58
Project Closeout & Termination
10 days

Figure 5.15 Sample bridge project bar chart, summary activities only.

features of Figure 5.14 are the addition of an identification (ID) number and duration for each activity and the exclusion of connecting arrows from this version of the chart.

The duration of a summary activity may be confusing. This software considers a summary activity's duration to be the total length of time between the start of the first activity in that summary's branch and the completion of the last activity in the branch, even though there may be gaps when no work is taking place. An example is the summary for 'pier', which shows a 90-day duration. 'Pier piling' is scheduled to begin on 3 May 2004 and 'pier structure' is scheduled to be completed on 3 September 2004, 90 working days later. Note that the total number of working days expected for the pier work is 4 + 7 + 2 + 10 + 20 = 43.

The software allows the filtering and grouping of activities in many different ways. Figures 5.13 and 5.14, based on the same dataset, have already been discussed. Figure 5.15 shows only the six elements in the second level of the WBS. Such a summary chart might be helpful in a synopsis prepared for one's client or upper management, although it may be misleading because gaps are scheduled to occur that do not appear on the element summaries.

The software is also capable of producing a network diagram depicting the results of the schedule analysis. For our bridge project, the network is given in Figure 5.16. Note that the layout is similar to that in Figure 5.12. This version shows the description, ID number, duration, early start date and early finish date for each activity, with critical activities shaded (shown in red in Microsoft Project) and non-critical unshaded (shown in blue in Microsoft Project). There is considerable flexibility in the design of the activity node templates; if desired, one could include the slack value for each activity, for example.

Our final output from the basic schedule analysis is a tabular report giving early and late start and finish dates and slack for each activity, similar to that prepared for our footing project and shown in Table 5.1. The order in which the activities appear is selected by the analyst and we have chosen to order them in Table 5.2 by increasing value of early start date.

Computer applications: some software features

The development of the computer during the past 40 years has made the analysis of large network schedules practical. For large projects, it is not unusual for a network totalling several thousand activities to be used to plan, schedule and monitor project progress. In the previous section, we have seen the results of a basic analysis of a very small network. Here, we describe the most important features and uses of modern project scheduling software. No attempt is made to cite or evaluate the several hundred available packages; a helpful source of such information is published by the Project Management Institute (1999).

Data input and error checking

The program prompts the user for various types of input. For an activity-on-node network to be processed by a typical program such as Microsoft Project 2000®, the minimum required input for each activity consists of an ID number, a description, a duration and the ID numbers of all of the activity's immediate predecessors, if any. The user also provides calendar information such as the project start date and the number of working days per week, plus general information such as project name, manager and location. While input is being provided, or as the first step in the calculation process, the program checks for such errors as the designation of an activity's predecessor without having defined that predecessor or the configuration of a set of activities that would imply a loop (for example, C follows B, B follows A, A follows C; that's not logical!).

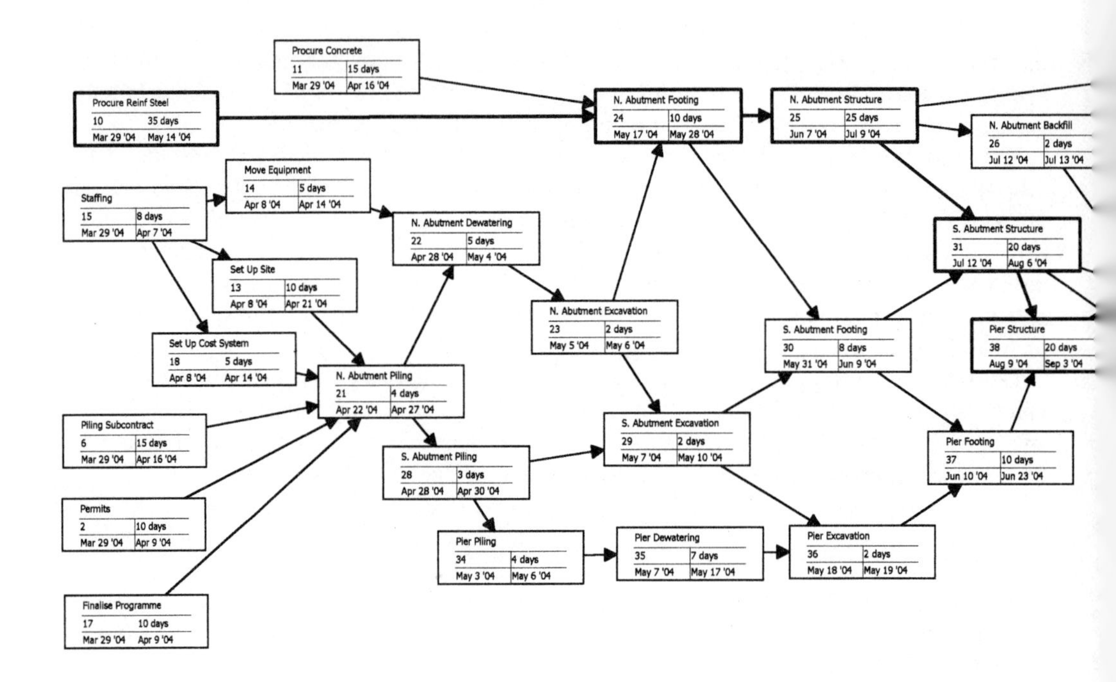

Figure 5.16 Bridge project schedule network, including analysis results.

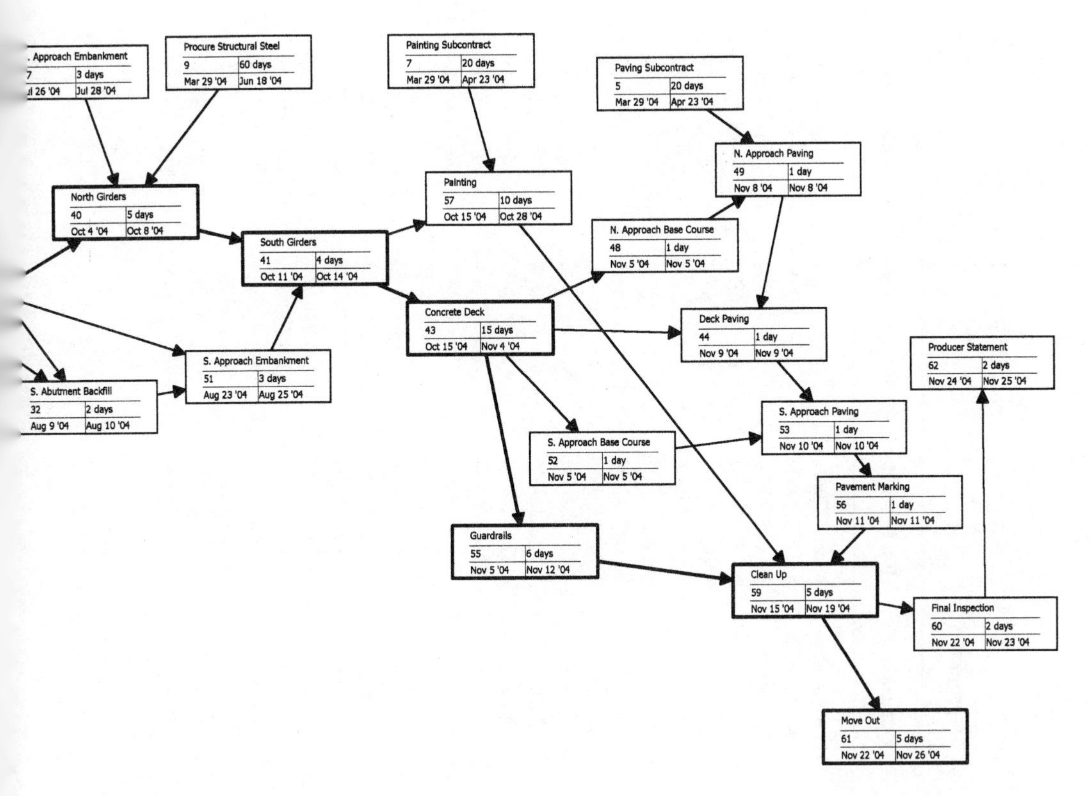

Figure 5.16 Continued.

Table 5.2 Activity table for bridge construction project

ID	Description	Duration	Early start	Early finish	Late start	Late finish	Slack
15	Staffing	8 days	Mon 3/29/04	Wed 4/7/04	Tue 4/6/04	Thu 4/15/04	6 days
2	Permits	10 days	Mon 3/29/04	Fri 4/9/04	Fri 4/16/04	Thu 4/29/04	14 days
17	Finalise programme	10 days	Mon 3/29/04	Fri 4/9/04	Fri 4/16/04	Thu 4/29/04	14 days
6	Piling subcontract	15 days	Mon 3/29/04	Fri 4/16/04	Fri 4/9/04	Thu 4/29/04	9 days
11	Procure concrete	15 days	Mon 3/29/04	Fri 4/16/04	Mon 4/26/04	Fri 5/14/04	20 days
5	Paving subcontract	20 days	Mon 3/29/04	Fri 4/23/04	Mon 10/11/04	Fri 11/5/04	140 days
7	Painting subcontract	20 days	Mon 3/29/04	Fri 4/23/04	Fri 10/1/04	Thu 10/28/04	134 days
10	Procure reinforced steel	35 days	Mon 3/29/04	Fri 5/14/04	Mon 3/29/04	Fri 5/14/04	0 days
9	Procure structural steel	60 days	Mon 3/29/04	Fri 6/18/04	Mon 7/12/04	Fri 10/1/04	75 days
14	Move equipment	5 days	Thu 4/8/04	Wed 4/14/04	Thu 4/29/04	Wed 5/5/04	15 days
18	Set up cost system	5 days	Thu 4/8/04	Wed 4/14/04	Fri 4/23/04	Thu 4/29/04	11 days
13	Set up site	10 days	Thu 4/8/04	Wed 4/21/04	Fri 4/16/04	Thu 4/29/04	6 days
21	N. abutment piling	4 days	Thu 4/22/04	Tue 4/27/04	Fri 4/30/04	Wed 5/5/04	6 days
28	S. abutment piling	3 days	Wed 4/28/04	Fri 4/30/04	Wed 6/16/04	Fri 6/18/04	35 days
22	N. abutment dewatering	5 days	Wed 4/28/04	Tue 5/4/04	Thu 5/6/04	Wed 5/12/04	6 days
34	Pier piling	4 days	Mon 5/3/04	Thu 5/6/04	Wed 6/30/04	Mon 7/5/04	42 days
23	N. abutment excavation	2 days	Wed 5/5/04	Thu 5/6/04	Thu 5/13/04	Fri 5/14/04	6 days
29	S. abutment excavation	2 days	Fri 5/7/04	Mon 5/10/04	Mon 6/21/04	Tue 6/22/04	31 days
35	Pier dewatering	7 days	Fri 5/7/04	Mon 5/17/04	Tue 7/6/04	Wed 7/14/04	42 days
24	N. abutment footing	10 days	Mon 5/17/04	Fri 5/28/04	Mon 5/17/04	Fri 5/28/04	0 days
36	Pier excavation	2 days	Tue 5/18/04	Wed 5/19/04	Thu 7/15/04	Fri 7/16/04	42 days
30	S. abutment footing	8 days	Mon 5/31/04	Wed 6/9/04	Wed 6/23/04	Fri 7/2/04	17 days
25	N. abutment structure	25 days	Mon 6/7/04	Fri 7/9/04	Mon 6/7/04	Fri 7/9/04	0 days
37	Pier footing	10 days	Thu 6/10/04	Wed 6/23/04	Mon 7/19/04	Fri 7/30/04	27 days

Table 5.2 Continued

ID	Description	Duration	Early start	Early finish	Late start	Late finish	Slack
26	N. abutment backfill	2 days	Mon 7/12/04	Tue 7/13/04	Mon 9/27/04	Tue 9/28/04	55 days
31	S. abutment structure	20 days	Mon 7/12/04	Fri 8/6/04	Mon 7/12/04	Fri 8/6/04	0 days
47	N. approach embankment	3 days	Mon 7/26/04	Wed 7/28/04	Wed 9/29/04	Fri 10/1/04	47 days
32	S. abutment backfill	2 days	Mon 8/9/04	Tue 8/10/04	Mon 10/4/04	Tue 10/5/04	40 days
38	Pier structure	20 days	Mon 8/9/04	Fri 9/3/04	Mon 8/9/04	Fri 9/3/04	0 days
51	S. approach embankment	3 days	Mon 8/23/04	Wed 8/25/04	Wed 10/6/04	Fri 10/8/04	32 days
40	North girders	5 days	Mon 10/4/04	Fri 10/8/04	Mon 10/4/04	Fri 10/8/04	0 days
41	South girders	4 days	Mon 10/11/04	Thu 10/14/04	Mon 10/11/04	Thu 10/14/04	0 days
57	Painting	10 days	Fri 10/15/04	Thu 10/28/04	Fri 10/29/04	Thu 11/11/04	10 days
43	Concrete deck	15 days	Fri 10/15/04	Thu 11/4/04	Fri 10/15/04	Thu 11/4/04	0 days
48	N. approach base course	1 day	Fri 11/5/04	Fri 11/5/04	Mon 11/8/04	Mon 11/8/04	1 day
52	S. approach base course	1 day	Fri 11/5/04	Fri 11/5/04	Tue 11/9/04	Tue 11/9/04	2 days
55	Guardrails	6 days	Fri 11/5/04	Fri 11/12/04	Fri 11/5/04	Fri 11/12/04	0 days
49	N. approach paving	1 day	Mon 11/8/04	Mon 11/8/04	Tue 11/9/04	Tue 11/9/04	1 day
44	Deck paving	1 day	Tue 11/9/04	Tue 11/9/04	Wed 11/10/04	Wed 11/10/04	1 day
53	S. approach paving	1 day	Wed 11/10/04	Wed 11/10/04	Thu 11/11/04	Thu 11/11/04	1 day
56	Pavement marking	1 day	Thu 11/11/04	Thu 11/11/04	Fri 11/12/04	Fri 11/12/04	1 day
59	Clean up	5 days	Mon 11/15/04	Fri 11/19/04	Mon 11/15/04	Fri 11/19/04	0 days
60	Final inspection	2 days	Mon 11/22/04	Tue 11/23/04	Tue 11/23/04	Wed 11/24/04	1 day
61	Move-out	5 days	Mon 11/22/04	Fri 11/26/04	Mon 11/22/04	Fri 11/26/04	0 days
62	Producer statement	2 days	Wed 11/24/04	Thu 11/25/04	Thu 11/25/04	Fri 11/26/04	1 day

Basic time calculations

We have shown, in our simple footing project network, how to perform the basic time calculations. The computer program goes through that same process to determine early and late start and finish times and slack for each activity. Because calendar information has been provided, a calendar date is assigned to each calculated start and finish time. Once the basic data have been provided, the calculation process is so rapid that it is a simple matter to change a few pieces of input and perform the calculations again, thus permitting the simulation of a number of 'what if' scenarios.

Tabular output

It is possible to filter and order the tabular output in a great variety of ways. We have shown one example in Table 5.2. Activities might be coded by craft or by location, whether or not the WBS is structured in that way. A report could be confined to all of the carpenter activities and made available to the lead carpenter. A look-ahead report could produce the listing of all activities expected to start within six weeks. A report could list only those activities that are critical or nearly so by selecting those with, say, two weeks or less slack. The possibilities are nearly endless!

Resource analysis

Activities cannot be carried out without personnel, equipment or materials. Many network scheduling programs allow the user to assign such resources to project activities to determine the overall impact on resource needs. If welders are in short supply, we could get some idea of the effect on our project duration of being able to hire no more than five, for example. Can we shift some non-critical activities to alleviate peaks in the curve of this labour requirement versus time? Suppose we plan to utilise a barge to support several of the water-related activities on our bridge project. Will one barge be enough? Analysis of this resource requirement resulted in the curve of barges versus time for 4 months of the project duration shown in Figure 5.17; here we have assumed the availability of two barges and have assumed that all activities start and finish at their early times. Note that, under this assumption, there are 2 weeks, or portions thereof, when three barges would be needed. Further analysis showed that shifting the schedules of some non-critical activities, so that they occur later than their early times, could reduce the maximum requirement to two barges. The results of this *resource levelling* are shown in Figure 5.18. Note that we have dealt with only one resource here, whereas often several resources are in short supply. Shifting activities to alleviate one resource problem may impact another resource adversely. Most modern project scheduling software permits the analysis of many resources concurrently.

Cost analysis and cost control

If the project budget is apportioned to each activity in the project network, many network scheduling programs have the capability to produce graphs of projected cash flow over the duration of the project. In a later section on development of the project budget, we shall demonstrate such a graph. Furthermore, one method of accounting for project costs as the project proceeds is to base cost reports on the actual costs of the individual activities and then compare the overall actual costs with those projected as of the same point in time.

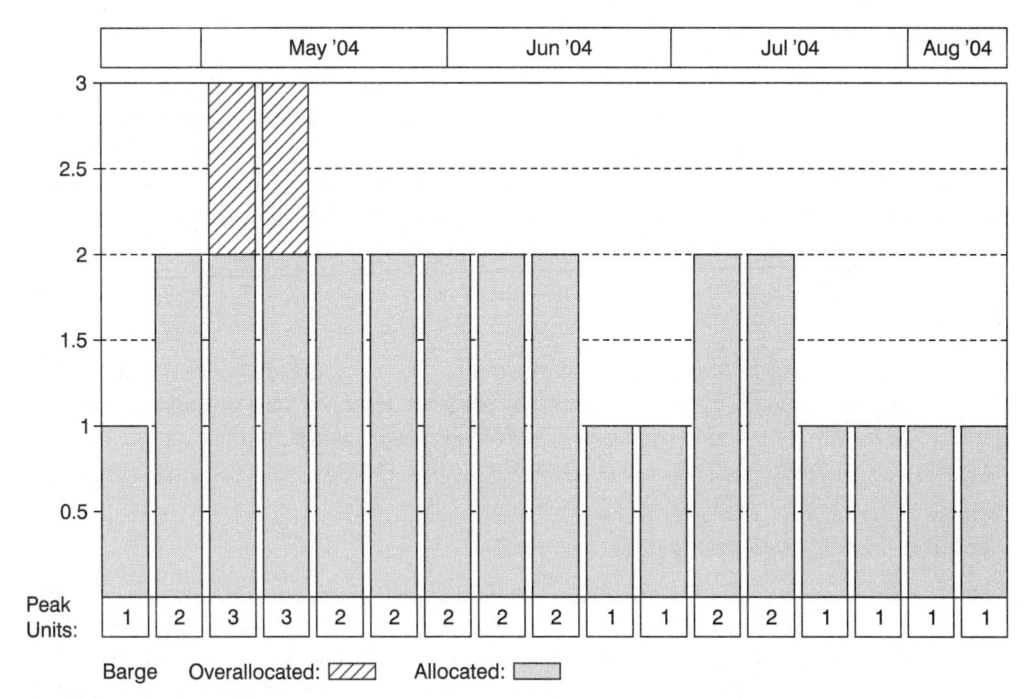

Figure 5.17 Portion of bridge project resource curve for barges, with over-allocation.

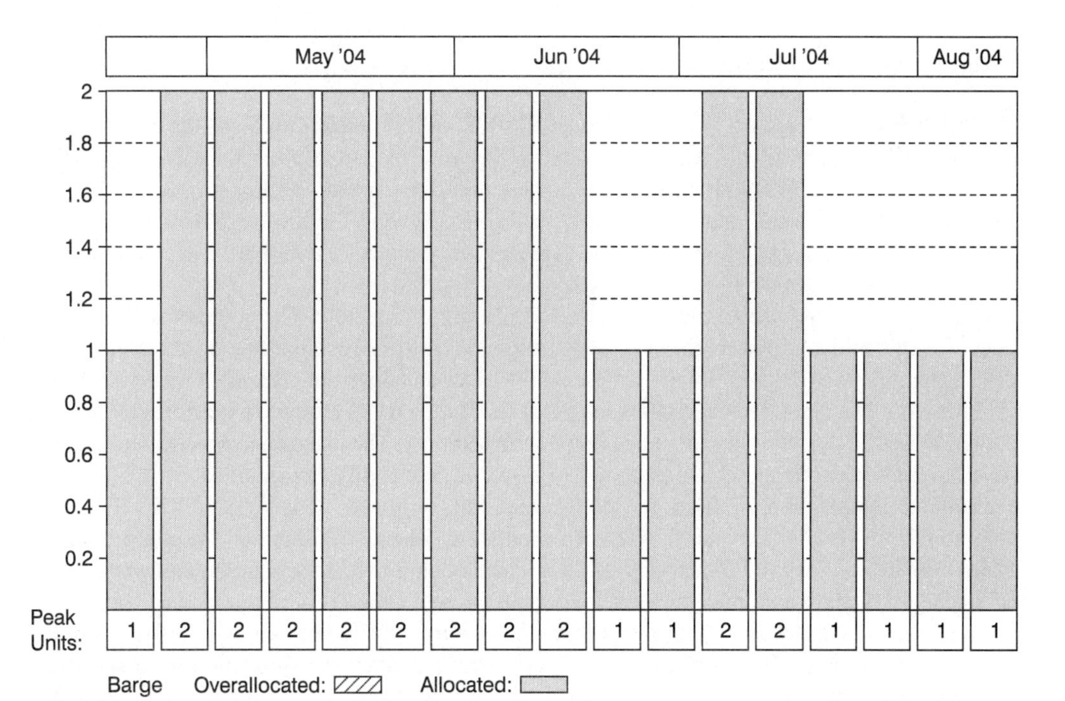

Figure 5.18 Portion of bridge project resource curve for barges, after resource levelling.

Schedule monitoring and updating

The tabular and graphical reports we have developed so far show the proposed project schedule. The software also provides an opportunity to monitor the project schedule by comparing actual with planned progress. Often a 'baseline' schedule can be saved to provide a lasting record of the original anticipated plan; then, the actual start and finish dates of each activity are recorded and compared against the baseline. It is common to update the schedule and report actual cash flow earnings concurrently, say, at the end of each month.

Graphical output

Modern computers have extensive graphical capabilities. We have already seen examples of bar charts, network diagrams and resource curves and we have described cash flow diagrams. Such graphical presentations are effective means of conveying large amounts of information visually; the walls of the project office are often adorned with current versions of these reports.

Electronic communication of input and output

When it is time to gather updating information about the project schedule, many programs allow the user to query members of the project team via e-mail or the Internet to seek their contributions to the input, with responses provided by electronic forms as well. After the updating analysis is completed, the new reports can be distributed electronically via e-mail or via the Internet. In the next chapter, we shall discuss the use of project websites and the inclusion of schedule information for use by the project team.

Operations modelling

The programming of project activities described so far in this chapter involves the coordination and timing of a multitude of tasks whose totality comprise the entire project. To do so, we have considered all of the WBS elements at the lowest end of each branch. Sometimes it is helpful to analyse in more detail some of the individual tasks, beyond the lowest WBS elements. By analysing individual operations, we can evaluate the impact of alternative techniques and resource commitments on their productivity.

Consider the manufacture, transportation and placing of concrete. The contractor, in making plans for several large placement operations, might want to know the effect on productivity of different numbers of delivery trucks, different crew sizes and different placement methods. A key characteristic of a repetitive operation such as this is that the times required to perform individual components of the operation cannot be known with certainty. The usual approach is to treat these times as random variables subject to known or assumed probability distributions.

Computer simulation is used to model such an operation. The CYCLONE (CYCLic Operations NETwork) software is one such modelling scheme (Halpin and Woodhead, 1998; Halpin and Riggs, 1992). Graphical modelling elements are used to represent such work items as loading, travelling and unloading, as well as queues and counters. The duration of each work item is simulated and the program then determines the total cycle time for an operation and the point in time when each item starts and finishes. After repeating the simulation for a specified number of runs, the program reports the proportion of time that each element is active. Thus, the analyst can identify places in the scheme where delays are occurring and where there are

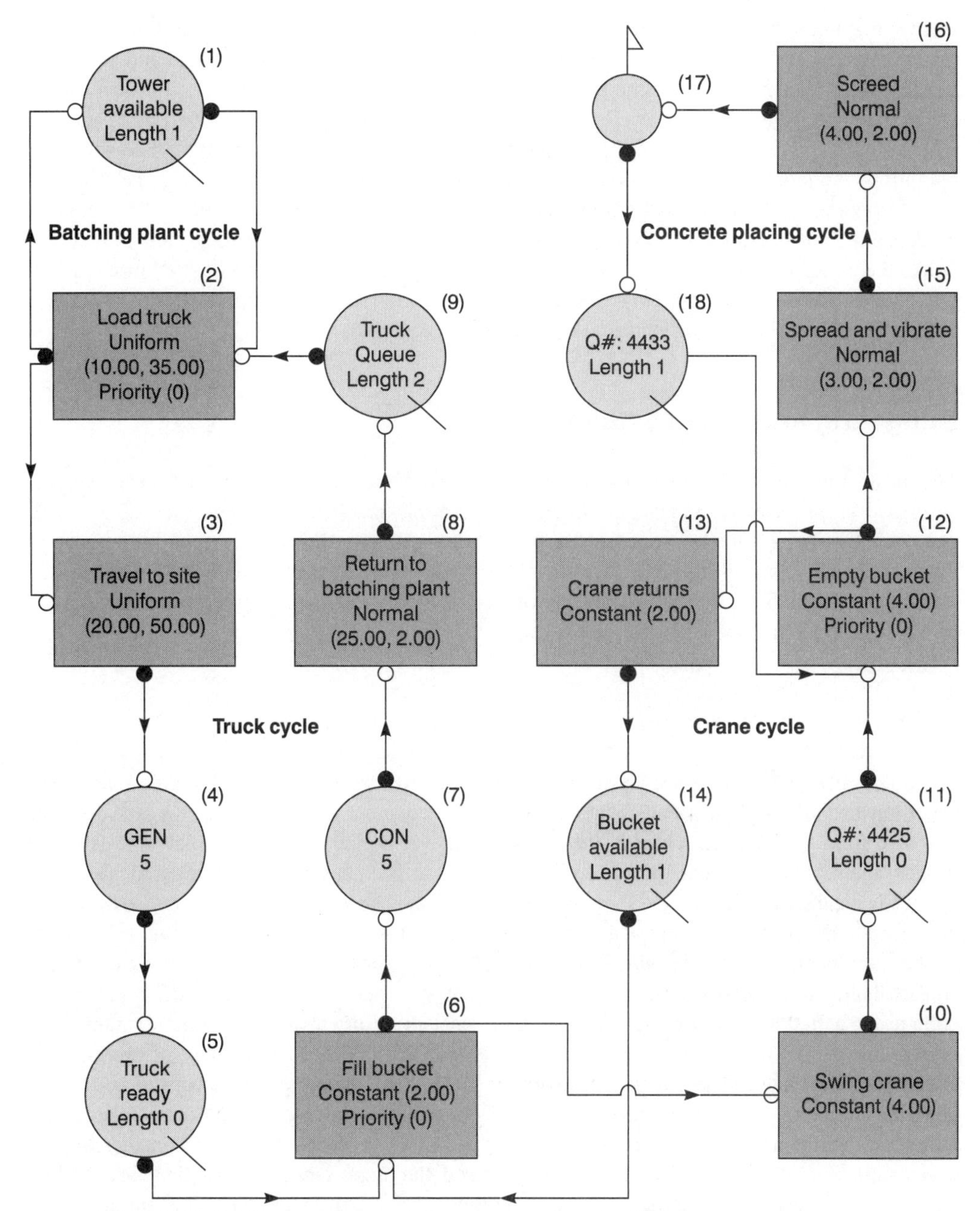

Figure 5.19 CYCLONE operations simulation network (Hewat, 2000).

bottlenecks due to lack of resources. Alternative schemes can be studied by changing sequences, numbers of pieces of equipment and average time required for the various elements.

A simple example of the use of CYCLONE in the analysis of a concrete batching and placing operation is shown in the network in Figure 5.19 (Hewat, 2000; Bennett, 2000). Four cyclical components are identified and analysed: batching plant, truck delivery, placement with crane and final spreading, vibrating and screeding. The analysis allows different numbers of trucks (shown here as 2) and buckets (shown here as 1), travel times that can be changed and variations in crew sizes that impact the time requirements for various components. The purpose of the study is to determine the total cycle time for any set of input variables, as well as to identify where delays and thus possible inefficiencies, are occurring. The Simphony suite of construction operations modelling programs (University of Alberta Construction Engineering and Management Programme, 1999) includes both general programs such as CYCLONE and specialised models for such operations as aggregate production and earthmoving. In addition to concrete work, the simulation approach to construction planning has been applied to such processes as tunnel excavation, roadway paving, the renovation of train tracks without interruption of service and the installation of large floating caissons (Bennett, 2000).

Budgeting and cost systems

The basis for the contractor's priced proposal was the quantity survey and cost estimate described in the previous chapter. Now that the contract has been awarded, the cost estimate becomes the basis for the contractor's project budget and the system that will assist in monitoring and controlling costs. We describe in this section the process of converting the estimate into a budget and the preparation of cash flow curves that show the estimated pattern of expenses over the project's life.

From estimate to budget

If the cost estimate is produced properly, the conversion from estimate to budget should be accomplished with little difficulty. The various work items that are the basis of the cost estimate will also form the basic structure of the budget, which is a plan for the financial aspects of the project against which actual financial performance can be measured and compared. Once the budget is established, the process of cost monitoring can be carried out. This process will be described in the following chapter, as we consider several issues related to operating the project. In the mobilisation stage, the budget and the framework for cost control will be established.

The three purposes of construction cost control systems are (1) to provide a means for comparing actual with budgeted expenses and thus draw attention, in a timely manner, to operations that are deviating from the project budget, (2) to develop a database of productivity and cost performance data for use in estimating the costs of subsequent projects and (3) to generate data for valuing variations and changes to the contract and potential claims for additional payments. We have discussed the second of these purposes in our consideration of cost estimating in Chapter 4 and have emphasised the importance of cost databases, while admitting the need for judgment for every unique project estimate. The first purpose, cost control and the change, variation and claims processes will be considered in Chapter 6.

In the previous chapter, we illustrated the use of an estimating software package called WinEst®, by including two tables (Tables 4.5 and 4.6), produced by that program. Many cost accounting programs provide the contractor with a means to track project costs, using the estimate as a basis. For illustration, we use Peachtree Complete Accounting 2002®, available from Peachtree Software Inc. (2002). The first step in using this package for project cost accounting is to export the WinEst® estimate to Peachtree and to structure the project budget in a form that will be used for cost control. Table 5.3 contains the project direct-cost budget for

Table 5.3 Direct cost budget for concrete wall project, after export to Peachtree Complete Accounting 2002

Bennett Construction
Direct Cost Budget
As of Apr 5, 2004

Filter Criteria includes: 1) IDs from 2004-611 to 2004-611. Report order is by ID.

Job ID Job description For customer	Phase ID	Phase description	Cost code description	Est. exp. units	Est. expenses
2004-611	1.505	Mobilisation	Equipment	1.00	120.00
Concrete wall			Labour	1.00	550.00
Meridian shippers	1.710	Clean up and move out	Equipment	1.00	120.00
			Labour	1.00	320.00
	2.110	Clearing and grubbing	Equipment	400.00	84.00
			Labour	400.00	132.00
	2.220	Silt excavation	Subcontractor	12.10	508.20
	2.221	Rock excavation	Subcontractor	5.30	707.55
	2.224	Backfill	Subcontractor	13.80	327.06
	3.115	Continuous edge forms	Labour	12.20	382.10
			Material	12.20	200.69
	3.135	Wall formwork	Labour	151.00	7429.20
			Material	151.00	2672.70
	3.211	Footing reinforcement	Labour	0.14	69.98
			Material	0.14	85.19
	3.213	Wall reinforcement	Labour	1.34	676.98
			Material	1.34	824.10
	3.310	Footing concrete	Equipment	2.40	47.40
			Labour	2.40	150.00
			Material	2.40	265.92
	3.318	Wall concrete	Equipment	15.00	228.00
			Labour	15.00	396.00
			Material	15.00	1900.50
	5.050	Anchor bolts	Labour	40.00	176.00
			Material	40.00	160.80
2004-611	Total				18 534.37

Courtesy Peachtree Software Inc. (2002).

our concrete wall project. Note that the work items and their values correspond to the direct-cost portion of our estimate given in Tables 4.1, 4.5 and 4.6. This table is confined to direct-cost items because that is the work whose costs must be controlled by management staff on the project worksite. We could include site overhead costs on this cost control scheme as well.

The *cost accounting code system* must be simple enough to be understood and used in the field and also sufficiently comprehensive to acquire the data necessary to fulfil the three purposes listed above. Too much detail is costly, may be misunderstood and misused and can result in information that is unneeded. On the other hand, a code simply representing 'concrete' will hardly identify problems with the placement of footing concrete for machinery foundations. Judgment is required to establish a code system that balances these considerations.

It is generally believed that a single cost-coding system should be used for all projects throughout the contractor's organisation. Most such companies tailor systems for their own use and there are many methods of classification 'represented by combinations of numbers and letters, both upper and lower case, full stops, colons, hyphens, etc., all of which have taken their place in the system' (Pilcher, 1992). A simple system might be based on alphabetical notation broken down by trade. Two or three letters or letter/number combinations might be sufficient. For example, Pilcher (1992) shows the use of BEM to represent laying engineered bricks in a manhole and DC3 for drainlaying 300 mm diameter concrete pipes; in each case, the first letter designates the trade (B for bricklaying and D for drainlaying), with subsequent letters or numbers indicating details within the trade.

Many contractors use a numerical coding system and often the coding is linked directly to the work breakdown structure, such as that in Figure 5.7. If the cost estimate is structured on the basis of the technical specifications, a cost-coding system may use the master specification format as its foundation. The following represents one of many possible coding layouts:

076 2004 08 06136.15 2, in which

- 076 = the 76th project begun this year
- 2004 = year of project start
- 08 = location in project; this might be the floor of a building or a kilometre location on a highway project
- 06136.15 = work type code, where 06 = wood and plastics from the 16 divisions of the CSI master format, 136 = heavy timber trusses as a subdivision of that work and 15 provides further details, such as elements of a certain size
- 2 = distribution code, representing materials, with other numbers for labour, equipment and the like.

Cash flow projection

In addition to establishing the overall project budget and the cost accounting scheme, the contractor must be concerned with the *timing* of expenses and revenues. To project these cash flows over the project's life will give the owner information about when to expect payment requests and in what amounts. It will also allow the contractor to plan its cash requirements and will permit planning for interim financing if revenues are not expected to cover expenses at certain points during the construction period. The preparation of a *cash flow projection curve* also provides the basis for one means of tracking overall project costs as the project proceeds.

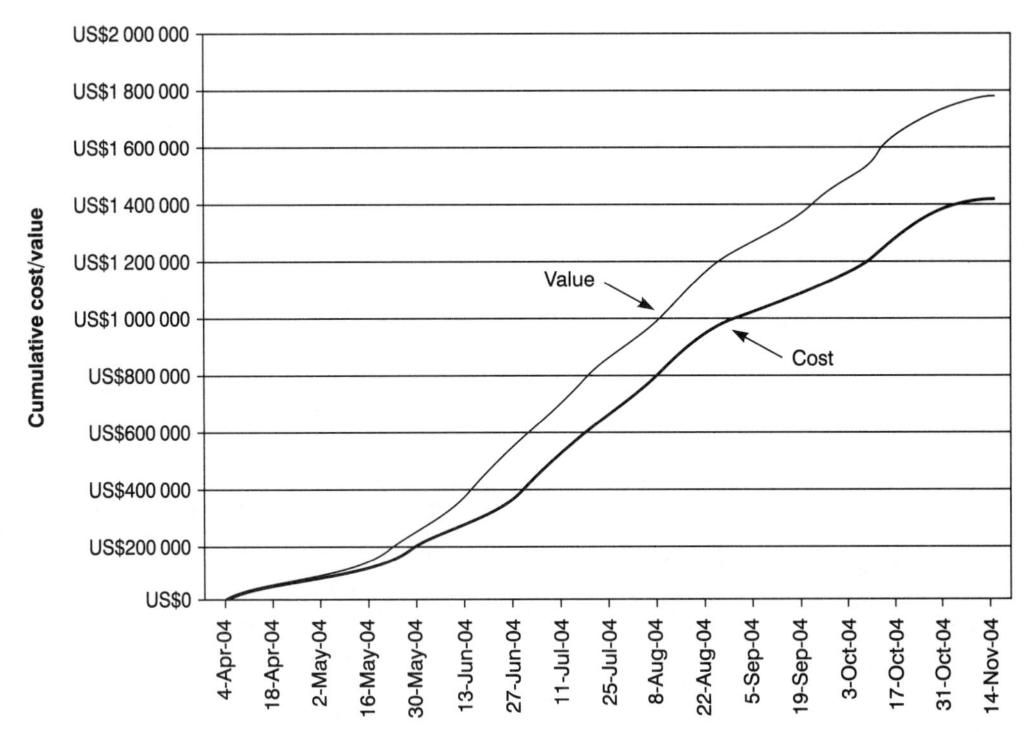

Figure 5.20 Project progress curves for bridge project, based on early start–early finish schedule.

An example of a financial budget in graphical form is the cash flow curve shown in Figure 5.20. This graph of cumulative cash flow as a function of time in months is based on the project schedule for the bridge project analysed earlier in this chapter. To develop this graph, activities were assigned their costs and values and, assuming that each activity will proceed on its early start–early finish schedule, these amounts were prorated across their respective durations and then summed for each week. The resulting 'lazy S' curve is a typical representation of project progress as a function of time. The project begins slowly and gradually increases its rate of progress, with the greatest rate occurring during the middle of the project, followed by slower progress toward the end. An empirical rule for a typical project is that approximately 50% of project progress occurs during the middle third of the project, with 25% occurring during each of the first and last thirds (Pilcher, 1992). In the cost curve in Figure 5.20, the total cost budget is US$ 1 420 230; the anticipated cost for the first 12 weeks (approximately a third of the 35-week project duration) is US$ 330 920 or 23% of the total cost. The estimated cost as of the end of the 24th week (approximately two-thirds of the total duration) is US$ 980 330, or 69% of the total cost.

Note that the graph in Figure 5.20 contains two curves. The curve of 'costs' forecasts estimated expenses over the life of the project, based on the quantities of work in each item and their unit costs as developed in the estimate. The second curve, of 'values', is a forecast of the value of the work in place, or contractor's gross earnings, at the end of each month. In this example, we have assumed a margin of 25% above costs to determine the amount the contractor will have earned as of the end of each period. At project completion, for example, the contractor will have earned US$ 1 420 230)(1.25) = US$ 1 775 266, if there are no changes to the contract.

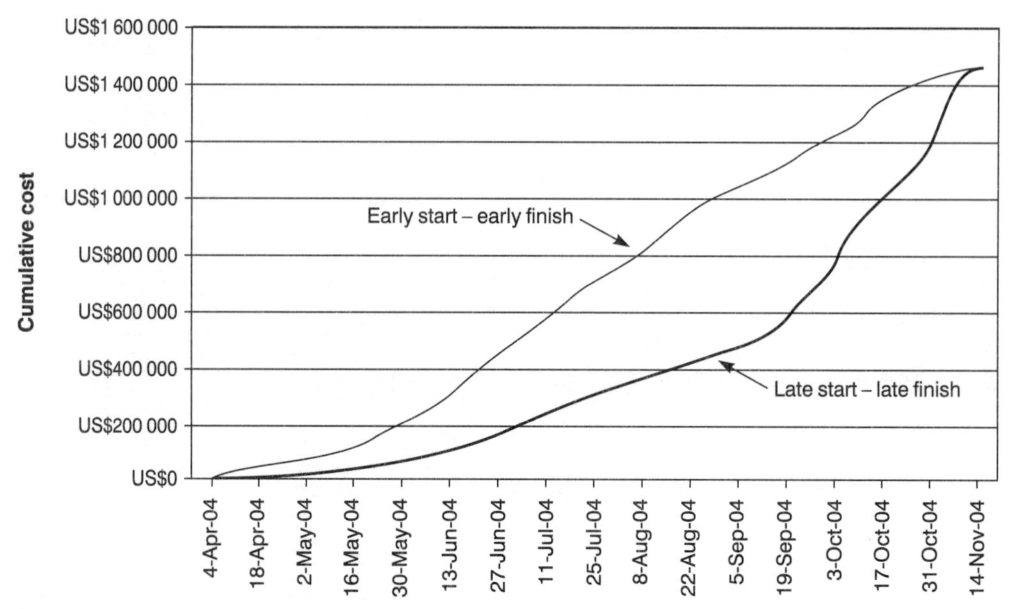

Figure 5.21 Cumulative cost, early and late curves.

This 'value' curve and the cost curve, taken together, are important in assisting the contractor to plan its cash flow needs. In actuality, this planning is more complicated than this figure implies, because there is usually a delay between the time an amount is earned by the contractor and the time of receipt of payment. This delay, which can be one or more months, and the fact that the owner may retain a portion of the earnings as an incentive for the contractor to keep producing, lead to situations in which the contractor's costs to date have exceeded revenues by a considerable amount, especially early in the project. We deal with an illustration of this situation in a sample exercise at the end of this chapter.

The cost curve in Figure 5.20 is a convenient basis for comparing actual total project costs to date with the estimated costs on the curve. The contractor can simply plot the actual costs and compare with the estimates. However, this approach is probably too simplistic for true cost control, as we shall discover in Chapter 6. At first glance, it would seem that if *actual* costs at a certain point in time are less than the *estimated* costs as of that same point in time, the project is in control, cost wise. But delays may have caused smaller-than-planned progress and the work actually accomplished may have cost more than estimated. Furthermore, the curve has been developed on the assumption that all activities will start and finish at their early times. We know that non-critical activities have some amount of slack that allows them to be delayed without impacting project progress adversely. Such delays can result in actual total costs to date that are lower than those shown on the curve, with no unfavourable consequences, provided that critical activities are on schedule.

In Figure 5.21 we have added a second cost curve to the early start–early finish curve in Figure 5.20, to show the cumulative costs at the end of each week under the assumption that all activities proceed as late as possible. The combination of the two curves is an envelope that gives a better indication of what the overall cost status ought to be throughout the project. Even this technique can be faulty in trying to track and control overall project costs, as will be explained in Chapter 6. One can record the actual overall costs at the end of each period on a graph such as this, but more detailed methods are needed to determine the status of individual cost items and then take action if their cost status is unfavourable.

Organising the worksite

An important activity as the contractor begins work in the field is to set up the site in a manner that will allow the work to proceed efficiently and effectively. Every job is unique, so guidelines for efficiency and effectiveness are difficult to spell out in general terms. However, a brief discussion of the various elements of worksite organisation will be useful. A well-organised jobsite can have a positive influence on the productivity of the entire project, from the delivery of materials to the handling of these materials as they are installed and from the manner in which workers move about the site to their personal needs and safety. An excellent reference is Chapter 5 in *Construction Jobsite Management* (Mincks and Johnston, 1998), from which many of the ideas in this section are taken.

Temporary services and facilities

The following are among the various temporary services and facilities that will likely be needed.

Offices

The jobsite office 'ranges from a clipboard in a pickup truck to multi-storey office facilities, depending on the scope of the project' (Mincks and Johnston, 1998). The office must be lockable and secure, have adequate desk space and areas for accessing drawings and other contract documents, provide wall space to display the project programme, have meeting and storage space and, if possible, offer a direct view of the worksite. Types of jobsite office facilities include existing buildings, modular office units, trailers and site-built offices. The contractor may be responsible for providing office space for design professional and owner representatives or subcontractors; even if not, locations for such quarters must be provided in the site layout plan.

Workshops and indoor storage

On a construction project of even modest size, it is likely that some on-site fabrication will be required and often such work is performed at inside locations protected from the weather. Examples include the preparation of concrete formwork and the pre-assembly of piping systems. Also, repairs of equipment and machinery may require such protected locations. In addition, indoor storage will be needed for various kinds of tools and for materials requiring weather protection that have been delivered and are awaiting incorporation into the project. Like the project office, various types of facilities can be used, including mobile structures and site-built buildings. Two considerations are of prime importance: (1) they must be located conveniently to the facility being constructed and close to delivery locations in the case of material storage and (2) they must be secure from theft and pilferage.

Dry shacks

These temporary facilities provide places for workers to store and eat their lunches and to change their clothes. They need not be elaborate, but they ought to be located conveniently to the work and be clean, adequately lit and relatively comfortable.

Temporary housing and food service

The 'construction camp' is an important feature of many projects in remote locations. When a project is located such that it is unreasonable for workers to return home every night, sleeping quarters and food service will be provided. Collective bargaining contracts may specify the nature and size of such accommodations. Possible facilities range from modular units and site-built buildings for projects of long duration to harbour-docked ships (see the case study of the Manapouri, New Zealand power project at the end of this chapter). In some cases, the contractor may simply provide a location for workers to park their trailers or recreation vehicles, which then allow them to sleep overnight and fix their own meals.

Temporary utilities

The office and shops will need the usual kinds of utilities, such as electric power, water and communication. In addition, power is likely to be needed throughout the building site and water will be needed from time to time. Until permanent utilities are available, some sort of temporary arrangements will be required. In the case of both power and water, it is important that the on-site source be as central as possible, that the lines do not interfere with the work in progress or delivery activities, that underground lines are not broken during excavation work and that the installations comply with code requirements.

Sanitary facilities

Drinking water, washing water and toilets must be provided for employees, including those of subcontractors. Drinking water is usually provided at the work location, often in 25 litre insulated containers, along with paper cups. Until permanent toilet facilities are available, wash water near temporary toilets is needed. To avoid excessive travel time, toilets should be placed adjacent to the work area. Thus, it is common to scatter these temporary portable chemical units at several locations. One health regulation requires that toilets be located within 200 feet, horizontally, of all workers and that they be available on every third floor of multi-storey structures. In addition to the sanitary needs of workers, the contractor must provide for the collection and disposal of construction waste. The importance of recycling requires that provisions be made for sorting various kinds of waste, such as timber, metal, plastics and drywall and accumulating them in separate bins.

Medical and first aid facilities

If the project is sufficiently large, the contractor may need to provide a separate building, staffed with medical personnel, to care for the injuries and other health issues that are likely to arise. Or a portion of the project office may be set aside for such purpose. In any case, the facility must be adequately stocked with proper supplies and equipment, be located close to the work activity and be clearly identified.

Access and delivery

Like any other community, the worksite must be planned so that it has adequate access to it and within it for the movement of people and the delivery and handling of materials. Access matters therefore are closely related to the planning of temporary facilities and storage and laydown areas. Access roads to the site must be planned so that they give direct access to storage areas, installation areas and entries to worksites. Especially in busy urban areas, such planning also must consider the effect on existing traffic patterns, with the potential for developing undesirable waiting lines of delivery vehicles and disruptions due to oversize trucks and loading equipment. All roadways, both to and on the site, must be built sufficiently to withstand expected loads and sizes of vehicles, they must be planned so that relocating them will not be necessary as the project proceeds and they must have adequate provision for dust control. Parking areas located near access roads but also close to the work will be required. In planning the locations of access roadways and other features, the contractor must also consider site drainage and water control, to prevent disruption to the work and possible environmental damage. Note that a drainage permit may be required, as indicated early in this chapter.

Delivery of materials to the jobsite is an activity that lasts throughout most of the project. An effective delivery plan will therefore be an important part of the overall management scheme. Among the considerations are the following.

- The route from main roads and streets to the jobsite must be clearly marked and adequate to handle the delivery vehicles.
- A smooth traffic pattern for delivery vehicles will allow ease of both access and egress.
- Material should be delivered as close to its installation location as possible. Thus, access to these locations will be important. An example is the delivery of pallets of roofing materials directly to the roof from the delivery vehicle.
- In congested areas, special arrangements may be needed to close streets or to make deliveries at night.
- If the contractor is responsible for unloading materials, provision of proper unloading equipment and personnel will be required.

Storage/laydown areas

The systematic storage of materials ensures savings in time when it is time to install the materials. These storage areas are frequently referred to as *laydown* areas on the jobsite. An area should be established for the delivery of each major material. Sorting of materials as they are delivered will reduce confusion later. The construction programme will be a helpful tool to determine when and in what quantities the various materials will need to be stored. Sufficient space must be provided not only for the materials themselves but also for handling equipment to manoeuvre in aisles or roadways. We have already indicated the possible need for inside storage in buildings, but outside storage also will often require weather protection from heat or cold, wind, rain or snow. Most materials will be set on pallets or skids to keep them off the ground and will be wrapped in tarps or other coverings. Drainage is another consideration for open-air storage. As always, the needs of subcontractors must be considered as plans for laydown areas are developed.

Security and signage

Jobsite security must be considered in planning the layout and staffing of the worksite. Any construction project is of interest to the public. While the contractor probably appreciates this interest, the public must nonetheless be excluded from the site, for reasons of both safety of the public and security of the completed works, the construction equipment and the stored materials. Most construction sites will require some sort of perimeter fence as part of the site layout plan, with decisions about the size, type, number and location of gates for personnel as well as delivery vehicles. In some special types of projects, such as the construction of an airport terminal adjacent to an active runway, security will include restricting access to some parts of the project only to certain authorised project personnel. Other site security issues may include night watchpersons, guard dogs and security patrols, as well as intrusion alarms on both permanent and temporary facilities. Adequate lighting is important, especially along perimeter fences.

Signs and barricades contribute to site safety and security and also provide general information and instruction and assist with smooth traffic flow. Safety signs indicate general danger, notify of the requirement for hard hats and specify particular hazards such as radiation or dust. General information signs identify the project and its team, designate routes to be followed by personnel and vehicles and indicate which parties may use various gates. Barricades prevent individuals from entering designated areas and divert traffic into desired areas. Figure 5.22 contains a collection of various project signs from several locations.

Quarries and borrow areas

Worksite planning may involve the contractor in acquiring and developing sites that will provide rock, gravel and other materials. Depending on the nature of the project, these sites may be on the worksite itself or some distance away. The owner may specify the use of certain locations or the contractor may have to locate them. On many projects, the development of such sites and the loading and hauling of the materials represent major proportions of the total project cost. Thus, access to rock quarries and borrow areas, the movement of vehicles and loading equipment within them and the routes used to haul to the areas where the materials will be deposited deserve detailed and expert planning.

The site layout plan

All of the worksite organisation considerations discussed above will converge in a plan that will be described in writing and shown on a drawing or series of drawings. According to Mincks and Johnston (1998), the jobsite layout plan includes the following: jobsite space allocation, jobsite access, material handling, worker transportation, temporary facilities, jobsite security and signage and barricades. Helpful references for establishing the plan include the construction documents, the programme, technical data on lifting and conveyance equipment, local codes, safety standards, size and weights of the large anticipated lifting loads and information from subcontractors and their storage needs. A helpful basis for the site layout drawing will be the project's site plan. The contractor can superimpose aspects of the site layout plan on this design drawing. Figure 5.23 shows one example of a project site layout plan, based on the site plan from the construction contract documents (Mincks and Johnston, 1998).

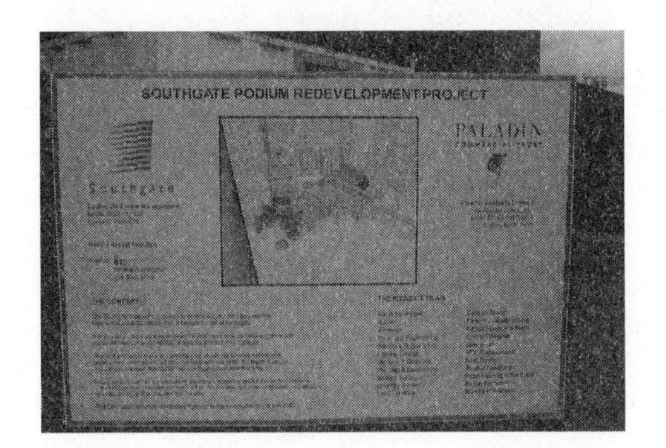

Figure 5.22 Project signage: a collection of pictures.

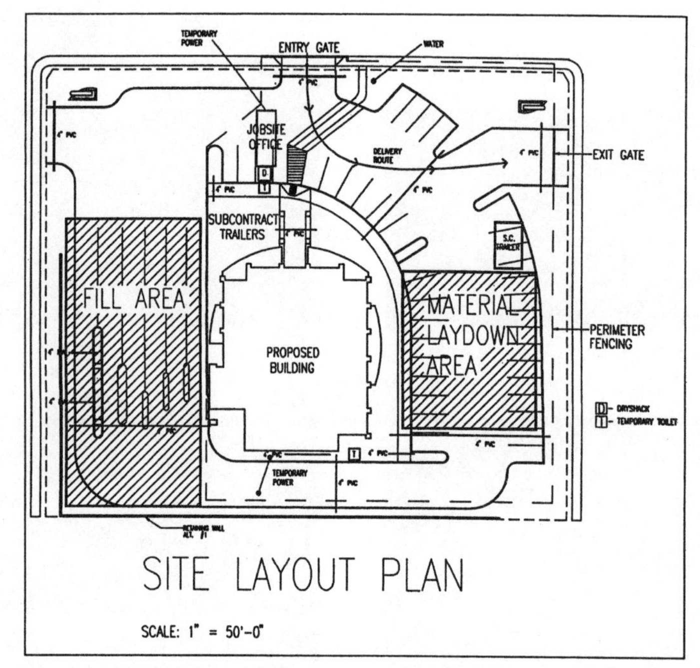

Figure 5.23 Jobsite layout plan. From *Construction Jobsite Management*, 1st edition, by W.R. Mincks and H. Johnston © 1998. Reprinted with permission of Delmar, a division of Thomson Learning. Fax 800 730–2215.

Buying out the job

The term *buyout*, in the context of construction project mobilisation, refers to procuring the materials and equipment that will be installed in the project and arranging subcontracts. It includes both selecting suppliers and subcontractors and finalising their purchase orders or subcontracts. The process begins during the tender preparation stage, as the contractor solicits and evaluates offers in the process of assembling the cost estimate. If the contractor is successful with its proposal and is awarded the contract, it then attempts to contract with the firms that submitted the most favourable offers. We discuss material procurement and subcontracting as the two distinct parts of the buyout process, even though both may be the responsibility of a single office or individual.

Material procurement

A construction company's purchasing policy is based on a combination of five overall objectives (Palmer et al., 1995): (1) price, (2) quality assurance, (3) delivery schedule, (4) vendor responsibility and (5) quantity control. Some of these factors relate directly to the material suppliers, while others, such as quantity control, are matters the contractor's firm must manage. Tradeoffs among these objectives are a constant challenge. For example, it may be best to pay a higher price for some item in exchange for the assurance of no delays in delivery or higher quality.

One interesting policy issue affecting both the individual project and the contracting firm is whether material procurement should be a function that is centralised in the home office or carried out entirely at the jobsite. A contractor with a regional or nationwide range of operations that takes on relatively large projects will establish large project offices capable of conducting most or all of the procurement function. The advantages of carrying out local, or jobsite, buying are the positive relationships such purchases are likely to engender in the local community and the probable better service in case of mistakes or emergencies. On the other hand, centralised buying tends to result in lower prices because larger quantities can be procured at one time, long-term, high-volume customers tend to be favoured in times of shortages, excess materials on one project can be more easily utilised on other projects and better financial control of the purchasing function can be achieved (Palmer et al., 1995).

The *materials procurement process* consists of several interrelated steps, as follows: receipt and evaluation of offers, purchasing or placement of the purchase order, approval by the owner or owner's representative, expediting or contact with the supplier to assure timely delivery, fabrication, shipping, delivery and inspection. For relatively minor items, many of these steps may be extraneous: the contractor may purchase small quantities of nails or bolts by sending a truck driver to the local hardware store with a purchase authorisation; provided these materials meet the technical specifications, the procurement process will be completed when they arrive back at the jobsite. For most materials, however, especially those manufactured specifically for the particular project, all of the listed steps will be followed. The first two take place during or prior to project mobilisation; we shall describe the remaining steps during our consideration of the project operations phase in Chapter 6.

In the process of compiling its cost estimate for the tender, the contractor can expect to receive, by post, fax, telephone or e-mail, offers from material suppliers. The management of the receipt and evaluation of the offers requires organisation and skill in ascertaining exactly what is being offered. Among the questions the astute estimator will want to ask are the following: Does the offer meet the plans and specifications? Is a substitution being offered? Will these materials come from inventory or from special order? What is the supplier's manufacturing capability? Can the supplier meet the required delivery schedule? Are delivery costs included in the offered price? Are taxes included? Does the offer include the cost of loading and unloading the materials?

It is likely that offers will not have been received for all materials during the tendering process; in that case, the estimator will have made a 'best guess' of the cost, based on prior experience and judgment. Now, during mobilisation, *all* materials must be purchased. A process for soliciting offers for materials not previously offered and for confirming the most favourable offers previously received will lead to decisions to purchase. Except for the smallest of projects, whose purchase arrangements may be made by telephone or by simply visiting the supplier, the terms of each purchase are incorporated in a *purchase order*, which is a contract between the contractor and the supplier. The simplest purchase might use a form like that from a mail order catalogue; for more complex materials and equipment, the purchase order will incorporate detailed plans and technical specifications. Purchase orders are intended to do four things (Palmer et al., 1995):

1 Establish the responsibility for buying the things that are needed and only those things.
2 Tell the supplier in writing exactly what has been ordered; how, when, and where it is to be shipped and billed; how much the company expects to pay for it; and any other information necessary to fill the order promptly and accurately.

3 Provide an internal check on purchasing, on receiving, and on the payment of money for items purchased.
4 Under the Uniform Commercial Code [in the US, and similar laws in other countries] fix certain of the terms of the contact of purchase and sale.

No matter how complex the transaction, a purchase order will contain five basic elements (Halpin and Woodhead, 1998): quantity or number of items required, item description, ranging from a standard description and stock number from a catalogue to a complex set of drawings and specifications, unit price, special instructions and signatures of agents empowered to enter into a contractual agreement. Purchase orders are numbered sequentially, with numbers strictly controlled. In addition to the original that is sent to the supplier, the contractor keeps copies for use by those who prepare vouchers for payment, who receive the order at the jobsite and who are responsible for expediting. Many of these 'copies' may be electronic, as many contracting firms conduct much of the communication regarding the procurement function via e-mail and the Internet. A typical purchase order is shown in Figure 5.24, which shows information about each item ordered plus instructions about how the shipment is to be marked for shipping, as well as various conditions governing the order.

One important aspect of any purchase that must be clearly defined is the point at which title of the purchased materials passes to the contractor. Consider a purchase made in Seattle, Washington, USA for use at a construction project in Fairbanks Alaska, 2450 km north by air and longer by sea or road. If the materials are damaged during their transport from Seattle to Fairbanks, which party will initiate a claim for compensation? The answer is the party who owns the goods at the time they are damaged. The purchase order must answer that question clearly. An extensive body of law has developed over questions related to shipping. We mention only a few aspects.

If delivery and payment take place concurrently, the transaction is called a cash sale and title passes when payment and delivery occur. If the purchase order requires the seller to deliver the materials personally to the buyer's location, the seller owns the materials until they are delivered. The most complicated situation is that in which the materials are shipped by common carrier; in that case, the general rule is that title passes to the buyer when the seller delivers the goods to the carrier for subsequent transport. As with most general rules, there are exceptions! A purchase order that specifies a 'free on board' (FOB) location means that the *seller* is required to deliver the materials to a common carrier and pay for transportation costs to the FOB location at no cost to the buyer, with title passing to the buyer only when delivery is completed. Thus, our purchase order might specify 'FOB Jobsite Fairbanks Alaska', in which case the materials would be owned by the seller until delivered to the Fairbanks project location. A variant is the 'cost, insurance, freight' (CIF) arrangement, under which title passes when the materials are delivered to the common carrier but the price paid includes the cost of the materials, usual insurance and freight to the buyer's location. Further discussion of title of purchases in construction can be found in Clough and Sears (1994). The general ideas apply to shipping of all kinds of goods throughout the world; a standard reference on international shipping is *CIF and FOB Contracts: British Shipping Laws* by Sassoon and Merron (1995).

Subcontracting

The contractor's other major responsibility under 'buying out the job' is to arrange subcontracts for those portions of the work the contractor will not perform itself. First, the contractor must

PURCHASE ORDER

GHEMM COMPANY, Inc.
P. O. BOX 507 – FAIRBANKS, ALASKA 99707

MARKING INSTRUCTIONS FOR SHIPPING

PURCHASE ORDER
Show number of this order on ALL invoices, bills of lading, packages, etc.

№ 1598

Date ________

Contract No. ________

Acknowledge receipt of this order AIRMAIL and advise shipping date.

By ________ Date ________

Please furnish the following subject to conditions above:

VIA ________ TERMS ________

ITEM	QUANTITY	DESCRIPTION	UNIT	UNIT PRICE	ITEM PRICE

INSTRUCTIONS PLEASE

1. Include packing list in each box or crate.
2. Show above P. O. Number on both packing list and invoices.
3. ________

By ________
Title

GHEMM COMPANY, Inc.

INTERIOR GRAPHICS & COPY

Figure 5.24 Sample purchase order (courtesy of Ghemm Company Contractors, Inc.).

decide which work will be subcontracted, a decision that is made, for most subcontracts, before the project mobilisation phase. Then, proposals will be received and analysed. Finally, negotiations lead to finalising the subcontracts.

To subcontract or not?

Subcontracting has its advantages and disadvantages. Subcontractors maintain organisations of personnel with special skills that often allow the work to be performed better and less expensively than the general contractor could do it. In addition, it may not be economical for the general contractor to acquire and maintain the special equipment needed for certain work. The skills provided by electricians, for example, are more likely to be found in an electrical subcontractor's organisation than in that of a general contractor. Likewise, general contractors often subcontract earthmoving work on building projects so as to avoid having to provide the required equipment. Furthermore, subcontracting allows the contractor to unload some of the risks and financial burdens of a large project onto other organisations. It is common practice among some general contractors, when preparing tenders, to add smaller margin percentages to work that will be subcontracted than to that which they will do themselves, in recognition of this reduced risk.

On the other hand, the contractor relinquishes some degree of control over the work by subcontracting. If delays occur or other difficulties arise, the contractor's task of responding and making corrections is more complicated because of this lack of direct control. Although the contractor has potentially less risk in such an arrangement, a subcontractor that proves to be financially unstable can leave unpaid obligations for which it has already been paid and the contractor may face a legal process and loss of considerably more than the profit expected from the project.

It is common practice that certain types of work will be subcontracted. General contractors that are preparing tenders and are on the planholder's list will automatically receive proposals from some speciality contractors without soliciting them. In building construction, subcontracts are likely to include electrical, mechanical and painting work. In highway construction, steel bridge erection, painting of roadway markings and installation of guiderails are commonly subcontracted. Nonetheless, the contractor will still have to weigh the advantages and disadvantages of subcontracting as it decides which portions of the work to turn over to these specialists.

Subcontract proposals

As noted in the previous section on material procurement, most initial proposals for materials and subcontracts are received by the contractor during the estimating and tendering process. Proposals for subcontracts, especially the larger ones, are likely to be received within a few days or even a few hours before the time when tenders are due and the natural activity and confusion inherent in tender day makes it a challenge to evaluate these proposals properly. Just as for material proposals, the contractor must be prepared to ask a series of questions of subcontractors, but it will want to know about both the subcontractor's organisation (unless that information is already known) and the proposed work. The contractor will want to know the subcontractor's experience, expertise, personnel, current workload and financial status. With regard to the proposal to provide subcontracting services, the contractor will require information about the particular specification sections that would be covered, with any proposed exceptions, whether the proposal is to furnish, unload and/or install, whether freight, insurance and taxes on furnished materials and equipment are included, how long the performance of the work is expected to take and whether the effort can fit into the master programme, what lead times will be required for shop drawings, manufacturing and delivery and whether the subcontractor intends to utilise its own equipment, storage facilities and office space or rely on the general contractor for these support services.

The following is a direct quotation from a proposal from a mechanical subcontractor; to save space, several of the exclusions are omitted here.

> Quotation for: Division 15 – Mechanical complete except for items specifically excluded below.
>
> Exclusions:
>
> Excavation/backfill/compaction
> Exterior utilities
> Pipe bedding/dewatering/shoring
> Concrete sawing/removal/core drilling/patching
> Debris removal
> ... ...
> Cost of bond/builders risk insurance
> Division 13900 Sprinkler
>
> Inclusions:
>
> Materials and equipment
> Installation labor
> Supervision
> One year warranty
> Freight to jobsite
> Red-lined as-built drawings
> Fuel gutter
>
> Total price ... ...

With statements such as these and answers to such questions as suggested above, the contractor must compare the offers and decide which is the most favourable to include in its tender price. As one example, suppose proposals are received from two electrical subcontractors. Electrical subcontractor A proposes to install foundations for electrical equipment with its own forces, subcontractor B states that it will not do this foundation work. Both proposals are otherwise equal in terms of work to be provided and the proposed price from contractor A is US$ 12 000 higher than that from contractor B. In evaluating the two proposals, the general contractor must decide whether it can do the foundation work itself for no more than US$ 12 000 in direct costs and other expenses and impacts.

Like material procurement, some subcontracted work may not be solicited until after the contract is awarded. The contractor may not have received subcontract proposals during the tendering process for work it intended to subcontract or it may decide after being awarded the contract that it will subcontract some work it had initially intended to perform with its own forces. In these cases, the solicitation, receipt and evaluation of proposals will take place during the project mobilisation phase, using the same guidelines suggested above.

Subcontract negotiation and award

Until the contractor is awarded the job, there is no need to negotiate and award subcontracts. Now, in the project mobilisation phase, these tasks will be carried out. The negotiation process will be conducted by principal officers of the contractor and subcontractor who are skilled in paying attention to the 'fine print' in contracts. The contractor's personnel will be primarily interested in whether the apparent successful subcontractor can do the work, has sufficient financial capability and will be able to meet schedule, safety, environmental and related responsibilities. Until the subcontract is signed, several topics are 'on the table' for discussion and resolution; some of these may not have even been thought of during the initial proposal submittal and evaluation. Palmer et al. (1995) provide a checklist of 39 to be considered by those who negotiate and award construction subcontracts. These questions range from 'Are the names and addresses of the parties correctly stated?' and 'To what extent does the subcontractor assume the responsibility to protect the work, the public and the owners of adjacent property from loss or damage?' to questions involving insurance coverage, jobsite cleanup and payment of legal expenses in case of disputes.

The form and content of the subcontract document varies as much as the work does itself. Standard forms are available from various organisations, such as the Associated General Contractors of America and the Fédération Internationale des Ingénieurs-Conseils. Many general contractors have developed their own subcontract forms that incorporate their subcontracting policies. Various owner organisations, especially in the public sector, have their own subcontract forms. The subcontract will incorporate by reference the drawings, general conditions, technical specifications and other documents, just as the prime contract does. The standard short form subcontract published by the Associated General Contractors of America is shown in Figure 5.25 (Associated General Contractors of America, 2000). It will be noted that, in addition to various matters pertaining to the particular project on page 1, the document includes three more pages of conditions ranging from safety, time and payment to claims and disputes. Four pages of instructions for using the form are also part of the full document. The wording is typical contract language, whether in a prime contract or a subcontract. In most cases, the subcontract will be signed by all parties prior to the commencement of any work covered by that contract. If such an arrangement is not possible, the general contractor can issue a letter of intent and notice to proceed, to get the work started and to assure the subcontractor that a completed subcontract will be forthcoming.

Project staffing

To carry out the work in the field requires people and those people must be organised in effective relationships. This section considers the organisation structure at the worksite and the sources of the labourers whose hands, minds, tools and equipment will assemble the final product in the field. We consider both collective bargaining agreements that provide union labour and the open, or merit, shop arrangement where no union is involved.

Worksite organisation structure

No matter what kind of project delivery system – traditional design–tender–build, design–build, construction manager, or others – is used for structuring the way the project is delivered to the owner, the worksite will typically feature an organisation based on the need to deploy specialty tradespeople and subcontractors to install the pieces of the project, plus a number of support specialists in such areas as engineering, safety and office management. The terminology used to designate the various responsibilities varies with different contractors, countries and regions. The size and complexity of the management organisation is dependent on the size of the project itself; many functions performed in the field for large projects are carried out by home or area office personnel when a project is small.

Figure 5.26 (see page 184) is a sample organisation chart showing the on-site staff for a relatively large project. It uses typical terminology and groupings of responsibility, although other configurations are far from rare. It is important to realise that the personnel assigned to any particular project are unlikely to be together ever again on another project in exactly the same organisation. A project, by definition, happens only once, after which its staff are assigned to other projects within or outside the company. The following quotation by Ricketts (1996) is an apt characterisation of construction project organisations:

> Each facility that the construction team produces, it produces only once; the next time its work will be done at the new location, to a new pattern, and under new, although often similar, specifications. Furthermore, from the very inception of each construction project, contractors are wholly devoted to completion of the undertaking as quickly and economically as possible and then moving out.

The *project manager* is the contractor's employee who is the focal point of the project and thus has overall responsibility for completing the project satisfactorily. In the organisation shown, we assume that the project manager (1) is the contractor's primary contact with the owner, (2) assures that the home or area office staff provide the necessary support for the project in such matters as purchasing, expediting and payment requests and (3) gives overall supervision and direction to the project superintendent. Because of the varied nature of this person's responsibilities, they will spend only part of their time at the worksite. If the company's projects are small, a project manager may assume this role for more than one project simultaneously. It is important to distinguish between the *contractor's project manager* and the *project manager* who might be engaged by the *owner* to manage the entire project from initial planning to the end of construction, as we have described in Chapter 2.

The *project superintendent* devotes their full time to the coordination and supervision of work at the worksite. As shown in Figure 5.26, this person supervises a *general foreperson* whose several crews install the various project elements (a small part of the diagram but a large percentage of the on-site staff!). *Subcontractors* also fall within the project superintendent's purview, often with an assistant to provide day-to-day direction. The various support roles include *project engineering*, with a variety of responsibilities ranging from field engineering and layout to cost and schedule control and other technical aspects of management. The *office manager* might be called the office engineer if thay shared some of the technical tasks with the project engineer in addition to the personnel and other business functions shown. The specialised functions of safety, environmental monitoring and control, equipment maintenance and industrial relations are, in our example, worksite activities that fall within the overall responsibility of the project superintendent.

Several persons may be assigned to any of the support functions in our organisation chart; on some projects, one or more scheduling engineers may devote their full time to monitoring the schedule and tracking project progress, for example. On the other hand, a small project might expect the project engineer to coordinate subcontractors and act as safety engineer in addition to the duties shown on the chart. We shall have occasion to describe many of the activities of

Job No.: ______________________ Account Code: ______________________

THE ASSOCIATED GENERAL CONTRACTORS OF AMERICA
AGC DOCUMENT NO. 603
STANDARD SHORT FORM AGREEMENT
BETWEEN CONTRACTOR AND SUBCONTRACTOR
(Where Contractor Assumes Risk of Owner Payment)

This document is endorsed by The Associated Specialty Contractors, Inc. (ASC), an umbrella organization composed of the following seven specialty contractor groups: Mechanical Contractors Association of America, National Electrical Contractors Association, National Insulation Association, National Roofing Contractors Association, Painting and Decorating Contractors of America, Plumbing-Heating-Cooling Contractors National Association, and Sheet Metal and Air Conditioning Contractors' National Association.

This Agreement is made this ______ (Day) day of ______ (Month), ______ (Year), by and between

CONTRACTOR, ______________________ (Name and Address) and

SUBCONTRACTOR, ______________________ (Name and Address).

PROJECT: ______________________ (Description of Project and Location).

OWNER: ______________________ (Name and Address).

ARCHITECT/ENGINEER: ______________________ (Name and Address).

1 SUBCONTRACT WORK To the extent terms of the agreement between Owner and Contractor (prime agreement) apply to the work of Subcontractor, Contractor assumes toward Subcontractor all obligations, rights, duties, and redress that Owner assumes toward Contractor. In an identical way, Subcontractor assumes toward Contractor all obligations, rights, duties, and redress that Contractor assumes toward Owner and others under the prime agreement. In the event of conflicts or inconsistencies between provisions of this Agreement and the prime agreement, this Agreement shall govern. Subcontractor shall perform Subcontract Work under the general direction of Contractor and shall cooperate with Contractor so Contractor may fulfill obligations to Owner. Subcontractor shall provide Subcontract Work for the Project in accordance with the Progress Schedule to be prepared by Contractor after consultation with Subcontractor, and as it may change from time to time. Subcontractor shall give timely notices to authorities pertaining to Subcontract Work and shall be responsible for all permits, fees, licenses, assessments, inspections, testing and taxes necessary to complete Subcontract Work. Subcontractor to provide ______________________ (Brief Description of Subcontract Work) ______________________ as more fully described in Exhibit A.

2 SUBCONTRACT AMOUNT Contractor agrees to pay Subcontractor for satisfactory and timely performance and completion of Subcontract Work: ______________________.

Retainage shall be ______________________ percent (______%), which is equal to the percentage retained from Contractor's payment by Owner for Subcontract Work.

3 INSURANCE Subcontractor shall purchase and maintain insurance that will protect Subcontractor from claims arising out of Subcontractor operations under this Agreement, whether the operations are by Subcontractor, or any of Subcontractor's consultants or subcontractors or anyone directly or indirectly employed by any of them, or by anyone for whose acts any of them may be liable. Subcontractor shall maintain coverage and limits of liability as set forth in Exhibit E.

4 BONDS Subcontractor ☐ shall ☐ shall not furnish to Contractor, as Obligee, surety bonds in a form as set forth in Exhibit F to this Agreement, and through a surety mutually agreeable to Contractor and Subcontractor, to secure faithful performance of Subcontract Work and to satisfy Subcontractor payment obligations related to Subcontract Work.

5 EXHIBITS The following Exhibits are incorporated by reference and made part of this Agreement:

EXHIBIT A: Subcontract Work, ______________________ pages.
EXHIBIT B: Prime agreement, Drawings, Specifications, General, Special, Supplementary, and other conditions, and addenda. (Attach a complete listing by title, date and number of pages.)
EXHIBIT C: Progress Schedule, ______________________ pages.
EXHIBIT D: Alternates and Unit Prices, include dates when alternates and unit prices no longer apply, ______ pages.
EXHIBIT E: Insurance Provisions, ______________________ pages.
EXHIBIT F: Bonds, ______________________ pages.
EXHIBIT __: Other ______________________ pages.

AGC DOCUMENT NO. 603 • STANDARD SHORT FORM AGREEMENT BETWEEN CONTRACTOR AND SUBCONTRACTOR
(Where Contractor Assumes Risk of Owner Payment)

Figure 5.25 Sample Subcontract Form. Reproduced with the express written permission of the Associated General Contractors of America under License No 0070. To order AGC contract documents, phone 1-800-AGC-1767 or fax your request to 703-837-5406, or visit the AGC website at www.agc.org.

6 **SAFETY** To protect persons and property, Subcontractor shall establish a safety program implementing safety measures, policies and standards conforming to (1) those required or recommended by governmental and quasi-governmental authorities having jurisdiction and (2) requirements of this Agreement. Subcontractor shall keep project site clean and free from debris resulting from Subcontract Work.

7 **ASSIGNMENT** Subcontractor shall not assign the whole or any part of Subcontract Work or this Agreement without prior written approval of Contractor.

8 **TIME**

8.1 **TIME IS OF THE ESSENCE** Time is of the essence for both parties. The parties agree to perform their respective obligations so that the Project may be completed in accordance with this Agreement.

8.2 **SCHEDULE** In consultation with Subcontractor, the Contractor shall prepare the schedule for performance of Contractor's work (Progress Schedule) and shall revise and update such schedule, as necessary, as Contractor's work progresses. Subcontractor shall provide Contractor with any scheduling information proposed by Subcontractor for Subcontract Work and shall revise and update as Project progresses. Contractor and Subcontractor shall be bound by the Progress Schedule. The Progress Schedule and all subsequent changes and additional details shall be submitted to Subcontractor reasonably in advance of required performance. Contractor shall have the right to determine and, if necessary, change the time, order and priority in which various portions of Subcontract Work shall be performed and all other matters relative to Subcontract Work.

9 **CHANGE ORDERS** When Contractor orders in writing, Subcontractor, without nullifying this Agreement, shall make any and all changes in Subcontract Work, which are within the general scope of this Agreement. Any adjustment in the Subcontract Amount or time of performance shall be authorized by a Change Order. No adjustments shall be made for any changes performed by Subcontractor that have not been ordered by Contractor. A Change Order is a written instrument prepared by Contractor and signed by Subcontractor stating their agreement upon the change in Subcontract Work. If commencement and/or progress of Subcontract Work is delayed without the fault or responsibility of Subcontractor, the time for Subcontract Work shall be extended by Change Order to the extent obtained by Contractor, and the Progress Schedule shall be revised accordingly.

10 **PAYMENT**

10.1 **SCHEDULE OF VALUES** As a condition of payment, Subcontractor shall provide a schedule of values satisfactory to Contractor not more than fifteen (15) days from the date of this Agreement.

10.2 **PROGRESS AND FINAL PAYMENTS** Progress payments, less retainage, shall be made to Subcontractor, for Subcontract Work satisfactorily performed, no later than seven (7) days after receipt by Contractor of payment from Owner for Subcontract Work. Final payment of the balance due shall be made to Subcontractor no later than seven (7) days after receipt by Contractor of final payment from Owner for Subcontract Work. These payments are subject to receipt of such lien waivers, affidavits, warranties, guarantees or other documentation required by this Agreement or Contractor. If payment from Owner for such Subcontract Work is not received by Contractor, through no fault of Subcontractor, Contractor will make payment to Subcontractor within a reasonable time for Subcontract Work satisfactorily performed.

10.3 **PAYMENTS WITHHELD** Contractor may reject a Subcontractor payment application or nullify a previously approved Subcontractor payment application, in whole or in part, as may reasonably be necessary to protect Contractor from loss or damage caused by Subcontractor's failure to (1) timely perform Subcontract Work, (2) properly pay subcontractors and/or suppliers, or (3) promptly correct rejected, defective or nonconforming Subcontract Work.

10.4 **PAYMENT DELAY** If Contractor has received payment from Owner and, if for any reason not the fault of Subcontractor, Subcontractor does not receive a progress payment from Contractor within seven (7) days after the date such payment is due, or if Contractor has failed to pay Subcontractor within a reasonable time for Subcontract Work satisfactorily performed, Subcontractor, upon giving seven (7) days' written notice to Contractor, and without prejudice to and in addition to any other legal remedies, may stop work until payment of the full amount owing to Subcontractor has been received. Subcontract Amount and time of performance shall be adjusted by the amount of Subcontractor's reasonable and verified cost of shutdown, delay and startup, and shall be effected by an appropriate Change Order.

10.5 **WAIVER OF CLAIMS** Final payment shall constitute a waiver of all claims by Subcontractor relating to Subcontract Work, but shall in no way relieve Subcontractor of liability for warranties, or for nonconforming or defective work discovered after final payment.

10.6 **OWNER'S ABILITY TO PAY**

10.6.1 Subcontractor shall have the right upon request to receive from Contractor such information as Contractor has obtained relative to Owner's financial ability to pay for Contractor's work, including any subsequent material variation in such information. Contractor, however, does not warrant the accuracy or completeness of information provided by Owner.

10.6.2 If Subcontractor does not receive the information referenced in Subparagraph 10.6.1, Subcontractor may request information from Owner and/or Owner's lender.

Figure 5.25 Sample Subcontract Form (continued).

11 INDEMNITY To the fullest extent permitted by law, Subcontractor shall defend, indemnify and hold harmless Contractor, Contractor's other subcontractors, Architect/Engineer, Owner and their agents, consultants, employees and others as required by this Agreement from all claims for bodily injury and property damage that may arise from performance of Subcontract Work to the extent of the negligence attributed to such acts or omissions by Subcontractor, Subcontractor's subcontractors or anyone employed directly or indirectly by any of them or by anyone for whose acts any of them may be liable.

12 CONTRACTOR'S RIGHT TO PERFORM SUBCONTRACTOR'S RESPONSIBILITIES AND TERMINATION OF AGREEMENT

12.1 FAILURE OF PERFORMANCE Should Subcontractor fail to satisfy contractual deficiencies or to commence and continue satisfactory correction of the default with diligence or promptness within three (3) working days from receipt of Contractor's written notice, then Contractor, without prejudice to any right or remedies, shall have the right to take whatever steps it deems necessary to correct deficiencies and charge the cost thereof to Subcontractor, who shall be liable for such payment, including reasonable overhead, profit and attorneys' fees. In the event of an emergency affecting safety of persons or property, Contractor may proceed as above without notice, but Contractor shall give Subcontractor notice promptly after the fact as a precondition of cost recovery.

12.2 TERMINATION BY OWNER Should Owner terminate the prime agreement or any part which includes Subcontract Work, Contractor shall notify Subcontractor in writing within three (3) days of termination and, upon written notification, this Agreement shall be terminated and Subcontractor shall immediately stop Subcontract Work, follow all of Contractor's instructions, and mitigate all costs. In the event of Owner termination, Contractor liability to Subcontractor shall be limited to the extent of Contractor recovery on Subcontractor's behalf under the prime agreement. Contractor agrees to cooperate with Subcontractor, at Subcontractor's expense, in the prosecution of any Subcontractor claim arising out of Owner termination and to permit Subcontractor to prosecute the claim, in the name of Contractor, for the use and benefit of Subcontractor, or assign the claim to Subcontractor.

12.3 TERMINATION BY CONTRACTOR If Subcontractor fails to commence and satisfactorily continue correction of a default within three (3) days after written notification issued under Paragraph 12.1, then Contractor may, in lieu of or in addition to Paragraph 12.1, issue a second written notification, to Subcontractor and its surety, if any. Such notice shall state that if Subcontractor fails to commence and continue correction of a default within seven (7) days of the written notification, the Agreement will be deemed terminated. A written notice of termination shall be issued by Contractor to Subcontractor at the time Subcontractor is terminated. Contractor may furnish those materials, equipment and/or employ such workers or subcontractors as Contractor deems necessary to maintain the orderly progress of Contractor's work. All costs incurred by Contractor in performing Subcontract Work, including reasonable overhead, profit and attorneys' fees, costs and expenses, shall be deducted from any monies due or to become due Subcontractor. Subcontractor shall be liable for payment of any amount by which such expense may exceed the unpaid balance of the Subcontract Amount. At Subcontractor's request, Contractor shall provide a detailed accounting of the costs to finish Subcontract Work.

12.4 TERMINATION BY SUBCONTRACTOR If Subcontract Work has been stopped for thirty (30) days because Subcontractor has not received progress payments or has been abandoned or suspended for an unreasonable period of time not due to the fault or neglect of Subcontractor, then Subcontractor may terminate this Agreement upon giving Contractor seven (7) days' written notice. Upon such termination, Subcontractor shall be entitled to recover from Contractor payment for all Subcontract Work satisfactorily performed but not yet paid for, including reasonable overhead, profit and attorneys' fees, costs and expenses. However, if Owner has not paid Contractor for the satisfactory performance of Subcontract Work through no fault or neglect of Contractor, and Subcontractor terminates this Agreement under this Article because it has not received corresponding progress payments, Subcontractor shall be entitled to recover from Contractor, within a reasonable period of time following termination, payment for all Subcontract Work satisfactorily performed but not yet paid for, including reasonable overhead and profit. Contractor's liability for any other damages claimed by Subcontractor under such circumstances shall be extinguished by Contractor pursuing said damages and claims against Owner, on Subcontractor's behalf, in the manner provided for in Paragraph 12.2 of this Agreement.

13 CLAIMS AND DISPUTES

13.1 CLAIMS RELATING TO CONTRACTOR Subcontractor shall give Contractor written notice of all claims within seven (7) days of Subcontractor's knowledge of facts giving rise to the event for which claim is made; otherwise, such claims shall be deemed waived. All unresolved claims, disputes and other matters in question between Contractor and Subcontractor shall be resolved in the manner provided in this Agreement.

13.2 DAMAGES If the prime agreement provides for liquidated or other damages for delay beyond the completion date set forth in this Agreement, and such damages are assessed, Contractor may assess a share of the damages against Subcontractor in proportion to Subcontractor's share of responsibility for the delay. However, the amount of such assessment shall not exceed the amount assessed against Contractor. Nothing in this Agreement shall be construed to limit Subcontractor's liability to Contractor for Contractor's actual delay damages caused by Subcontractor's delay.

3

AGC DOCUMENT NO. 603 • STANDARD SHORT FORM AGREEMENT BETWEEN CONTRACTOR AND SUBCONTRACTOR
(Where Contractor Assumes Risk of Owner Payment)

Figure 5.25 Sample Subcontract Form (continued).

13.2.1 CONTRACTOR CAUSED DELAY Nothing in this Agreement shall preclude Subcontractor's recovery of delay damages caused by Contractor.

13.3 WORK CONTINUATION AND PAYMENT Unless otherwise agreed in writing, Subcontractor shall continue Subcontract Work and maintain the Progress Schedule during any dispute resolution proceedings. If Subcontractor continues to perform, Contractor shall continue to make payments in accordance with this Agreement.

13.4 MULTIPARTY PROCEEDING The parties agree, to the extent permitted by the prime agreement, that all parties necessary to resolve a claim shall be parties to the same dispute resolution proceeding. To the extent disputes between Contractor and Subcontractor involve in whole or in part disputes between Contractor and Owner, disputes between Subcontractor and Contractor shall be decided by the same tribunal and in the same forum as disputes between Contractor and Owner.

13.5 NO LIMITATION OF RIGHTS OR REMEDIES Nothing in Article 13 shall limit any rights or remedies not expressly waived by Subcontractor which Subcontractor may have under lien laws or payment bonds.

13.6 STAY OF PROCEEDINGS In the event that provisions for resolution of disputes between Contractor and Owner contained in the prime agreement do not permit consolidation or joinder with disputes of third parties, such as Subcontractor, resolution of disputes between Subcontractor and Contractor involving in whole or in part disputes between Contractor and Owner shall be stayed pending conclusion of any dispute resolution proceeding between Contractor and Owner.

13.7 DIRECT DISCUSSION If a dispute arises out of or relates to this Agreement, the parties shall endeavor to settle the dispute through direct discussion.

13.8 MEDIATION Disputes between Subcontractor and Contractor not resolved by direct discussion shall be submitted to mediation pursuant to the Construction Industry Mediation Rules of the American Arbitration Association. The parties shall select the mediator within fifteen (15) days of the request for mediation. Engaging in mediation is a condition precedent to any form of binding dispute resolution.

13.9 OTHER DISPUTE PROCESSES If neither direct discussions nor mediation successfully resolve the dispute, the parties agree that the following shall be used to resolve the dispute.

(Check one selection only.)

☐ **Arbitration** Arbitration shall be pursuant to the Construction Industry Rules of the American Arbitration Association, unless the parties mutually agree otherwise. A written demand for arbitration shall be filed with the American Arbitration Association and the other party to the Agreement within a reasonable time after the dispute or claim has arisen, but in no event after the applicable statute of limitations for a legal or equitable proceeding has run. The arbitration award shall be final. This agreement to arbitrate shall be governed by the Federal Arbitration Act, and judgment upon the award may be confirmed in any court having jurisdiction. ◆

☐ **Litigation** Action may be filed in the appropriate state or federal court. ◆

13.10 COST OF DISPUTE RESOLUTION The cost of any mediation proceeding shall be shared equally by the parties participating. The prevailing party in any dispute that goes beyond mediation arising out of or relating to this Agreement or its breach shall be entitled to recover from the other party reasonable attorneys' fees, costs and expenses incurred by the prevailing party in connection with such dispute.

14 JOINT DRAFTING The parties expressly agree that this Agreement was jointly drafted, and that they both had opportunity to negotiate terms and to obtain assistance of counsel in reviewing terms prior to execution. This Agreement shall be construed neither against nor in favor of either party, but shall be construed in a neutral manner.

CONTRACTOR: ______________________ ◆

BY: ______________________ ◆

PRINT NAME: ______________________ ◆

PRINT TITLE: ______________________ ◆

SUBCONTRACTOR: ______________________ ◆

BY: ______________________ ◆

PRINT NAME: ______________________ ◆

PRINT TITLE: ______________________ ◆

5/00
4

AGC DOCUMENT NO. 603 • STANDARD SHORT FORM AGREEMENT BETWEEN CONTRACTOR AND SUBCONTRACTOR
(Where Contractor Assumes Risk of Owner Payment)

Figure 5.25 Sample Subcontract Form (continued).

INSTRUCTIONS FOR COMPLETION OF
AGC DOCUMENT NO. 603
STANDARD SHORT FORM AGREEMENT BETWEEN CONTRACTOR AND SUBCONTRACTOR
(Where Contractor Assumes Risk of Owner Payment)

This document is endorsed by The Associated Specialty Contractors, Inc. (ASC).

2000 EDITION

The Standard Short Form Agreement Between Contractor and Subcontractor (Where Contractor Assumes Risk of Owner Payment), AGC Document No. 603 (AGC 603), has been premised in substantial part on language and concepts found in the 1998 Edition of AGC Document No. 650. Both of these documents, AGC 603 and AGC 650, have been approved and endorsed by ASC, an umbrella organization composed of the following seven specialty contractor groups: Mechanical Contractors Association of America, National Electrical Contractors Association, National Insulation Association, National Roofing Contractors Association, Painting and Decorating Contractors of America, Plumbing-Heating-Cooling Contractors' National Association, and Sheet Metal and Air Conditioning Contractors' National Association. AGC 603 is intended as a convenient, short-form subcontract presented in a format in which transaction-specific information is provided or referenced at the beginning of the agreement. Further, the format of this short form agreement is distinctive from the formats of AGC long form agreements to minimize document length. In many instances, articles, such as "the," deliberately have been left out of the text of the Agreement for the purposes of brevity.

In AGC 603, payment to the Subcontractor is not conditioned on the Contractor having received from the Owner payment for Subcontract Work satisfactorily performed. See the discussion in these Instructions, below. AGC Document No. 604, Standard Short Form Agreement Between Contractor and Subcontractor (Where Contractor and Subcontractor Share Risk of Owner Payment), 2000 Edition, can be used when conditioned payment is valid in the jurisdiction and elected by the parties.

This document replaces AGC Document No. 603, Short Form Subcontract, which was published in 1987. AGC 603, 2000 Edition, benefited from an inclusive development process in which contractors, subcontractors and others offered comments and constructive feedback on its language.

AGC DOCUMENT NO. 603 • STANDARD SHORT FORM AGREEMENT BETWEEN CONTRACTOR AND SUBCONTRACTOR
(Where Contractor Assumes Risk of Owner Payment)

Figure 5.25 Sample Subcontract Form (continued).

GENERAL INSTRUCTIONS

Standard Form

These instructions are for the information and convenience of the users of AGC 603, 2000 Edition. They are not part of the Agreement nor a commentary on or interpretation of the contract form. It is the intent of the parties to a particular agreement that controls its meaning and not that of the writers and publishers of the standard form. As a standard form, this Agreement has been designed to establish the relationship of the parties in the standard situation. Recognizing that every project is unique, modifications may be required. See the following recommendations for modifications.

Legal and Insurance Counsel

THIS DOCUMENT HAS IMPORTANT LEGAL AND INSURANCE CONSEQUENCES, AND IT IS NOT INTENDED AS A SUBSTITUTE FOR COMPETENT PROFESSIONAL SERVICES AND ADVICE. CONSULTATION WITH AN ATTORNEY AND AN INSURANCE ADVISER IS ENCOURAGED WITH RESPECT TO ITS COMPLETION OR MODIFICATION. FEDERAL, STATE AND LOCAL LAWS AND REGULATIONS MAY VARY WITH RESPECT TO THE APPLICABILITY AND/OR ENFORCEABILITY OF SPECIFIC PROVISIONS IN THIS DOCUMENT.

AGC SPECIFICALLY DISCLAIMS ALL WARRANTIES, EXPRESS OR IMPLIED, INCLUDING ANY WARRANTY OF MERCHANTABILITY OR FITNESS FOR A PARTICULAR PURPOSE. PURCHASERS ASSUME ALL LIABILITY WITH RESPECT TO THE USE OR MODIFICATION OF THIS DOCUMENT, AND AGC SHALL NOT BE LIABLE FOR ANY DIRECT, INDIRECT OR CONSEQUENTIAL DAMAGES RESULTING FROM SUCH USE OR MODIFICATION.

COMPLETING THE AGREEMENT

Completing Blanks

Diamonds (♦) in the margins indicate provisions requiring the parties to fill in blanks with information.

Modifications

Supplemental conditions, provisions added to the printed agreement, may be adopted by reference. It is always best for supplements to be attached to the agreement. Provisions in the printed document that are not to be included in the agreement may be deleted by striking through the word, sentence or paragraph to be omitted. It is recommended that unwanted provisions not be blocked out so that the deleted materials are illegible. The parties should be clearly aware of the material deleted from the standard form.

It is a good practice for both parties to sign and date all modifications and supplements.

Photocopying the Completed Agreement

The purchaser of this copyrighted document may make up to nine (9) photocopies of a completed document, whether signed or unsigned, for distribution to appropriate parties in connection with a specific project. Any other reproduction of this document in any form is strictly prohibited, unless the purchaser has obtained the prior written permission of The Associated General Contractors of America.

OBTAINING ADDITIONAL INFORMATION

To obtain additional information about AGC standard form contract documents and the AGC Contract Documents Program, contact AGC at 333 John Carlyle Street, Suite 200, Alexandria, VA 22314; phone (703) 548-3118; fax (703) 548-3119, or visit AGC's web site at www.agc.org.

AGC 603

The provisions of this standard form Agreement are organized so that information relating to the Subcontract Work, Subcontract Amount, Insurance, Bonds, and other information or Exhibits are included or referenced on the first page of the subcontract.

Blanks are provided at the top of the first page for the Contractor's Job Number and Account Code for convenience and quick reference.

The date of the Agreement and identification of the parties and the Project are essential information to be accurately inserted.

Article 1 SUBCONTRACT WORK

Here the parties insert a brief description of the Subcontract Work.

Article 2 SUBCONTRACT AMOUNT

Here the parties are to insert the amount to be paid for the Subcontract Work. Retainage is to be a percentage equal to the percentage retained from the Contractor's payment by the Owner for Subcontract Work.

Article 3 INSURANCE

The parties should determine the specific Subcontractor coverages and limits of liability required for the project, and attach this information as Exhibit E.

Figure 5.25 Sample Subcontract Form (continued).

Article 4 BONDS

Here the parties indicate if the Subcontractor shall furnish surety bonds for the Project. The form of the bond shall be provided in Exhibit F and through a surety mutually agreeable to the Contractor and Subcontractor.

Article 5 EXHIBITS

Here specific exhibits are incorporated by reference. The parties are to indicate the number of pages for each exhibit. The Subcontract Work represents Exhibit A. The Prime agreement, Drawings, Specifications, General, Special, Supplementary and other conditions represent Exhibit B. It is recommended that the complete listing by title, date and number of pages be provided in this Exhibit. Progress Schedule represents Exhibit C. Alternates and Unit Prices represent Exhibit D. Insurance provisions represent Exhibit E. Bonds represent Exhibit F. A space is provided to reference an additional exhibit, if needed by the parties.

Article 6 SAFETY

The Subcontractor is to establish a safety program that conforms to the requirements of governmental authorities and this Agreement. The Subcontractor is required to keep the project site clean and free from debris.

Article 7 ASSIGNMENT

The Subcontractor is required to receive the Contractor's written approval before assigning the Subcontractor Work.

Article 8 TIME

Time is of the essence for both parties to the Agreement.

Article 9 SCHEDULE

The Subcontractor provides the Contractor with scheduling information and updates. Parties are bound by the Progress Schedule. The Progress Schedule, its details and changes are submitted to the Subcontractor reasonably in advance of performance. The Contractor can determine and, if necessary, change the time, order and priority in which the Subcontract Work shall be performed. In consultation with the Subcontractor, the Contractor will prepare the schedule for performance of the Contractor's work (the Performance Schedule) and will revise and update as necessary.

Article 9 CHANGE ORDERS

When the Contractor orders in writing, the Subcontractor, without nullifying this Agreement, shall make any and all changes in Subcontract Work. Any adjustment in the Subcontract Amount shall be authorized by Change Order, which is defined here. If commencement and/or progress of Subcontract Work is delayed without the fault of or responsibility of the Subcontractor, the time for Subcontract Work shall be extended by Change Order.

Article 10 PAYMENT

10.1 SCHEDULE OF VALUES The Subcontractor is required to submit a schedule of values satisfactory to the Contractor not more than 15 days from the date of this Agreement.

10.2 PROGRESS AND FINAL PAYMENTS This Paragraph does not condition the Subcontractor's payments on receipt of payments by the Contractor from the Owner. Progress payments to the Subcontractor are to be made within seven days after the Contractor receives payment from the Owner for the Subcontractor's Work. If, through no fault of the Subcontractor, the Owner does not pay the Contractor for the Subcontractor's Work, the Contractor assumes that liability "within a reasonable time." A "reasonable time" enables the Contractor to attempt to secure the payment from the Owner; what is "reasonable" for a particular project will depend on a variety of project-specific factors.

10.[illegible] PAYMENT WITHHELD The Subcontractor's application for payment may be rejected for any of the three reasons indicated.

10.5 WAIVER OF CLAIMS This Paragraph states that final payment constitutes a waiver of all claims by the Subcontractor. The Subcontractor is still liable for warranties, or for nonconforming or defective work discovered after payment.

10.6 OWNER'S ABILITY TO PAY Upon request, the Subcontractor has the right to receive from the Contractor information the Contractor obtained relative to the Owner's financial ability to pay for Contractor's work. The Contractor does not warrant the accuracy of the Owner's information.

Article 11 INDEMNITY

The Subcontractor shall indemnify and hold harmless Contractor, Contractor's subcontractors, Architect/Engineer, Owner and their agents, consultants, employees and others from all claims for bodily injury and property damage that may arise from the performance of the Subcontract Work.

Article 12 CONTRACTOR'S RIGHT TO PERFORM SUBCONTRACTOR'S RESPONSIBILITIES AND TERMINATION OF AGREEMENT

12.1 FAILURE OF PERFORMANCE This Paragraph governs the Contractor's recourse when the Subcontractor fails to perform, including notice to cure. If the Contractor performs work under these provisions or subcontracts its performance, it has the right to charge the cost thereof to the Subcontractor, who is liable for such payment including reasonable overhead, profit and attorney's fees.

Figure 5.25 Sample Subcontract Form (continued).

12.2 TERMINATION BY OWNER This Paragraph provides that, in the event the Owner terminates the prime agreement, the Contractor, after providing written notification, may terminate the Subcontract.

12.3 TERMINATION BY CONTRACTOR This Paragraph addresses termination of the Subcontractor by the Contractor for cause.

12.4 TERMINATION BY SUBCONTRACTOR This Paragraph details the circumstances under which the Subcontractor may terminate the Agreement and the recoverable costs.

Article 13 CLAIMS AND DISPUTES

13.1 CLAIMS The Subcontractor shall give the Contractor written notice of all claims within seven days of the Subcontractor's knowledge of the fact giving rise to the event.

13.2 DIRECT DISCUSSION The parties are encouraged to settle their disputes through direct discussions. If these discussions are not successful, the parties must attempt mediation as a condition precedent to any other form of binding dispute resolution (Paragraph 13.8). Any disputes not resolved by mediation are to be decided by the dispute resolution process selected in Paragraph 13.9.

13.3.1 OTHER DISPUTE PROCESSES The parties select either arbitration or litigation to solve a dispute that neither direct discussions nor mediation resolved.

Article 14 JOINT DRAFTING

This Article states that the Agreement shall be construed in a neutral manner, since the Agreement has been adopted by the parties as an expression of their intent.

5/00
iv

AGC DOCUMENT NO. 603 • STANDARD SHORT FORM AGREEMENT BETWEEN CONTRACTOR AND SUBCONTRACTOR
(Where Contractor Assumes Risk of Owner Payment)

Figure 5.25 Sample Subcontract Form (continued).

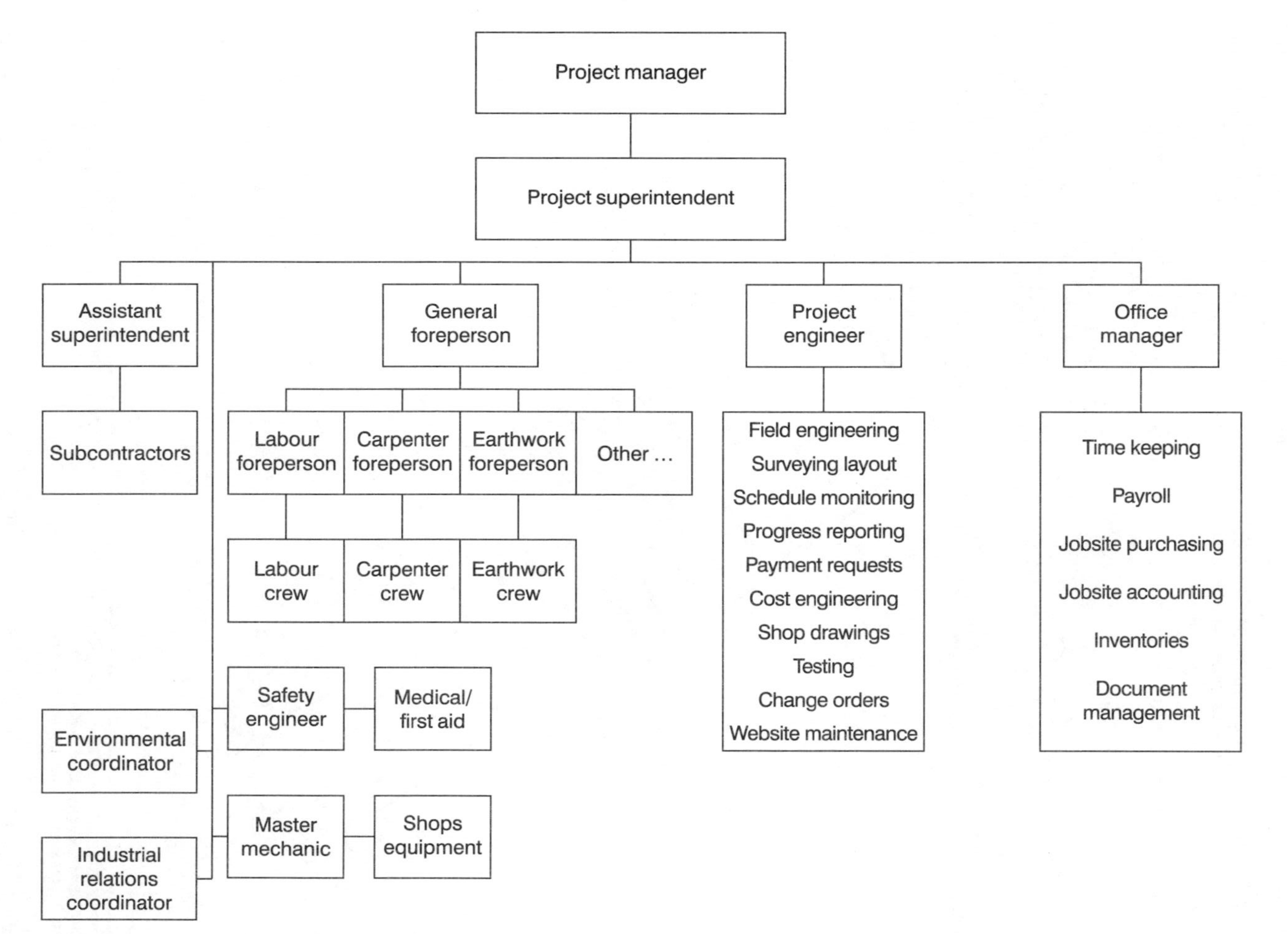

Figure 5.26 Worksite project organisation for a large project.

the various functions in the following chapter, when we discuss the operation of the construction team on the site. Our purpose in the present chapter is to indicate that an organisation structure must be established and that the roles must be filled, either from within the contractor's organisation or by hiring from outside. Although the contractor's personnel shown in Figure 5.26 will comprise the majority of staff on site, some non-contractor personnel will also be part of the team, including owner and design professional representatives and others, such as material suppliers and public agency inspectors, will visit the site from time to time.

Union labour

In addition to establishing and staffing a management structure, the contractor will need to staff the project with *operating personnel* (or *operatives*) who will provide the direct labour. If the contractor has negotiated collective bargaining agreements with construction *trade unions* (also called *craft unions*), either by itself or as part of a multi-employer bargaining unit, the source of the tradespeople will be the local labour unions. Whether or not a union contractor was directly involved in negotiating the union agreement, the contract, when signed, is binding upon both parties, and the relations between contractor and worker are largely governed by that contract. The contractor requests from the local union office the personnel needed for the project, for whatever duration they may be needed, rather than contacting the employees directly.

When a construction tradesperson is first employed on a project, no negotiation takes place about such matters as rates of pay and working conditions, because those issues have been finalised in the collective bargaining agreement. Such agreements are usually made between the contractor or an association of contractors, on one side and either individual craft unions on behalf of their members in a local or regional area or with alliances of unions, on the other. Examples of individual craft unions are the Canadian affiliates of the International Brotherhood of Electrical Workers or the United Union of Roofers, Waterproofers and Allied Workers in the USA. A local or regional Building Trades Council is an example of an alliance of unions. In some cases, a single union may represent all construction crafts; in Australia, for example, the Construction Division of the Construction, Forestry, Mining and Energy Union is comprised of bricklayers, carpenters, plasterers, painters, crane drivers, steel fixers, operators, construction labourers and trades assistants (Construction, Forestry, Mining, and Energy Union of Australia, 2002). A typical collective bargaining agreement covers a period of 1 or 2 years and contains a long list of agreed-upon provisions covering wages and fringe benefits, hiring procedures, hours of work and overtime pay, holidays, vacations and leaves of absence, apprentice programmes, work rules stipulating types of work that will be assigned to members of a certain union, number of forepersons required and tools and transportation allowances, rules governing strikes and work slowdowns, the geographical area of jurisdiction covered by the agreement, substance abuse, safety and accident provision and non-discrimination issues, dispute resolution procedures and a subcontractor clause that prohibits the contractor from engaging a non-union subcontractor to do specified types of work.

Large national contractors may sign collective bargaining agreements with the national bodies of labour unions covering all work within that country. In the recent past international agreements have emerged that cover some of the issues found in more local agreements. In 2000, the Germany-based international construction group, HOCHTIEF, signed an agreement with the German Workers' Union and the International Federation of Building and Wood Workers agreeing to a 'social charter' imposing minimum standards of work conducted on its projects. In its code of conduct, the charter includes sections on 'freedom of employment, ban on child labour, freedom of association and collective bargaining, adequate wages, working time and working conditions' (International Federation of Building and Wood Workers (IFBWW) Press Release, 2000).

In Chapter 6, we shall discuss human resource management during project operation, including the role of union labour. But here it is important to recognise the value and also the limitations of unions in construction. The primary advantage enjoyed by the union contractor is the easy and ready access it has to a pool of skilled labour; apprentice programmes for training new entrants into the crafts are a major source of expertise in construction. In addition, the worker comes to the job with already-defined rules governing their employment. These rules may also be a disadvantage, because all workers of a certain classification are provided the same benefits and therefore there exists no opportunity to reward merit. In addition, because negotiations are conducted with the union and not with the employee and also because there is no guarantee of long-term employment, the worker's loyalty usually lies more with the union than with the employer.

One of three possible membership requirements can exist on a unionised project. One is a *closed shop*, of which an employee must already be a member in order to be employed. In the USA, current labour law prohibits closed shops. The most common form is the *union shop*; under this arrangement, a new employee is not required to be a member of the union but, if not, must join within a certain number of days, defined as 7 days in the USA. Finally, in an *agency shop*, there is a union and a collective bargaining agreement. However, union membership is voluntary; non-members have the same privileges as members (including the 'privilege' of paying union dues!), but have no voice in union matters.

A *Project Labour Agreement* (PLA) is a special collective bargaining agreement which governs labour relations between all labour unions and all contractors, including subcontractors, working on a particular project, for the duration of the project. A report on the use of such an agreement for a US$ 130 million bridge repair and rehabilitation project carried out in the 1990s in the eastern USA (Report on Rockland/Westchester County Labor Analysis, no date) stated the following:

> All parties to be involved in the construction – unions, contractors, owners (often through a construction manager) – are required to be signatories to the PLA, which supersedes any prior-existing agreements. The PLA provides for standardised work practices; hours; holidays; grievance, dispute and arbitration procedures; and overall labor/management harmony for the duration of the project. Most importantly, the PLA precludes any strikes, lock-outs, work stoppages and/or any other disruption of work for any reason during the term of the PLA. The term 'Pre-Hire' is derived from the agreement by the parties, prior to construction, to hire all workers through the respective union hiring halls.

Such agreements have been a part of private sector construction for many decades. An agreement of this type was a key to completing the Trans Alaska Oil Pipeline on schedule in 1977; in negotiating with the unions, the pipeline owners agreed to lucrative wages, benefits and working conditions for union workers in exchange for the assurance work would not be stopped by labour disputes. More recently, PLAs have entered the public sector, but not without legal challenges from non-union contractors and employee associations who fear loss of work and from parties who believe they violate public tendering statutes because the costs involved might preclude awarding the contract to a less expensive, non-union contractor. In most cases, the agreements have been declared legal, provided there is a showing of responsibility to the public fisc (Mara, 1999).

Open shop and merit shop – non-union contracting

If the contractor operates a non-union shop, it hires its own employees and arranges its subcontracts without regard to their unionised status. The contractor establishes its own rates of pay and fringe benefits and is responsible for recruiting, hiring, training, evaluating, promoting and discharging its employees. The decline in union membership throughout the civilian work force in the USA, from a high of about 28% in the late 1950s to about 14% in 2000, has been reflected in a similar trend in the construction industry. In 1981, 33% of construction workers in the USA were members of unions or were covered by union contracts; in 2000, that figure was 20% (Allen, 1994; Bureau of Labor Statistics, 2002). Thus, although US construction has been heavily unionised in the past, these statistics indicate that about 80% of all such work is currently performed by non-union workers.

Two types of non-union operations are used in the construction industry. A completely 'unorganised' contractor has no affiliation with a national organisation and sets all of its rules regarding employment of its workers; such an arrangement is called an *open shop*. The other approach is called a *merit shop*; in this case the contractor is a member of a regional or national contractors' association and agrees to abide by employment practices established by that organisation. Workers can easily accept employment with any member of the association, but hiring is done locally, directly between the contractor and the worker (Hendrickson and Au, 1989).

Despite the popularity of non-unionised work forces, the contractor must face two challenges: the recruiting of its workers and the assurance that they will have adequate skills. Recruiting takes place through advertising in local media, by way of open shop hiring halls, and, in the case of merit contracting, through the use of registers maintained by contractors' associations. Whereas apprenticeship programmes are a popular and beneficial advantage of unionised contracting, the burden of training is upon the contractor when there is no union. Merit shop associations have undertaken to fulfil this skill development need.

Non-union contracting allows the contractor more flexibility in the composition and assignment of work crews, permits the employer to pay workers based on their ability and performance and generally results in less expensive projects. Although wage rates tend to be somewhat lower than in the unionised sector, employment is more continuous and workers' annual wages are often at least as high as for those with similar jobs working for union contractors. Because they are employed directly by the contractor, workers tend to develop more loyalty toward their employers and take an interest in the company's financial well-being.

Special considerations in mobilising for remote projects

If the contractor has obtained a new project in the Canadian Arctic or the Middle Eastern desert, preparation to carry out that project involves considerations that would not be required if the project were located in downtown Stockholm or suburban Adelaide. Although an extensive discussion is not possible here, we list below some issues the contractor operating in such remote regions must incorporate into its mobilisation planning. For further details, see McFadden and Bennett (1991), Changsirivathanathamrong (1998), Charaf (1999) and Cuetara (1999).

- Because the project is located a long way from sources of supply, plans for acquiring and delivering materials and equipment must be carefully developed. Included is the fact that the site may be accessible during only parts of the year. With no local hardware store, forgotten items will be time consuming and costly to obtain.

- Health and safety concerns may warrant establishing more extensive medical facilities than would be needed on a more centrally located project.
- The jobsite will require facilities for housing and feeding workers and staff to provide these support services.
- Incentives, such as extra-pay increments and transportation privileges, may be needed to attract sufficient numbers of qualified workers.
- In developing the programme, the contractor may be able to plan on 7-day-per-week work effort, because workers will be living at the site.
- Protection of the work due to extreme environmental conditions may be required. Examples are cooling of massive concrete works, sheltering of work from intense sunlight and rain and temporary enclosures and heating for projects in cold regions.
- When environmental conditions are extreme, reduced worker productivity must be expected.
- Likewise, construction equipment performance may be adversely affected by extreme temperatures (hot or cold), snow and ice and lack of daylight.
- The contractor must plan to provide extra considerations for worker safety and comfort, such as large quantities of drinking water, shade protection and extra rest breaks in hot regions and cold weather clothing in the Arctic.
- Pre-fabrication and modularisation of project components off site will be potential means for cost savings.
- A remote site may be located in a country different from that of the contractor's main office. If the languages are different, plans must be developed for effective communication with employees from the local area.

At the end of this chapter, we describe some of the features of the remote construction camps that were an essential part of the 800-mile Trans-Alaska Oil Pipeline, constructed during the 1970s.

Discussion questions

1. Interview a general contractor in your area to determine what permits, consents and licences are required for a typical project carried out by this contractor.
2. Find a tender (bid) bond and a performance or payment bond for a project in your area and compare their wordings to those found in the text. What are the significant differences?
3. Develop a work breakdown structure for the concrete wall project shown in Figure 4.5. Your answer is not required to correspond to the work items in this 'project's' cost estimate.
4. For the work breakdown structure in Figure 5.7, select either the north abutment footing or the concrete deck element and create more detail by adding two more levels to the hierarchy.
5. In the footing construction schedule in Figure 5.10, suppose we could arrange to procure the reinforcing steel in 1 day, thus reducing that activity's duration by 3 days. How much would this modification change the project duration if nothing else were changed? Why?
6. The schedule in Figure 5.13 is rather inefficient because it shows that no activity will take place while the pier concrete cures. Discuss several ways to make this schedule more efficient.

7 The construction of a sidewalk project consists of the activities given below.

Activity	Duration (working days)	Immediate predecessor(s)	Value (US$)	Number of labourers
Move-in	2		500	3
Excavate	3	Move-in	1500	2
Grade surface	2	Excavate	800	2
Order and deliver forms	6		900	0
Order and deliver reinforcing	10		2200	0
Pre-fabricate forms	4	Order and deliver forms	2600	4
Install forms	3	Grade surface; prefab-ricate forms	2100	3
Place and compact base	2	Install forms	1100	2
Place reinforcing	1	Place and compact base; order and deliver reinforcing	600	2
Place concrete	1	Place reinforcing	4500	5
Strip forms	3	10 working days after completion of place concrete	1650	2
Cleanup and move-out	2	Strip forms	900	3

Note that there is a lag of 10 working days between the completion of 'place concrete' and the beginning of 'strip forms'.

(a) Draw an activity-on-node schedule network diagram for this project.
(b) Calculate the early and late start and finish time and the slack for each activity.
(c) Identify the critical path.
(d) If the scheduled project start date is 9 May 2005 and the work will proceed 6 days per week, find the expected project completion date.
(e) Assume that the value of each activity is spread evenly across its respective duration. Determine the amount that will have been earned at the end of each working day, based on an early start and late start schedule. Draw two cash flow projection curves, one for an early start schedule and one for a late start schedule.
(f) The indicated number of labourers is required for each working day the activity is underway. Assuming an early start schedule, prepare a histogram showing the number of labourers required for each project day.

8 The sidewalk construction project in Question 7 is a very linear project, with very few parallel sequences of activities. Thus, its schedule often has only one activity underway at any one time. A somewhat shorter programme might be achieved by phasing the work in the following way. Divide the following activities into two phases each: excavate, grade surface, install forms, place and compact base and place reinforcing. Each phase will have a duration of half of the activity's original duration (half-day durations are okay!). After the first phase

has been completed, its second phase *and* the first phase of its immediate successor can begin. Thus, when Phase 1 of excavate (1.5-day duration) has been completed, Phase 2 of excavation and Phase 1 of grade surface can begin.

Leave all other activities except the five given above as single activities and draw a revised network diagram. Repeat parts (b)–(f) for this revised diagram. How many days does this revision save compared to the original programme? How does this change affect the maximum labourer requirement?

9 What actions might be taken to save even more time than the result of your revised analysis in Question 8? How much time do you think you could save?

10 Observe a construction project and note the various repetitive operations. Write a brief report on your findings. Do you believe a modelling analysis could improve this operation? Why or why not?

11 A contractor is planning the budget and cash flow for a 12-month project with the following estimated costs per month.

Month	Estimated cost (US$)
1	60 000
2	120 000
3	185 000
4	265 000
5	305 000
6	350 000
7	310 000
8	270 000
9	215 000
10	190 000
11	145 000
12	85 000

(a) Calculate and plot the cumulative cost per month.

The 'value' of the project works will be 130% of the costs. That is, the margin is 30% in addition to the costs given above.)

(b) Calculate, and plot on the same graph, the cumulative value per month.

The contractor will submit a request for payment at the end of each month, for the value of works installed in that month. Payment will be received 2 months later. The owner will withhold 10% of each payment request as retainage for all months 1–11, so that the contractor will receive 90% of its month 1–11 payment requests, each 2 months later. When the payment for month 12 is received, at the end of month 14, it will include all of the amounts retained and will complete payment of the entire project value.

(c) Calculate, and plot on the same graph, cumulative payments received by the contractor at their proper points in time.

(d) For several months, the cumulative costs will exceed cumulative payments received. Calculate this difference for each month and plot it on the same graph, labelling the amounts as 'working capital required'. Identify the maximum amount of required working capital and the month when it will be required. Assume that costs will be incurred at the end of each month.

(e) Suggest means by which the contractor might secure the required working capital.

12 For the bridge construction project whose cross-section is shown in Figure 5.6, sketch a possible site layout plan, incorporating ideas from the section on 'organising the worksite'. Because very little is given about this project, make some assumptions and state them in your answer.

13 List several arguments in favour of centralised purchasing of all materials for a construction project and several arguments in favour of having individual projects do all of their own purchasing. Which do you tend to prefer? Why?

14 List five other topics that might be considered during subcontractor negotiations, in addition to those given in the text.

15 Interview a local construction manager to determine the organisation structure for a current or recent project worksite. Does the project have a chart that depicts this organisation? If so, try to obtain a copy. If not, try to create one from the information you glean at your interview. How does this organisation differ from that given in the text? In what ways is it the same?

16 Find out about trends in construction collective bargaining in your country or region in the past 10 years. What proportion of construction work is done by merit or open shop contractors? How has this proportion changed? What are some reasons for the changes, if any?

17 List some of your ideas for assuring that the construction industry in your area will have adequately trained operatives to perform construction work for the workload expected in the next 5 years.

References

Allen, S.G. 1994. *Developments in Collective Bargaining in Construction in the 1980s and 1990s*. NBER Working Paper No. w4674. National Bureau of Economic Research. http:/papers.nber.org/papers/W4674.

Arizona State University. 2002. *Purchasing and Business Services*. http://www.asu.edu/purchasing/.

Associated General Contractors of America. 2000. *Standard Short Form Agreement between Contractor and Subcontractor (where Contractor Assumes Risk of Owner Payment)*. Associated General Contractors of America Document No. 603.

Babcock, D.L and L.C. Morse. 2001. *Managing Engineering and Technology*, 3rd edn. Prentice-Hall.

Bennett, F.L. 1996. *The Management of Engineering: Human, Quality, Organizational, Legal, and Ethical Aspects of Professional Practice*. John Wiley.

Bennett, F.L. 2000. Information technology applications in the management of construction: an overview. *Asian Journal of Civil Engineering (Building and Housing)*, **1**, 27–43.

Bureau of Labor Statistics. 2002. *Career Guide to Industries: Construction*. US Department of Labor. http://www.bls.gov/oco/cg/cgs003.htm.

Changsirivathanathamrong, A. 1998. Logistical support of remote construction. In *Construction in Extreme Environments. CEE 596 – Current Topics in Construction Management*. Cornell University School of Civil and Environmental Engineering.

Charaf, J. 1999. Management of construction equipment in hot regions. In *Management of Construction in Adverse and Culturally Diverse Environments. CEE 596 – Advanced Construction*. Cornell University School of Civil and Environmental Engineering.

Clough, R.H. and G.A. Sears. 1994. *Construction Contracting*, 6th edn. John Wiley.

Construction, Forestry, Mining, and Energy Union of Australia. 2002. *CFMEU Construction Online*. http://www.cfmeu.asn.au/construction/.

Contractor Magazine. 2001. New Zealand Contractors' Federation and Contrafed Publishing.

Cuetara, G. 1999. Construction site safety: special considerations for hot and cold regions projects. In *Management of Construction in Adverse and Culturally Diverse Environments. CEE 596 – Advanced Construction*. Cornell University School of Civil and Environmental Engineering.

Dorsey, R.W. 1997. *Project Delivery Systems for Building Construction*. Associated General Contractors of America.

Fisk, E.R. 2003. *Construction Project Administration*, 7th edn. Prentice-Hall.

Halpin, D.W. and L.S. Riggs. 1992. *Planning and Analysis of Construction Operations*. John Wiley.

Halpin, D.W. and R.W. Woodhead. 1998. *Construction Management*, 2nd edn. John Wiley.

Hendrickson, C. and T. Au. 1989. *Project Management for Construction: Fundamental Concepts for Owners, Engineers, Architects, and Builders*. Prentice-Hall.

Hewat, B. 2000. *Construction Operations Simulation Using CYCLONE*. University of Canterbury, New Zealand, ENCI 627 Construction Operations Analysis and Management Research Paper.

International Federation of Building and Wood Workers (IFBWW) Press Release. 2000. *IFBWW, IG BAU, and HOCHTIEF agree globally valid social standards*. http://home-pages.iprolink.ch/~fitbb/INFO_PUBS_SOLIDAR/Press_-_Hochtief.html.

Jamieson, B.K. 2001. *Partnering – Construction Project Management*. ContraxAsia. http://www.cxa.net/Articles/dtp1.htm.

Johnston, H. and G.L. Mansfield. 2001. *Bidding and Estimating Procedures for Construction*, 2nd ed. Prentice-Hall.

Mara, B. 1999. Project Labor Agreements: An Overview of Contractual and Legal Considerations. In *Contract Claims, Alternate Dispute Resolution, Collective Bargaining, and Labor Relations in the Construction Industry. CEE 596 – Advanced Construction*. Cornell University School of Civil and Environmental Engineering.

McFadden, T.T. and F.L. Bennett. 1991. *Construction in Cold Regions: A Guide for Planners, Engineers, Contractors, and Managers*. John Wiley.

Mincks, W.R. and H. Johnston. 1998. *Construction Jobsite Management*. Delmar.

Municipality of Nenana. 2002. *Contract for Installation of Five Underground Steel Pipe Casings in Nenana, Alaska. Nenana Phase II Water and Sewer Project*. Design Alaska.

New Zealand Building Business: Your Building Guide, South Island Edition. No date. Productive Promotions.

Palmer, W.J., W.E. Coombs and M.A. Smith. 1995. *Construction Accounting and Financial Management*, 5th ed. McGraw-Hill.

Peachtree Software Inc. 2002. Peachtree Complete Accounting 2002 Release 9.0.01. http://www.peachtree.com.

Pilcher, R. 1992. *Principles of Construction Management*, 3rd ed. McGraw-Hill.

Project Management Institute. 1999. *Project Management Software Survey*.

Report on Rockland/Westchester County Labor Analysis. No date. Prepared for project labor agreement covering certain construction and repair work on the Governor Malcolm Wilson Tappan Zee Bridge. New York State Thruway Authority.

Ricketts, J.T. 1996. Construction management. In Merritt, F.S., M.K. Loftin and J.T. Ricketts, ed. *Standard Handbook for Civil Engineers*, 4th edn. McGraw-Hill.

Sassoon, D.M and H.O. Merron. 1995. *CIF and FOB Contracts: British Shipping Laws*, 4th ed. Sweet and Maxwell.

Surety Association of Canada (2002). *Surety Defined.* http://www.surety-canada.com/index.html.

University of Alberta Construction Engineering and Management Programme. 1999. *Simphony: An Environment for Developing Special Purpose Construction Simulation Tools.* http://cem.civil.ualberta.ca.

Washington State Department of Transportation. 2001. *Operations Office Notice. August 29.* http://www.wsdot.wa.gov/TA/Operations/LAG/BidBondForm.html.

Case study: Wilmot Pass and Wanganella – logistical planning and personnel accommodation in a remote corner of New Zealand

Special thanks to Tony Savage, Plant Manager for Meridian Energy at the Manapouri Power Station, New Zealand, for providing background information and pictures used in this case study.

The West Arm Power Station, located at the end of the West Arm of New Zealand's Lake Manapouri, is that country's largest hydro-electric generating facility, producing about 15% of the New Zealand's electric power. The station is housed in a man-made underground cavern, draws water from Lake Manapouri and discharges it through two 10 km long outrace tunnels into Doubtful Sound at Deep Cove, 178 m lower. The station provides about 700 MW of power to a single user, an aluminum smelter located at Bluff, via a 171 km transmission line. Located in the isolated Fiordland region of New Zealand's South Island in an area that receives as much as 8.75 m of rainfall annually, the site before development of the power project was virtually inaccessible, save for water access across Lake Manapouri to the West Arm Station and, at the outfall of the tailrace, by water access at Deep Cove. A single tramping track connected the two locations over the steep and heavily forested Wilmot Pass.

Among the multitude of challenges facing the construction planners in the early 1960s was the need to provide delivery and personnel access to the West Arm site. Some 300 000 tonnes of materials and equipment, including heavy generators and other components typical of electrical power stations, would be required to be transported from foreign manufacturers. The nearest deep water port at that time was located at Bluff and a 200 km paved road connected that location with the east side of Lake Manapouri. A railway operated for the first 140 km of that distance. Thus, one option was to upgrade the existing land transport infrastructure to accommodate the heavy loads, convey them from Bluff to the east edge of the lake and then use waterborne craft to carry them to the west side for installation. The other primary option, and the one selected, was to construct an all new road from Deep Cove over the 700 m high Wilmot Pass to the West Arm site, a distance of 20 km. The general location of the project in New Zealand's Fiordland is shown in Figure 5.27, while the location of the road in

Figure 5.27 Manapouri power project location on New Zealand's South Island.

greater detail is shown in Figure 5.28. Among the considerations in this economic decision was the fact that load sizes would still be limited on the Bluff–Manapouri route and on-site costs would increase for both manufacturing and erecting smaller components of generating equipment at the site (Meridian Energy, 2002).

The route generally follows the original tramping track through some of New Zealand's most rugged country, climbing to the pass from Deep Cove at maximum grades of 10.5%, except near the top where the grade is 16%. Curves are as large as possible in the rough terrain to allow use by heavy equipment transporters. Stream crossings consist of culverts where possible, but several steel bridges with concrete

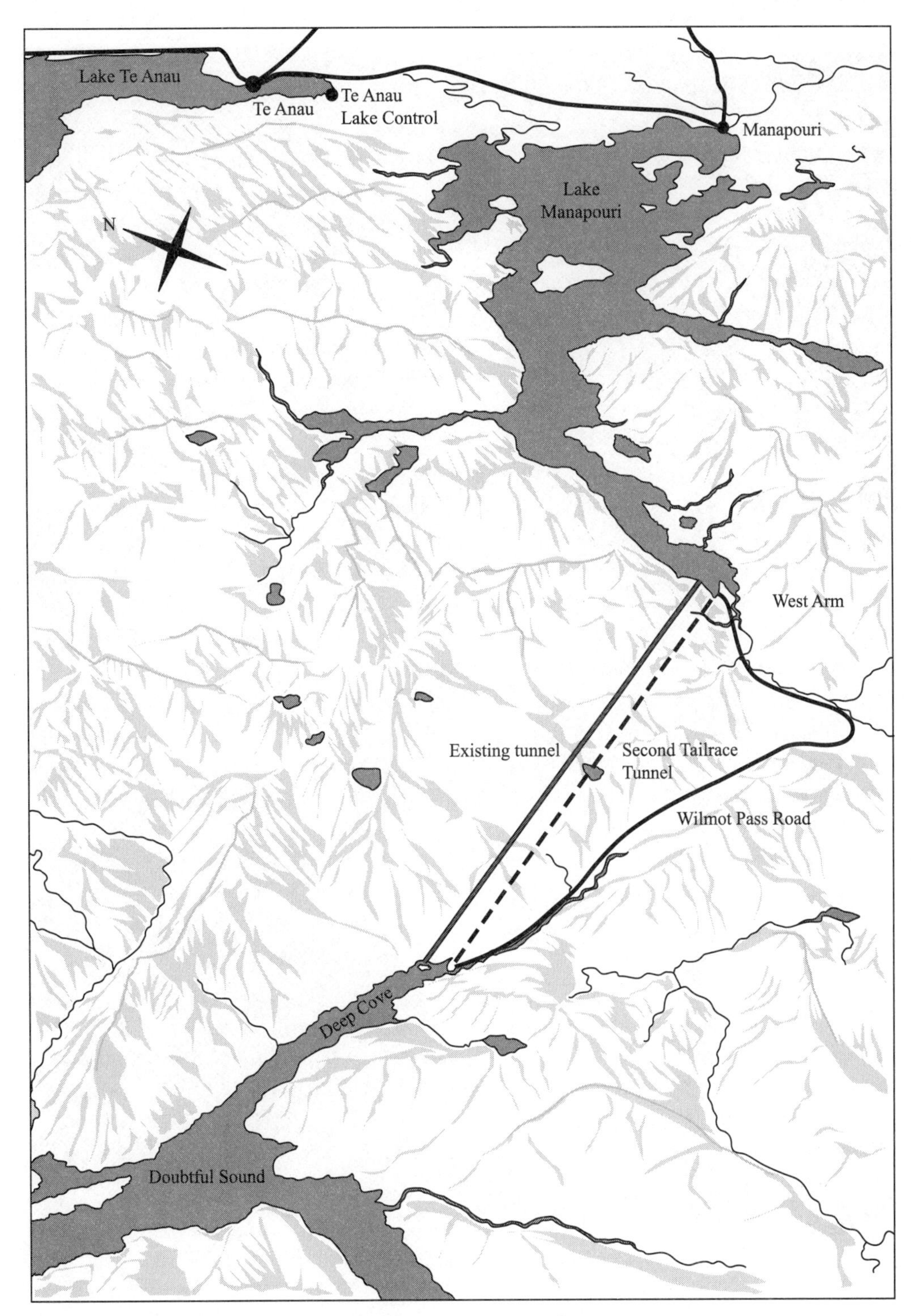

Figure 5.28 Wilmot pass road between West Arm and Deep Cove.

Figure 5.29 Bridge construction on Wilmot pass road, 1965.

Figure 5.30 Deep Cove side of access road over 700 m Wilmot Pass.

decks are used. One such bridge is shown under construction in Figure 5.29. The road surface is of shingle. Construction proceeded from both ends, commencing in October 1963 and lasting 2 years, an average completion rate of less than 1 km per month.

> The road work proved to be a formidable task. It was considered to be, by people involved in both undertakings, comparable to and worse in many respects than the construction of the famed Burma Road during World War II. The incessant rains, abominable swamp conditions on flat going and sheer rock faces that had to be negotiated over the mountains combined to produce extreme working conditions for both men and machines. (Langbein, 1971)

When the route was completed in 1965, vital supplies and equipment began to flow from the new port at Deep Cove to the site at West Arm and workers had much more convenient access as well. The road continues to operate today; Figure 5.30 shows an impressive photographic view of the road on the Deep Cove side of the pass. It was used extensively during the drilling of a second tailrace tunnel, completed in 2001, and it also provides a route for numerous buses that offer tourists breathtaking views of the area and access to waterborne tours of Doubtful Sound via Deep Cove.

Among the unusual aspects of the Manapouri power project was the use of a ship as a floating hostel. The *M.V. Wanganella* had seen service as a first-class trans-Tasman passenger liner, plying the waters between Australia and New Zealand, before being converted to a hospital ship during World War II. After the war, she returned to her

Figure 5.31 *M.V. Wanganella*, floating hostel for Manapouri project workers, 1963–1970.

original route until the early 1960s and was about to be scrapped, when those planning the Manapouri project chose her 'as an alternative to the difficult, expensive and time-consuming operation of building a construction workers' camp on mountainous, bush-clad sides of Doubtful Sound' (Bennett, 2002). From 1963 to 1970, the ship (shown in Figure 5.31) provided accommodation at the new wharf in Deep Cove for upwards to 500

> tunnellers and miners, drivers of locos, trucks and bulldozers, operators of cranes, mucking, concrete and other heavy machines, fitters, welders, electricians, carpenters, painters, cooks, clerks, inspectors, surveyors, engineers and countless other workers who would appreciate a comfortable bed in *Wanganella's* warm cabins. (Bennett, 2002)

A hospital was built aboard the ship and a movie theatre and library were provided for the

> tough, hardened men from all over the world – Americans, Germans, Spaniards, Poles, Greeks, Italians, Yugoslavs, Frenchmen, Hungarians, Latvians, Estonians, Irishmen, Scotsmen, Englishmen, Pacific Islanders, Australians and New Zealanders, both Maori and Pakeha – [who] came and went from the *Wanganella*. (Bennett, 2002)

Among the popular spots on board was the bar, from where, it was said, three thousand empty beer cans per day were discarded overboard, leading to the assurance that the ship was unsinkable because her hull rested on a mountain of sunken beer cans! Although other locations at Deep Cove were developed as wharfs, the vessel also served as a floating dock, with her derricks used to unload stores and equipment.

References

Bennett, S.W. 2002. *M.V. Wanganella History* (unpublished).
Langbein, J.A. 1971. The Manapouri Power Project, New Zealand. *Proceedings of the Institution of Civil Engineers*, **50**, 311–351.
Meridian Energy. 2002. *Technology and Teamwork: The Second Tailrace Tunnel, Manapouri Power Station.*

Case study: Temporary quarters for pipeline workers on a far-away project – Trans-Alaska Oil Pipeline construction camps

The Trans-Alaska Oil Pipeline was constructed between 1975 and 1977 to carry crude oil from the oil-rich North Slope of Alaska at Prudhoe Bay to the ice-free port of Valdez on the state's southern coast. It was a massive undertaking, costing approximately US$ 9 billion and was, at the time of its completion, the largest construction project ever undertaken with private funds. Today the pipeline is owned and operated by Alyeska Pipeline Service Company, a consortium of oil companies led by BP Pipelines (Alaska) Inc., Phillips Transportation Alaska, Inc. and ExxonMobil Pipeline Company, Inc. Part of the project included the installation of 800 miles (1290 km) of 48" (122 cm) diameter steel pipe, half of which is buried and the other half of which is supported above ground by cross beams and vertical support members driven into the ice-rich permafrost (permanently frozen ground). In addition, the system includes 12 pumping stations, a communication and control system, storage and loading facilities at Valdez and an extensive gathering system at Prudhoe Bay. At its peak flow in 1988, the pipeline transported 2.1 million barrels of crude oil per day; with predicted declining production, the flow was about 1.0 million barrels per day in 2002 (Alyeska Pipeline Service Company, 2002). The engineering and construction challenges in this land of extremes are a fascinating story for the student of construction management. Temperatures range from +90°F (+32°C) to −65°F (−54°C). In the extreme north, summertime brings continuous daylight, while wintertime brings continuous darkness. Moisture conditions range from an Arctic desert to an average annual snowfall of 540" (13.7 m) at Thompson Pass. For a construction management perspective on the entire project, see McFadden and Bennett (1991).

Much of Alaska, especially its northern half, is a vast wilderness, lacking any infrastructure of the type needed to support a large construction undertaking. Typical of these conditions is the view in Figure 5.32, located north of the Yukon River and south of Atigun Pass in the Brooks Range of mountains. It was necessary to build many temporary facilities, some of which later became permanent, owned and operated either by the pipeline operators or the Alaska government. For example, 14 airfields, ranging in length from 2500 to 5000 feet (0.75 to 1.5 km) were developed in northern Alaska to support the project; two of these continue to be used by Alyeska. A 358 mile (575 km) long haul road was pioneered from an existing roadway to Prudhoe Bay; the road includes the first and only bridge across the Yukon River in Alaska and is now part of the state's highway system.

Figure 5.32 Large portions of the Trans-Alaska Oil Pipeline were built in remote areas of the USA's largest state.

At the peak of construction, in October 1975, 28 072 people were employed on the pipeline project. The vast majority were scattered along its length, in what a writer has called an 800 mile Skinny City (Cole, 1997). It is estimated that about 70 000 different people worked on the project, from the first efforts at development in 1969 to project completion in 1977. Among the many challenges was the need to provide living accommodations for these workers, because 'going home at night' (to places like Texas and Oklahoma, home states of many of the pipeline workers) was not an option. A total of 29 construction camps were built, one at each of the 12 pumping stations, one at the Valdez terminal and 16 others located adjacent to the line. The map in Figure 5.33 shows the pipeline route, together with several camps and pumping stations. Camp names like Coldfoot, Prospect, Sourdough and Sheep Creek are apt designations for the harsh, rugged and typically Alaskan environment in which the workers worked and lived. The largest camp, at the Valdez Terminal, contained 3480 beds. Along the pipeline, the largest camp was at Isabel Pass, with 1652 beds, while the camp at Sourdough was the smallest, with 112 beds.

Built at a cost of US$ 6–10 million each, the camps were of standardised modular construction; basic housing units consisted of 28 two-person rooms. Room and board costs were paid by 'Uncle Al', as Alyeska was known, leading many workers to consider the life at Skinny City to be 'Fat City'. With few exceptions, the camps have been removed, in accordance with a stipulation of one of the permits granted by the government that allowed the pipeline to be constructed:

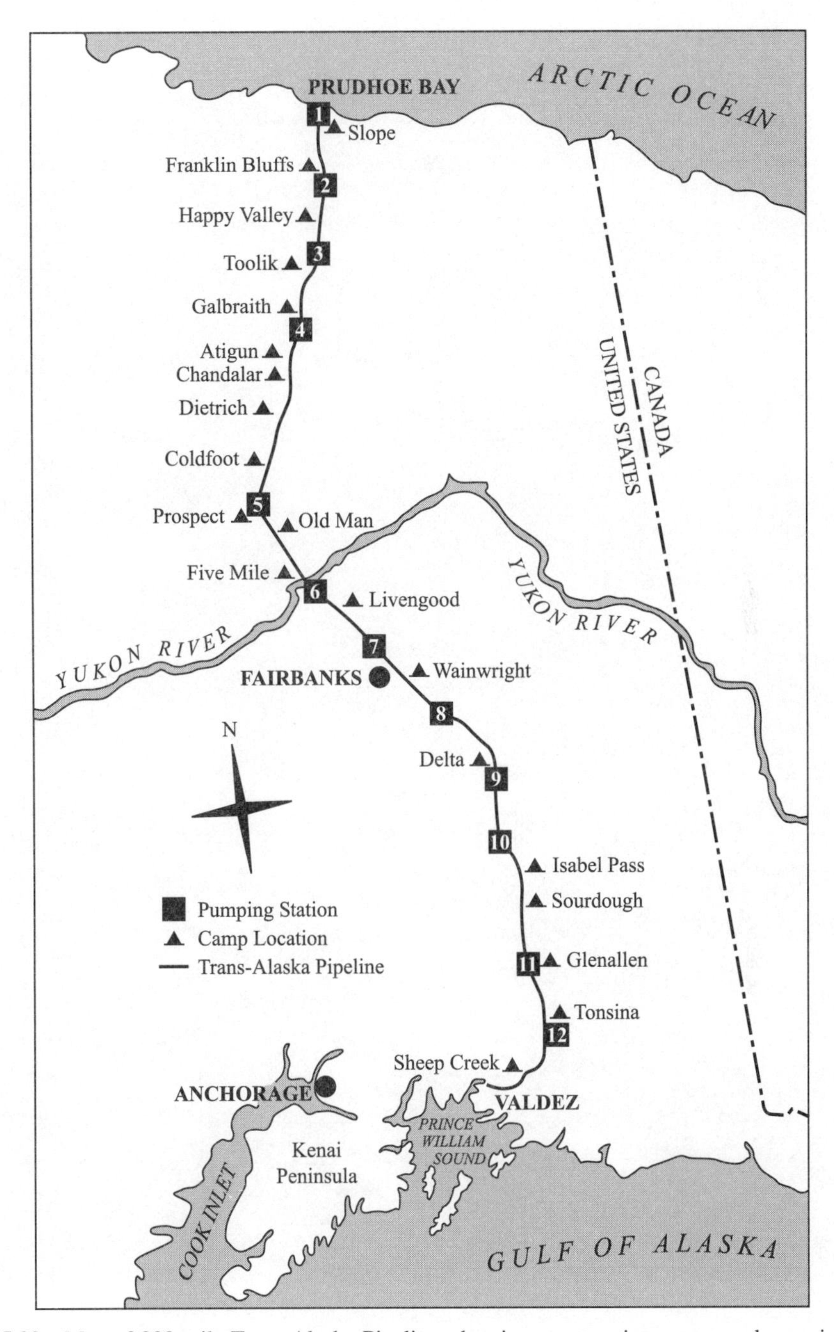

Figure 5.33 Map of 800-mile Trans-Alaska Pipeline, showing construction camps and pumping stations (Cole, 1997). Reproduced by permission of the author.

> In the pipeline camps, the normal social structure was replaced by a hybrid that revolved around working, eating, and sleeping, and two-to-a-room cubicles in a trailer . . .

Figure 5.34 Atigun construction camp, Trans-Alaska Pipeline project. © Alyeska Pipeline Service Company.

From the air, the typical camp looked like a cross between a sparse mobile home park and a heavy equipment yard. The long, rectangular pre-fab units would be lined up like building blocks and usually connected by enclosed, heated walkways.

Figure 5.35 Sourdough construction camp, Trans-Alaska Pipeline project. © Alyeska Pipeline Service Company.

> The camps, built on pilings or skids with above-ground water and sewer pipes, had a temporary look and feeling to them in keeping with their origins as transplants in the wilderness. Powerful generators hummed in the background, and the smell of diesel exhaust was in the air. The lights burned through the long winter nights in kitchens, warehouses, offices, and garages. (Cole, 1997)

Aerial views of two construction camps are shown in Figures 5.34 and 5.35.

Food service at the construction camps was famous for its high quality and abundance. In the early stages of construction, the caterers were paid on a cost-plus basis and there was little incentive to limit portions or types of food. Steak, frog legs, Beef Wellington, lobster, king crab and prime rib were standard fare, with a typical dinner menu consisting of soup du jour, New York steak, Shrimp Louie, garlic bread, French bread, Brussels sprouts with cheese sauce, fried potatoes, seven salads, peach pie, ice cream and cookies (Cole, 1997). Food was a morale booster, and there were complaints when a modicum of cost cutting was introduced; a culinary worker said the protesters were like seagulls, who 'eat, sleep, and squawk'.

Several of the camps published photocopied newspapers, including the *Atigun Times*, which called itself 'America's Farthest North Daily Newspaper', the *Coldfoot Chronicle* and the *Toolik Tribune*. Because the Toolik camp was about 35 miles north of Atigun, the *Tribune*'s masthead read 'Indisputably America's Farthest North Daily Newspaper'.

References

Alyeska Pipeline Service Company. 2002. *Pipeline Facts: Pipeline Construction.* http://www.alyeska-pipe.com.

Cole, D. 1997. *Amazing Pipeline Stories: How Building the Trans-Alaska Pipeline Transformed Life in America's Last Frontier.* Epicenter Press.

McFadden, T.T. and F.L. Bennett. 1991. *Construction in Cold Regions: A Guide for Planners, Engineers, Contractors, and Managers.* John Wiley.

6

Project operations phase

Introduction

With the start of construction operations in the field, the project takes on a different focus and the construction manager is called upon to perform a variety of responsibilities aimed at effective and efficient use of resources in the assembly of the several parts of the project. All of the previous planning, design, tendering and mobilisation have been the prelude to the actual performance of field operations. While some activities described in the previous chapter, such as procurement and staffing, are likely to continue during the early stages of the project operations phase, now the emphasis is on monitoring and controlling various aspects of the project, managing resources effectively and coordinating the vital documentation and communication activities.

The chapter begins by describing the monitoring and control of the project schedule, budget, quality, safety and environmental concerns. It continues with a discussion of resource management, in this case the management of personnel, materials and equipment. It then describes such communication and documentation issues as submittals, measurement and progress payments, variations, the overall document management responsibility and the role of electronic communications in the modern construction project. The chapter concludes with some legal issues, including the claims process, methods for dispute prevention and resolution and some particular legal matters that frequently arise during project operations.

Monitoring and control

The project programme represents the plan for the schedule of the work. Likewise, the budget is the plan for the cost aspects of the work. In a similar way, the contractor will have plans for the management of quality, safety and environmental concerns. During project operations, it is essential that actual performance be compared with planned performance in all of these areas and action taken to remedy any indicated deficiencies. This responsibility is termed *monitoring and control*, where *monitoring* refers to methods for comparing actual with planned performance and *control* denotes the actions taken to attempt to bring deficient aspects of the project into conformance.

Schedule updating

Periodically, perhaps monthly, the contractor will compare schedule progress with that shown on the project programme. Often this task is carried out in conjunction with requests for payment prepared for the owner; sometimes such a schedule updating is a condition precedent to the approval of the payment request. The purpose is to determine whether the various activities that were planned to be active during the previous period were actually active, the extent of their progress and, especially, the anticipated project completion date based on progress to date. Another purpose of the update is to incorporate any new information about already planned activities, to add information about any new work not previously planned and to determine their impacts on other activities and on the overall project completion date.

If the contractor is using only a simple bar chart for the programme, actual progress can be plotted underneath each work item bar indicating, for active work, when each item started and when it finished if it did. The use of bar charts for schedule updating is somewhat limited, as the ultimate purpose is to estimate the current anticipated project completion date; because the interrelationships among work items is difficult to depict, the impact of various work items on others and on project completion cannot be easily determined.

The network-based project schedule provides a convenient means for periodic monitoring and control. The process is rather straightforward. Each activity is examined to determine its status, as of the updating date. For those activities that have been completed since the last updating, their 'remaining durations' (the amount of time still required to complete the activity) are set to zero. For an activity currently underway at the time of the updating, the remaining duration will be the estimated time required to complete it. The monitoring and control process provides an important opportunity to evaluate the progress of each activity. Efforts to date may indicate that the original duration estimates were inaccurate and the updated schedule should reflect the best guess for each activity's remaining duration.

By way of illustration, consider an activity to install beams for a floor system. If the original estimated duration was 12 working days and if the work is 75% complete based on quantity installed, one should *not necessarily* conclude that (0.25)(12) = 3 working days remain to complete the activity. The remaining duration may be 5 working days, if the work is proving unusually difficult or it may be 2 days if the original duration estimate was, in actuality, too high.

The updating process also affords an opportunity to look at, and revise if appropriate, the durations of activities not yet started. During the preparation of the original programme, certain assumptions must be made: the method to be used to install tankage or cabling, the productivity of a carpenter crew, the mode of transport for delivery of materials. Even if an activity has not yet been started, new information that alters these assumptions may result in revised activity durations. Furthermore, revised activity sequencing may also be called for during the schedule update, based on more information or decisions to alter methods used. Scheduling is a dynamic process and the results of monitoring and control activity should reflect the most current experience, conditions and assumptions.

After activity duration and sequence data have been modified to reflect current conditions, the basic network calculations are performed, just as they were for the original schedule. And, as before, the results indicate the early and late start and finish times and the slack for each activity that has not been completed, the critical path and the current estimated project completion date. If the results are compared with the original programme, the manager can easily tell whether the project is on schedule. There may be legitimate reasons for a projected overrun of the original project completion date, such as owner requests for additional work or allowable delays due to weather or labour problems. (We should note that the network schedule is an excellent means of illustrating the time effects of such impacts.) But if the project is truly behind schedule, the contractor can employ a variety of methods to try to bring it back into conformance. Among these methods are the following (Mincks and Johnston, 1998): (1) altering the activity sequences so as to compress the schedule by carrying out some activities earlier than originally planned and shortening the critical path; (2) reducing the durations of some of the activities currently underway, by implementing such approaches as increasing crew sizes, changing methods or adding shifts; and (3) attempting to gain time later in the project by re-evaluating future activities. The third method may be the easiest to plan but the most difficult to carry out; it is easy to rationalise that things will get better!

In Figure 6.1, we show the results of a schedule updating for the hypothetical bridge construction project that was the subject of much discussion in Chapter 5. This figure corresponds to Figure 5.13, a bar chart indicating a project start date of 29 March 2004, with a target completion date of 26 November 2004. We assume the project has been underway for 4 weeks and thus we update the schedule as of 26 April 2004. The hatched bars represent the *baseline* or the original schedule as shown in Figure 5.13. The solid vertical line is placed at the 'current' date of 26 April 2002. Actual or anticipated progress is shown in black or dark checkerboard for critical activities (red in Microsoft Project) and grey or light checkerboard for non-critical activities (blue in Microsoft Office). The effects of several changes to the original schedule are shown, including delays in the piling subcontract, the programme finalisation and the cost system. For example, the piling subcontract and cost-system delays have impacted the start of north abutment piling installation and all activities that follow. The most significant change to the original schedule is the delay in procuring reinforcing steel. Because this activity is on the critical path, this 10-day delay will cause the project completion date to be delayed by 10 working days, from 26 November to 10 December 10 2004, as shown on the updated bar chart.

Monitoring is relatively easy! Control, or bringing the project programme back into conformance, is the challenge! Suggestions in the case of our bridge project might include a faster delivery method for the reinforcing steel, overtime to shorten the abutment installation or other activities, concurrent rather than sequential work on both north and south abutments and redesign of the concrete deck and possibly the use of pre-cast concrete or steel decking that might save installation time. Most if not all of these suggestions will cause increases in cost and the manager must weigh carefully the cost and schedule impacts of the various possible courses of action. The information that the project is 2 weeks behind after only 4 weeks is not pleasant, but at least the method has provided the information in a timely manner, when action can still be taken to try to bring the programme back into control.

Cost control

Recall that in Chapter 5 we stated the three purposes of construction cost systems as follows: (1) to provide a means for comparing actual with budgeted expenses and thus draw attention, in a timely manner, to operations that are deviating from the project budget, (2) to develop a database of productivity and cost performance data for use in estimating the costs of subsequent projects and (3) to generate data for valuing variations and changes to the contract and potential claims for additional payments. In this section, we focus on the first of these purposes, the monitoring and control of costs during the construction operations phase, based on whatever system was established during project mobilisation. Two related outcomes are expected from the periodic monitoring of costs: (1) identification of any work items whose actual costs are

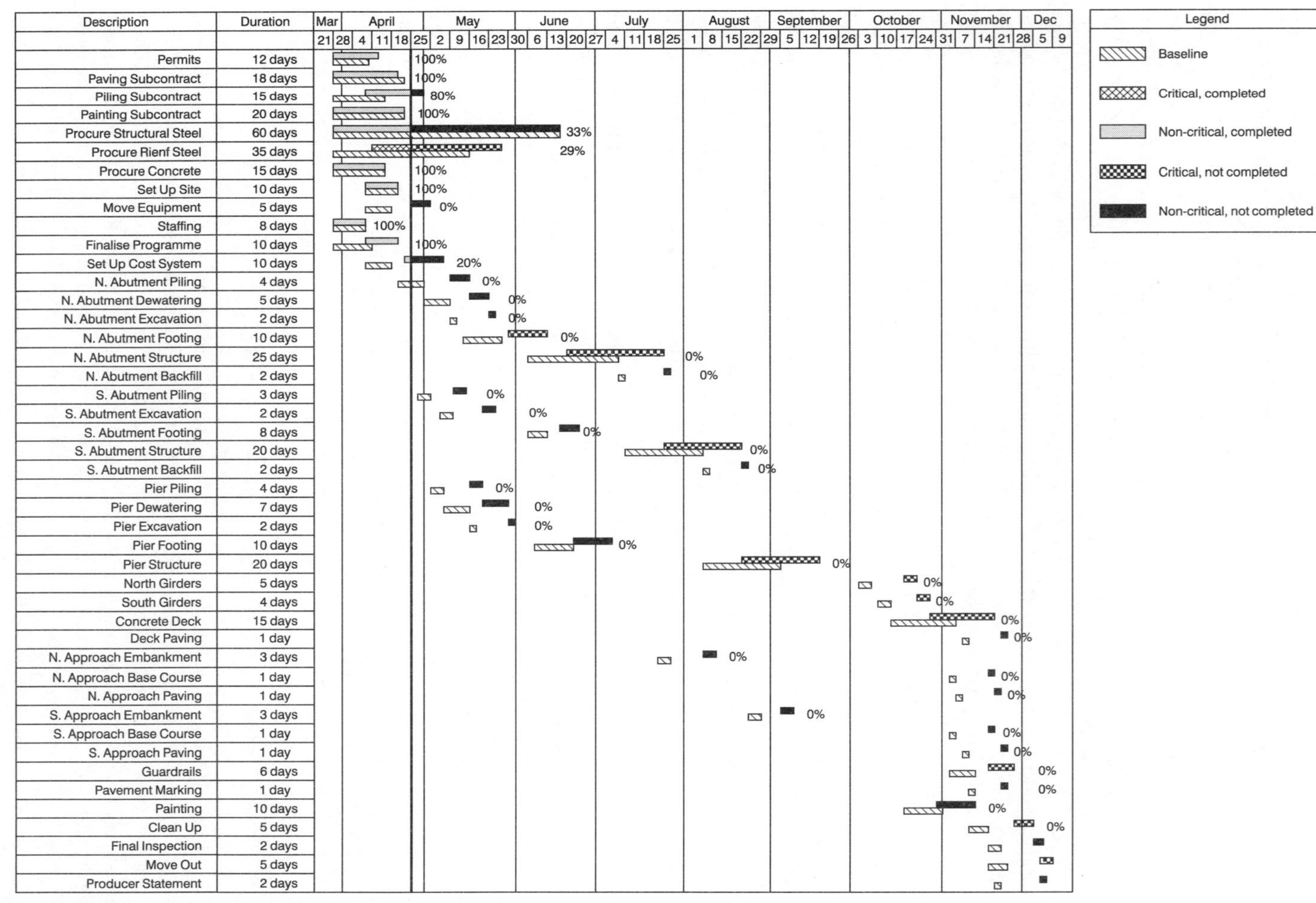

Figure 6.1 Sample bridge project bar chart, updated as at 26 April 2004.

exceeding their budgeted costs, with subsequent actions to try to bring those costs into conformance with the budget and (2) estimating the total cost of the project at completion, based on the cost record so far and expectations of the cost to complete unfinished items.

Data sources

The various elements of construction costs include labour, materials, equipment (plant) and subcontracts, as well as both project and general overhead costs. The control of labour (including the labour portions of overheads) and equipment costs is probably subject to more uncertainty than that for the other elements. Unless an error is made in the cost-estimating process, the cost of materials, subcontracts and non-labour portions of overhead are estimated with sufficient accuracy that, as the project proceeds, they usually do not vary greatly from their budgeted amounts. The successful construction contract, from a cost standpoint, is the one whose labour and equipment costs are controlled in a timely and effective manner.

Labour efforts for each worker are reported by the person's field supervisor on a time card, which shows the number of hours devoted to each work item. Typically, a single daily time card is completed by the foreperson for the entire crew under the supervisor's jurisdiction, with one or two lines for each employee and the numbers of hours are shown under the cost-accounting codes for the respective work items. Two lines per employee are used when overtime work is performed; one line is devoted to regular time and one to overtime, so that different pay rates can be applied. The raw data supplied by the time cards is summarised in two primary ways, a report for each employee that forms the basis of payroll accounting and a report for each work item that leads to summary reports of project costs. Most contractors use some form of automated data processing system to prepare all of these reports, except for the raw time cards that often arrive from the field replete with mud spatters and thumbprints.

The cost of materials must also be charged to each work item. If a certain material is used on only one work item, the charge process is straightforward; the invoice, when received, is coded with the appropriate cost-accounting code and the charge is added to the cost of other materials for that code. An extra step is required if a material is purchased in bulk for use on two or more work items. For example, 300 pieces of formwork plywood may be purchased to make forms for concrete walls, columns, beams and raised slabs. Even if there is only one cost-accounting code for each of these four structural concrete components (and there may be many more than four, to distinguish different locations within the project, for example), some system must be established to charge the various codes with the proper cost from the plywood inventory. In the case of formwork, an additional consideration involves the common practice of reusing the material one or more times and the proper allocation of the cost to each use. An additional complication arises if the material is reused on a future project; the company must have a procedure for charging the two projects with their respective relevant costs.

Equipment (plant) time is reported in much the same way as labour time, with daily equipment time cards assigning hours to the various cost accounting codes. The raw data is transformed into summaries for each code and for each piece of equipment, just as it is for labour. The appropriate hourly charge rate, however, is less straightforward than that for labour, because equipment costs may include ownership costs (first cost and interest), maintenance costs, direct operating costs (fuel, oil and the like) and the cost of transporting the equipment to the site. Usually the cost of equipment operators is included with the labour component of costs rather than the equipment component. If the equipment is rented by the contractor from a rental agency, the costs will be the rental rate, the transportation costs if any and the operating costs. If it is owned by the contractor, a system must be in place to calculate an hourly charge rate that includes ownership and maintenance costs. Often contractors establish separate equipment divisions, either simply on paper or as separate legal entities, to 'rent' their equipment to their various projects.

Subcontract charges are probably the simplest of all to charge to the job. The subcontract establishes the cost, usually on a lump sum basis and if not changed by contract modification all parties know in advance what the costs will be. As they are billed by the subcontractor and paid by the contractor, they are charged to the appropriate cost-accounting code.

The reporting of project overhead costs can be handled in various ways. Whatever the method, the purpose, like that for other types of expense, is to compare actual costs with those estimated in the project budget. If the cost estimate for project overhead was based on a schedule listing the various items of overhead, such as project manager, office rental and cell phones, with their respective quantities (often in months), estimated monthly costs and total costs, the actual expenses can be accumulated using that list as a basis, and a project overhead cost report can be prepared (Clough and Sears, 1994)

In the next section, we discuss the cost status reports that result from the data sources described above.

Cost reports

Cost reports are intended to provide timely information about the status of the project budget so that managers can take any actions indicated to attempt to bring the project into compliance with the original budget. They must be timely, being available soon after the end of the period they cover. They must be as accurate as possible while meeting the requirement for timeliness. They must be simple and concise so that they will be used and understood by those charged with interpreting and acting upon them. Clough *et al.* (2000) offer the following comment:

> Project cost accounting must strike a workable balance between too little and too much detail. If the system is too general, it will not produce the detailed costs necessary for meaningful management control. Excessive detail will result in the objectives of the cost system being obliterated by masses of data and paperwork, as well as needlessly increasing the time lag in making the information available.

We have discussed, in Chapter 5, the configuration of a cost accounting code system, which will be the basis of both the cost estimate for tendering purposes and the cost accounting procedures carried out during project operation. The level of detail provided by the cost accounting scheme ought to balance the considerations of concern in the above quotation. Another consideration is the frequency of report generation. If a report is produced only every several months, the opportunities to correct out-of-control items is diminished and the effort to produce such a report may result in its delay well beyond the end of the reporting period. On the other hand, a report produced every few days will be excessively costly and likely will be neglected as another piece of routine paperwork. Many companies have their project personnel prepare weekly cost reports with information about active work items in summary fashion, followed by monthly reports with more detail (Pilcher, 1992). The amount of detail shown on a cost status report is dependent on the people who will use the report; summary reports by major cost classification will suffice for company executive personnel, while as much detail as possible is important for operating personnel in the field.

No matter what the system, and every company has its own, the periodic cost reports will give, for each item, the quantity and/or percentage completed to date, the cost to date, the estimated cost remaining to complete the item and the cost estimate for the item at completion. Totals of these costs are also calculated for the entire project. To provide a means of comparing actual costs with budgeted costs, the budgeted cost may be shown or the unit cost of work to date can be calculated and compared with the budgeted unit cost. To illustrate, we show a cost report for the concrete wall project whose cost estimate was calculated in Chapter 4. The report is confined to the direct costs only, which totalled US$ 18 534 in our estimate. Whereas some reporting systems would report separate values for the various elements of cost (labour, materials, plant and subcontractors), our report deals only with the totals. We have also simplified the analysis by assuming that no changes have occurred in the originally estimated quantities. Even with these simplifications and a very simple hypothetical project, the general approach to meaningful cost status reporting can be learned from this example.

The spreadsheet in Table 6.1 contains the cost status report for our concrete wall project. We assume that the project has been underway for 2 weeks and that cost data represent the status at the end of the second week. The report is divided into four sections: (1) general information about each work item, (2) quantity information, (3) cost information and (4) unit cost information. A quick study of the quantity and cost information can yield important facts about the general status of the project. We can learn which parts of the work was completed in week one, which was started in week one and completed in week two, which has not yet been started and so forth.

The first three columns – item number, item and unit – are copied directly from the project budget (Table 4.1). Each of the columns in the quantity, cost and unit cost sections will be described in turn.

Quantity section

- The *total* values are the estimated quantities taken directly from the estimate sheet in Table 4.1.
- The *end-of-previous-week* values are the quantities completed during and reported at the end of the previous week, in this case the end of week one.
- The *this-week* values are those quantities installed during the second week, as determined by physical measurements and delivery data.
- The *to-date* values are calculated by adding the *this-week* values to the *end-of-previous-week* values, respectively, for each work item, to determine the total quantities in place as of the reporting date.
- The *% complete to date* for each item is its *to-date* quantity divided by its *total* quantity, with the result expressed as a percentage. Note that this value is a measure of *quantity* completed, not necessarily cost to date as a percentage of estimated total item cost nor time used as a percentage of time allocated to the item.

Cost section

- The *total-budget* values are the estimated cost of each item, taken directly from the estimate sheet in Table 4.1.
- *The end-of-previous-week* values are the actual costs of each item taken from the report for the end of week one.
- The *this-week* values are the actual costs of labour, material, equipment and subcontracts for each item for week two, as reported via time sheets, material invoices and similar documents.

Table 6.1 Concrete wall cost summary and forecast

Item number	Item	Unit	Quantity					Cost							Unit cost			
			Total	End of previous week	This week	To date	% complete to date	Total budget (US$)	End of previous week (US$)	This week (US$)	To date (US$)	Estimate to complete (US$)	Estimate at completion (US$)	Estimated variance at completion (US$)	Budget (US$)	End of previous week (US$)	This week (US$)	To date (US$)
1	Mobilisation	job	1	1	0	1	100	670	700	0	700	0	700	30	670.00	700.00		700.00
2	Clearing and grubbing	m^2	400	350	50	400	100	216	180	40	220	0	220	4	0.54	0.51	0.80	0.55
3	Silt excavation	m^3	12.1	11	1.1	12.1	100	508	435	50	485	0	485	(23)	42.00	39.55	45.45	40.08
4	Rock excavation	m^3	5.3	4	1.3	5.3	100	708	520	305	825	0	825	117	133.50	130.00	234.62	155.66
5	Backfill	m^3	13.8	0	0	0	0	327	0	0	0	327	327	0	23.70			
6	Footing Forms – Place and Remove	smca	12.2	2	10.2	12.2	100	583	95	462	557	0	557	(26)	47.77	47.50	45.29	45.66
7	Footing reinforcement	tonne	0.138	0	0.138	0.138	100	155	0	164	164	0	164	9	1124.42		1188.41	1188.41
8	Footing concrete	m^3	2.4	0	2.4	2.4	100	463	0	463	463	0	463	0	193.05		193.08	193.08
9	Wall forms – place and remove	smca	151	10	63	73	48	10 102	780	4515	5295	5590	10 885	783	66.90	78.00	71.67	72.53
10	Wall reinforcement	tonne	1.335	0	0.455	0.455	34	1 501	0	487	487	1014	1 501	0	1124.40		1070.33	1070.33
11	Anchor bolts	ea	40	0	10	10	25	337	0	95	95	230	325	(12)	8.42		9.50	9.50
12	Wall concrete	m^3	15	0	0	0	0	2 525	0	0	0	2525	2 525	0	168.30			
13	Cleanup and move-out	job	1	0	0	0	0	440	0	0	0	440	440	0	440.00			
							47.4	18 534	2710	6581	9291	10 126	19 417	883				
				Budgeted cost of work performed					2620		8783							

- The *to-date* values are calculated by adding the *end-of-previous-week* and *this-week* values for the respective items, to find the total cost to date for each item.
- The *estimate-to-complete* values represent the judgment of project management staff for the cost required to complete the remaining work on the item. If an item is 100% complete, of course, its *estimate to complete* will be US$ 0. For an item that is partially complete, it is possible to assume that the cost to complete the item will be in proportion to its cost to date and its current completion percentage, so that, for example, if an item is 25% complete and has cost US$ 4000 so far, its remaining 75% will cost (0.75/0.25)(US$ 4000) = US$ 12 000. However, it is essential that management personnel intervene in this estimate to judge all factors that will affect productivity on the remaining work, rather than this estimate being calculated automatically. Seldom will a simple proportionate estimate be appropriate. If an item has not been started, the *estimate to complete* is likely to be the *total budget*, although information about requirements or methods received after preparation of the estimate may require that this value be revised at this time.
- The *estimate-at-completion* values are calculated by adding the *to-date* and *estimate-to-complete* values. These values represent the best current estimates of the total cost of each item.
- The *estimated variance at completion* is the difference between the *estimate at completion* and the original *total budget*. In our system, a positive variance is unfavourable, meaning that the *estimate at completion* is greater than the *total budget*. Be aware that there is no general agreement among various practitioners and authors as to the sign (+ or –) of a variance; some systems calculate variance by subtracting *estimate at completion* from *total budget*, rather than the reverse as we have done and thus define a positive variance as favourable (Clough et al., 2000; Pilcher, 1992). Any system must be consistent in the manner in which variances are calculated and the user of variance reports must understand how to interpret positive and negative variances. Note in our example that, among the items partially complete at the end of week two, wall forms is expected to overrun the budget (positive variance), anchor bolts is expected to underrun the budget (negative variance) and wall reinforcement is anticipated to be even with its original budget.

Each cost column also shows a total. Among others, we note the total direct cost budget of US$ 18 534 and an estimate at completion for all direct costs of US$ 19 417, with a resulting unfavourable *variance at completion* of US$ 883. We also note the actual direct cost of all work to date is US$ 9291.

Unit cost section

In each of the four columns in the unit cost section, the unit cost is calculated by dividing the cost value by its respective quantity value. For example, the budgeted unit cost for clearing and grubbing is US$ 261/400 m^2 = US$ 0.54 m^2, whereas the unit cost for this week for this same item is US$ 40/50 m^2 = US$ 0.80 m^2. Unit cost information identifies trends that might otherwise be hidden. Note, for example, that the unit cost of wall forms at the end of week one (US$ 78.00) was well above the budget (US$ 66.90); week two experience resulted in a lower unit cost (US$ 71.67), bringing the unit cost to date down to US$ 72.53, still above the budget.

The two values at the bottom of Table 6.1 are calculated to indicate the value of the work completed so far based on the total budget for each item or the *budgeted cost of work performed*. Often this amount is called the project's *earned value*. It is useful in determining how much the

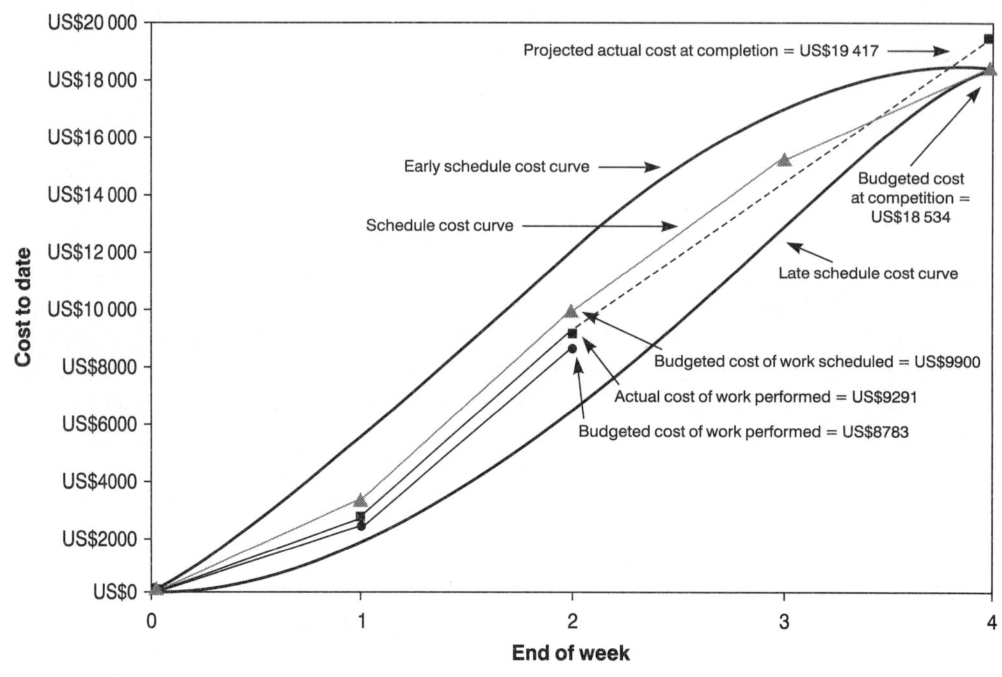

Figure 6.2 Concrete wall cost control curve.

completed work should have cost and is also useful as one measure of project progress, as we shall see. The *budgeted cost of work performed* is a simple weighted average, found by multiplying each item's total budget by its completion percentage and then summing these products for all items. The *earned value* of mobilisation completed so far, for example, is (100/100)($670) = $670, because this item is 100% complete. For Wall Forms, the Earned Value is (48/100)(US$ 10 102) = US$ 4849. If we perform these calculations for all items and find their sum, it is US$ 8783, the 'worth' or 'value' or *budgeted cost of work performed* for the entire project as of the end of the second week. Note that we have also included a corresponding value, US$ 2620, for the end of week one.

Finally, the report shows an overall completion percentage of 47.4%, found by dividing the *budgeted cost of work performed* (US$ 8783) by the *total budget* (US$ 18 534) and expressing the result as a percentage.

A report similar to that in Table 6.1 could be prepared for each of the elements of cost or they could all be shown in detail on a single report. Or a report with less detail might be prepared, perhaps for upper management, including only the completion percentage, budget, estimate at completion and variance for each item. The possibilities are numerous; this example attempts to set forth some basic ideas about the preparation and use of such documents.

We shall take some of the information in Table 6.1 one step further and illustrate the use of cost–time curves for displaying cost status information. Figure 6.2 contains five curves. In black we see an envelope consisting of two curves, an early start schedule cost curve and a late start schedule cost curve, similar to Figure 5.21 in the previous chapter, based on a network schedule analysis not shown here. Recalling that the contractor has flexibility on scheduling non-critical activities, we show, in grey, a *scheduled cost curve*, based on a firm plan for the schedule of all activities; some non-critical activities are planned to take place at their early times, while others are planned for later than their early times. This curve represents the cumulative flow of costs over time if the project proceeds in accordance with that firm plan.

Against the scheduled cost curve in Figure 6.2, we plot two other pieces of information. First, (black line with squares) we show the actual costs to date; in the case of week two, the US$ 9291 is taken directly from our summary and forecast in Table 6.1. Plotting actual costs to date against the scheduled cost curve provides an interesting comparison, but such a comparison does not provide much information about the cost or schedule status of the project. Because the actual cost of work performed (ACWP) is less than the budgeted cost of work scheduled (BCWS) (US$ 9900 from Figure 6.2), does that mean the project costs are under control or the project is on schedule? Is the project under control because the ACWP falls within the early and late schedule cost curves? The answers are that we cannot answer these questions with only the information presented so far, because we do not know how much it *should have cost* to do the work completed to date. Furthermore, we could have made much progress and spent lots of money on non-critical activities and neglected critical ones, resulting in a behind-schedule project whose schedule status can only be found by updating the network schedule. To know how much the project should have cost so far, we turn to the final figure, the *budgeted cost of work performed* (BCWP), a value of US$ 8783 as explained above and plotted on the curve (black line with circles) in Figure 6.2. This is the project's earned value so far, the amount that the work items that have made progress to date should have cost, in total, if their actual unit costs for the quantities installed were exactly equal to the unit costs in the estimate. We can calculate two helpful variances, a cost variance and a schedule variance, where

cost variance (CV) = ACWP – BCWP = US$ 9291 – US$ 8783 = +US$ 508.

schedule variance (SV) = BCWP – BCWS = US$ 8783 – US$ 9900 = –US$ 1117.

The sum of these two variances is called the total variance, where

total variance (TV) = CV + SV = (ACWP – BCWP) + (BCWP – BCWS) = ACWP – BCWS, which is simply the difference between the two points on the two lines (black line with squares and black line with circles) in Figure 6.2 or US$ 9291 – US$ 9900 = US$ 508 – US$ 1117 = –US$ 609, a figure that is not very helpful in assessing the project's status.

The *cost variance* for our project at the end of the second week is positive, meaning that the actual costs are higher than they should have been for the same amount of work; this positive variance is unfavourable and indicates that action needs to be taken. We knew that the cost status was not good, because our forecast in Table 6.1 had predicted an actual cost at completion of US$ 19 417, as plotted at week four in Figure 6.2, which led to an estimated variance at completion of +US$ 883.

The *schedule variance* is a less helpful value than the cost variance, because schedule status is so dependent on progress of critical activities. The fact that our schedule variance is negative, meaning that the budgeted cost of the work that was scheduled to have been performed by the end of the second week is greater than the value of the work that has actually been performed, is an indication that the project schedule is in trouble, although the –US$ 1117 value does not have any good meaning related to the project timeframe. (It is probably misleading to locate the plot of the projected actual cost at completion (US$ 19 417) at the end of week four, as we now expect the project to take longer than that, pending an update of the network schedule.)

Thus, our *earned value analysis*, using the BCWS, has led to an indication that the project is in trouble from both a cost and a schedule standpoint. If we had confined our analysis to a comparison of the ACWP with the BCWS by calculating only the total variance, these additional indicators would not have been available. Other types of variances can be calculated, such as those that measure performance of each individual work item or those about one category of expense such as labour, but the sample shown above should be sufficient to give a general understanding of variance analysis as applied to construction cost control.

Table 6.1 is an example of a spreadsheet-based cost control document that an individual contractor might develop specifically for its own use. Other options include software such as Peachtree Complete Accounting 2002® (Peachtree Software Inc., 2002), described in Chapter 5. In Table 6.2, we show a cost status report for our concrete wall project, as generated by this software. Note that this report, when compared with Table 6.1, shows (1) the same estimated and actual costs as of the report date, (2) considerably less detail overall and (3) breakdowns of item costs by labour, material, equipment and subcontractor. As one would expect, most cost accounting packages allow considerable flexibility in the format and amount of detail shown in such reports.

Other aspects of financial control

An organisation's accounting information is used for three primary purposes (Hendrickson and Au, 1989).

- Internal reporting to the company's project managers and other operating personnel for detailed, frequent planning, monitoring and control, of the type we have discussed in this section.
- Internal reporting to the company's more senior managers for aiding strategic, long-range planning.
- External reporting to shareholders, government agencies, regulatory bodies, lenders and other outside parties.

Preparation and analysis of external reports is referred to as *financial accounting* and involves such documents as balance sheets (also called statements of financial position), income statements (also called profit-and-loss statements) and cash flow statements. These important company-wide documents are beyond the scope of this book, but the person involved in company management, as opposed to project management, will quickly become familiar with them. *Cost accounting* (also called *managerial accounting*) comprises the sorts of analyses and reports covered earlier in this section that assist the organisation internally to determine status and make decisions.

Costs and revenues associated with projects are reflected in both the financial and cost accounting systems of the construction organisation. The accounting system includes the *general ledger*, where all expense transactions are recorded, *accounts payable* (creditors) and *accounts receivable* (debtors) *journals*, which record invoices as they are received and when paid and billings to clients and their subsequent payment, respectively, inventory records that provide up-to-date information on material availability and job cost ledgers that summarise costs of each project, arranged by cost account number. It can be seen that the company's financial accounting system basically records summary revenue and expense information from each of its

Table 6.2 Cost status report for concrete wall project, produced by Peachtree Complete Accounting 2002

Bennett Construction
Cost status report
As of Apr 23, 2004

Filter Criteria includes: 1) IDs from 2004-611 to 2004-611. Report order is by ID.

Job description Job ID For customer Percent complete	Phase ID	Phase description	Cost code description	Est. exp.	Act. expenses	Diff. expenses
Concrete wall	1.505	Mobilisation	Equipment	120.00	150.00	30.00
2004-611			Labour	550.00	550.00	
Meridian shippers	1.710	Cleanup and move-out	Equipment			
47			Labour			
	2.110	Clearing and grubbing	Equipment	84.00	86.00	2.00
			Labour	132.00	134.00	2.00
	2.220	Silt excavation	Subcontractor	508.20	485.00	–23.20
	2.221	Rock excavation	Subcontractor	707.55	825.00	117.45
	2.224	Backfill	Subcontractor			
	3.115	Continuous edge forms	Labour	382.10	350.00	–32.10
			Material	200.69	207.00	6.31
	3.135	Wall formwork	Labour	3591.60	4020.50	428.90
			Material	1292.10	1274.50	–17.60
	3.211	Footing reinforcement	Labour	69.98	81.00	11.02
			Material	85.19	83.00	–2.19
	3.213	Wall reinforcement	Labour	230.73	207.00	–23.73
			Material	280.87	280.00	
	3.310	Footing concrete	Equipment	47.40	50.00	2.60
			Labour	150.00	150.00	
			Material	265.92	263.00	–2.92
	3.318	Wall concrete	Equipment			
			Labour			
			Material			
	5.050	Anchor bolts	Labour	44.00	44.00	
			Material	40.20	51.00	10.80
2004-611	Total			8782.53	9291.00	508.47

Courtesy Peachtree Software Inc. (2002)

projects, in addition to general company expenses, for overall company use, whereas the cost accounting system is oriented more towards individual projects. Among the issues of special interest to construction organisations whose projects span more than one calendar year is whether the financial accounting system will record revenues on a *completed contract* basis, under which income for the entire project is recorded in the year when the project is completed or the *percentage of completion* method, wherein income is recorded in proportion to the relative amount of work completed each year.

Helpful discussions of this issue and many others related to construction accounting and financial management are found in Palmer et al. (1995) and Hendrickson and Au (1989).

Quality management

At numerous places in this book, including the introductory material in Chapter 1, we have emphasised the triple constraint, or the three primary objectives of schedule, cost and quality, that the project manager must balance in carrying out the project. If monitoring and control are about schedule and cost, they are also surely about the management of quality. Thus, this section is devoted to responsibilities the contractor must fulfil in managing quality on the project, as well as some other more general information about quality management in construction.

First, what is quality? There are many definitions. The Institute of Quality Assurance (2002) says this about quality:

> In its broadest sense, quality is a degree of excellence: the extent to which something is fit for its purpose. In the narrow sense, product or service quality is defined as conformance with requirement, freedom from defects or contamination, or simply a degree of customer satisfaction.

The American Society of Civil Engineers (ASCE), in its much-lauded *Quality in the Constructed Project: A Guide for Owners, Designers and Constructors* (American Society of Civil Engineers, 2000) defines quality as 'the fulfilment of project responsibilities in the delivery of products and services in a manner that meets or exceeds the stated requirements and expectations of the owner, design professional and constructor'. If we utilize this understanding of quality, then we see that the contractor's task in managing quality is to assure compliance with the technical requirements of the project, as described in the contract documents, through a series of steps that plan, execute, monitor and control the physical aspects of the work. The approach, if viewed in this way, is thus parallel to the management of the project's programme and budget.

The contractor's activities in assuring quality are said to involve both quality assurance and quality control. Although these terms are sometimes used synonymously, it is helpful to distinguish between them. Quality assurance may be defined as 'planned and systematic actions focused on providing the members of the project team with confidence that components are designed and constructed in accordance with applicable standards and as specified by contract.' Quality control, on the other hand, is 'the review of services provided and completed work, together with management and documentation practices, that are geared to ensure that project services and work meet contractual requirements' (American Society of Civil Engineers, 2000). Thus a contractor's *quality assurance* programme includes all activities that 'assure quality' including (but certainly not limited to) selection of subcontractors, training of the workforce, use of proper methodology and the various testing activities. *Quality control* is one aspect of quality assurance, the aspect that (1) 'reviews' work done to determine whether it conforms to requirements (the *monitoring* function) and (2) brings it into conformance if the examination identifies defects (the *control* function).

Total quality management has taken on increasing importance throughout all industries worldwide during the past twenty years. Many contractors subscribe to its philosophy and their quality assurance/quality control programmes derive directly from its basic tenets. Total quality management can be defined as 'a cooperative form of doing business that relies on the talents and capabilities of both labour and management to continually improve quality and productivity using teams with the ultimate goal of delighting both internal and external customers' (Bennett, 1996). Although total quality management is often thought of as a series of statistical techniques for measuring and controlling quality, it really includes a management philosophy and approach that goes well beyond these techniques, as the definition suggests.

Although we have chosen to treat quality management mainly in this chapter on project operations, it must be an issue throughout the project's life cycle. Indeed, the entire process is oriented toward assuring a quality product that conforms to user requirements, is free of defects and satisfies the customer, as the Institute of Quality Assurance definition given above suggests. We have already considered the initial programming and planning, the various steps in the design process, the development of technical specifications, various reviews including constructability and value engineering and the selection of the contractor and subcontractors; we will consider activities related to project termination and closeout in Chapter 7. Even though they do not occur during the project operations phase, these efforts, if performed effectively, can be thought of simply as elements of good, sensible management practice that contribute to a quality project. Specifications, for example, are constantly in use by the contractor. As we described in Chapter 3, they specify both work and materials and include general specifications and codes promulgated by various industry and professional organisations as well as those developed specifically for the project. Whereas traditionally specifications have been *descriptive*, stipulating the process the contractor must follow, *performance specifications* now govern many construction operations. The method utilised to achieve the required performance is left to the contractor. In the case of Portland cement concrete, for example, a descriptive specification stipulates such things as a required mix design, slump and aggregate size and quality; in contrast, a performance specification calls for strength, durability, colour and other end-result requirements and the contractor would have some choices in means to achieve the required performance level.

The Fédération Internationale des Ingénieurs-Conseils (FIDIC) has identified decreasing quality of construction as a worldwide problem. A 2001 report suggests that 'the squeeze on costs, both of construction and supervision, is having an adverse effect on construction quality, as can be predicted' (Fédération Internationale des Ingénieurs-Conseils, 2001). One interesting finding is that such modern delivery systems as design–build, which shift risk and responsibility from the owner to the contractor, may lead to reductions in both quality and cost effectiveness, because they reduce or eliminate the role of the owner's engineer. The report calls for such measures as more appropriate contract provisions, more rigorous quality processes and the use of consulting engineers independent from the project to serve as verifiers (Hirotani, 2001).

One means of attempting to alleviate construction quality problems and assuring quality performance by construction contractors is certification by the International Organisation for Standardisation. That organisation uses the designation ISO to refer to all of its standards, the term being derived from the Greek *isos*, meaning *equal*; thus its standards apply equally or universally for a given situation or condition. The ISO standards related to quality are known as the ISO 9000 series of standards. The ISO 9001 standards are for organisations that design, produce, install and service products, while the ISO 9002 standards apply to organisations that only produce, install and service products. The standards apply to all kinds of organisations. In the realm of construction, ISO 9002 certification is sought by contractors who do not perform design services, while design–build contractors fall under the ISO 9001 guidelines.

The ISO 9000 certification process requires an organisation to develop a quality system that meets guidelines in 19 (in the case of ISO 9002) or 20 (for ISO 9001) specified topics, including management responsibilities, contract review requirements, purchasing requirements, control of quality records and training requirements. After the quality system has been developed and implemented, an accredited external auditor evaluates the system's effectiveness. If found worthy, the system is approved and registered as *ISO 9000 certified*. Many owner organisations require their contractors to be ISO certified or to meet similar standards. Although improved quality is by no means a certain result, the rigorous process required to obtain ISO 9000 certification is an important and effective means of identifying and rewarding top-notch contractors. Increasingly, contractors around the world are seeking this designation. Publications by Hoyle (2001) and Chung (1999) are useful guides for contractors contemplating ISO 9000 certification.

Even before the project begins, the contractor is expected to have its company-wide quality assurance/quality control programme in place. *Quality in the Constructed Project: A Guide for Owners, Designers and Constructors* (American Society of Civil Engineers, 2000) lists the following elements as appropriate for such a programme:

- Recruiting and assigning a skilled workforce
- Quality control organisation
- Project progress schedule
- Submittal schedule
- Inspections
- Quality control testing plan
- Documentation of quality control activities
- Procedures for corrective action when quality control and/or acceptance criteria are not met.

Note that these procedures comprise an essential set of 'elements of good, sensible management practice', which we suggested above to be the foundation for effective quality management.

During the project operation phase, the contractor is usually required by the contract to furnish and abide by a quality plan. The Asian Development Bank (2000), in its guidelines for contractors doing business in India, advises

> The bidding documents require the contractor to submit a quality assurance plan, and show proof of having met with quality specification under previous contracts. The bid prices must reflect the quality requirements under ADB financed projects and include for extra staff and equipment for quality assurance . . .

Among the project-specific requirements of the contractor's quality assurance/quality control programme are the following (American Society of Civil Engineers, 2000):

- Use of qualified subcontractors
- Inspection, control, and timely delivery of purchased materials, equipment, and services

- Identification, inventory, and storage of materials, parts, and components pending incorporation into the project
- Control of measuring and test equipment
- Segregation and disposition of nonconforming materials, equipment, or components
- Maintenance of records specified by contract and required by the contractor's QA/QC programme to furnish evidence of activities affecting quality.

The purpose of the on-site testing and inspection activities required by the contract is to protect the public, the owner, the design professional and the contractor. Testing is normally conducted by independent outside agencies for such parts of the project as soils, Portland cement and asphalt concrete, welding and bolt torque, with these same organisations often inspecting horizontal and vertical alignment and the placement locations of reinforcing steel. Public officials representing the local code authority or municipal building office can be expected to visit the site to assure compliance with local building codes. These inspectors are not authorised to direct contractor personnel, but usually do have the power to issue *stop work orders* in the case of severe infractions.

Transit New Zealand is that country's national public agency responsible for the design, construction, operation and maintenance of public roads. For a contract deemed to have a 'high' quality assurance level, its *Quality Standard* (Transit New Zealand, 1995) requires contractors to become certified under a system based upon, but somewhat less formal than, the ISO 9000 process. In addition, such a contract requires a contract quality plan for the specific project. In the case of projects with 'normal' quality assurance levels, a contract quality plan is required, plus a simple quality system during the quality period, but certification is not required. The quality system elements in which the contractor's organisation must demonstrate competence in order to become certified include (1) management responsibility, (2) quality system, (3) contract review, (4) document and record control, (5) purchasing and subcontracting, (6) control and inspection of the work, (7) non-conforming work and quality system improvement, (8) internal quality audits, (9) training and (10) safety and resource management. As one example of the requirements, we show one of the sections of the purchasing and subcontracting element:

> 5.4 Incoming Materials
>
> All materials purchased by the contractor or alternatively supplied by the Principal shall be checked for compliance with the specified requirements prior to incorporation in the works (preferably on receipt) and verification shall be noted on the relevant 'Inspection Checklist'.
>
> All materials shall be handled and stored in a manner that prevents damage or deterioration and verification that none has occurred shall be noted on the relevant 'Inspection Checklist'.

The contractor is required to set forth a plan for meeting this stipulation, including samples of its proposed inspection checklists, as one part of its application for certification.

The contractor's quality performance may be evaluated and rewarded or penalised under a scheme originated in Singapore in 1989 (Building and Construction Authority, 2002). This Construction Quality Assessment System (CONQUAS), which has been applied to more than 1700 public and private building projects in Singapore, has resulted in measured improvement in construction quality in that country and is being used elsewhere, including the UK (Building Research Establishment, 2002). The assessment looks at four components: architectural works,

Table 6.3 Rates of reportable non-fatal injuries by industry, UK, 1999–2000

Industry	Rate[1]
Construction	2530
Agriculture	2520
Transport, storage and communication	2140
Manufacturing	2110
Public administration and defence	1550
Health and social work	1470
Distribution and repair	1430
Extraction and utility supply	1400
Hotels and restaurants	1400
Other community, social and personal service activities	1200
Education	810
Finance, real estate and business activities	570

[1] Reportable non-fatal injuries per 100 000 workers.
Adapted from Health and Safety Commission (UK), 2001.

mechanical and electrical works, structural works and whole life performance; within each component, various assessment items are evaluated through site inspections throughout the construction process. In Singapore, a bonus/default scheme applies to public-sector projects above a certain monetary value, wherein the contract value is adjusted depending upon the results of the quality assessment.

Safety management

Although they are not part of the project manager's 'big three' objectives of schedule, budget and quality, the monitoring and control of safety and environmental impact are major issues in all construction operations. If we think of time, cost and technical performance as objectives to be optimised, or at least balanced, then we could consider workplace safety and the impact of operations on the environment as constraints on the attainment of those objectives. If safety were ignored, projects might be completed more quickly; if the environment were not protected, costs might be lower, at least in the short run. But these constraints are very real in modern construction management and so we deal with each in turn, by providing some background information and then discussing important elements of a contractor's on-site monitoring and control programme.

Unfortunately, construction is a hazardous business. Despite active safety programmes on the part of contractors, unions and owners and many legal regulations, accident experience continues to place the construction industry among the most accident prone of all types of work. Table 6.3 summarises one piece of information from the UK, showing the reportable non-fatal injury rate for all industries for the 1999–2000 reporting year (Health and Safety Commission (UK), 2001). For construction, the rate is 2530 such injuries per 100 000 workers, highest of all industries and slightly greater than agriculture at 2520. Although there are other ways in which to cast such numbers (some subcategories of manufacturing (not shown) such as quarrying and

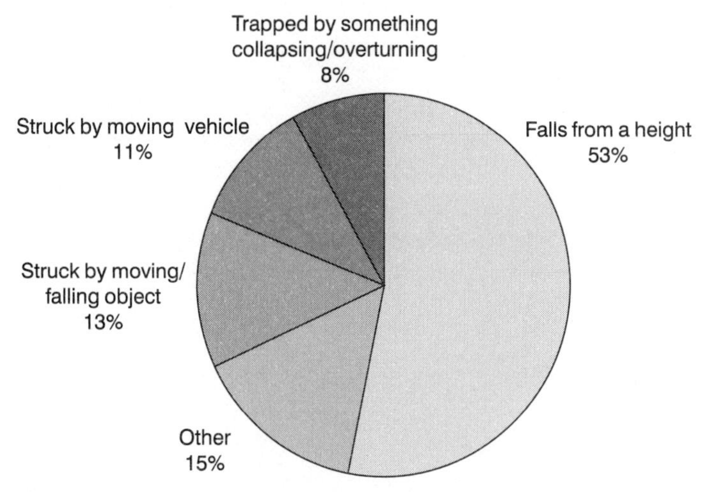

Figure 6.3 Types of accidents causing fatal construction accidents, UK, 1996/1997–2000/2001 (total = 422).

wood manufacture have higher rates than construction), it clear that construction non-fatal injury experience in the UK is less than exemplary. Similar data for fatal accidents lead to the same conclusion; data from other countries and regions indicate that safety is a major concern throughout the worldwide construction industry.

If the contractor intends to design a safety programme to minimise accidents, it is important to know the types of accidents that occur most frequently. In Figures 6.3 and 6.4, we show summaries of five recent years of construction accident statistics from the UK (Health and Safety Commission (UK), 2001). Of the 422 fatalities, more than half were caused by falling from a height, while other major causes were being struck by a moving or falling object, being struck by a moving vehicle and being trapped by something collapsing or overturning, as shown in Figure 6.3. For the 20 030 major non-fatal injuries in Figure 6.4 (where 'major' includes fractures, amputations, loss of sight and similar traumatic injuries), over half were caused by falling, tripping or slipping from a height or on the same level, while other major causes were from being struck by moving or falling objects and from handling, lifting or carrying.

The cost of accidents includes both their direct and their indirect costs. The term *cost iceberg* has been used to refer to the fact that the direct costs are often only a small fraction of the indirect costs (Rowlinson, 2000). A common suggestion is that the ratio of indirect to direct costs is 4:1, although various studies place the ratio at between 2:1 and 11:1, depending in part on the cost of medical care. Direct costs are those that are reimbursed by workers' compensation or other insurance, while indirect costs are other costs that impact the project budget due to the injured worker being off the job, the assistance given by others, the impact on crew productivity, damaged materials and equipment, supervisory assistance and the typical slowdown by many other workers on site from simply observing the accident and its aftermath (Hinze, 2000). These indirect costs have been calculated as averaging about US$ 15 000 for an accident that results in lost workdays and subsequent restricted work. The total of the contractor's costs related to safety can be thought of as the sum of (1) the costs of investing in safety, such as training, incentive programmes, staffing, drug testing and protective equipment and (2) the costs of lack of safety, or accidents. As investment in safety increases, the direct and indirect costs of accidents can be expected to decrease and vice versa. For the contractor, the economic challenge is to balance

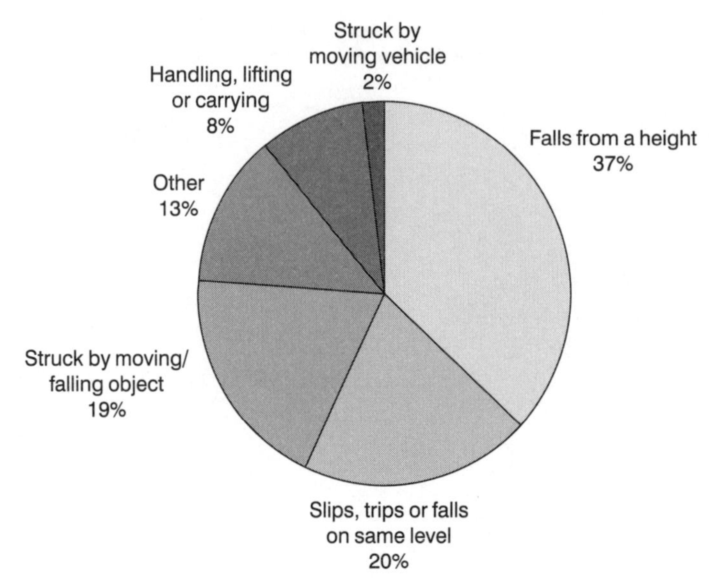

Figure 6.4 Types of accidents causing major non-fatal construction accidents, UK, 1996/1997–2000/2001 (total = 20 030).

these elements so that their sum is low, while meeting any applicable laws and regulations. To spend nothing on safety, with resulting high accident costs, is clearly unacceptable. At the other extreme, there is some expenditure on safety above which extra expense will result in minimal and unjustified decreases in accident costs.

The nature of the construction site is such that many hazards can exist. In Table 6.4 we list some of the common hazards. With the exception of ice and snow and the simplest of projects,

Table 6.4 Common construction site hazards

Ladders
Lifting equipment
Openings, holes and platform edges
Rigging, ropes and chains
Scaffolding
Trenches and other excavations
Heavy equipment and trucks
Machinery (woodworking, pipe fabrication, compaction)
Motor vehicles
Hazardous substances (asbestos, carcinogens)
Littered workplace (debris, liquids, tools)
Electricity
Fire (open flames, combustible liquids)
Welding and cutting
Ice, snow and mud
Noise

Based on Ahmed *et al.* (2001) and Halpin and Woodhead (1998).

all of the other hazards are likely to be present on the site, wherever in the world it is located, depending perhaps upon the type of project. An exercise at the end of this chapter asks the reader to think about the types of accidents and resulting injuries that could occur from each hazard and the action the contractor might take to prevent each.

On the construction site, the contractor is legally required to operate a safe workplace, from both a contractual standpoint and due to the various laws and regulations promulgated by public authorities. (Even if, for some strange reason, there were no contract, the laws and regulations would still apply.) Article 10 of the *General Conditions of the Contract for Construction* (American Institute of Architects, 1997) requires the following:

> 10.1.1 The Contractor shall be responsible for initiating, maintaining and supervising all safety precautions and programs in connection with the performance of the Contract.

Among the detailed provisions is the following:

> 10.2.1 The Contractor shall take reasonable precautions for safety of, and shall provide reasonable protection to prevent damage, injury or loss to:
>
> 1 (employees and other persons . . .)
> 2 (the work and materials and equipment . . .)
> 3 (other property at the site . . .).

While the word *reasonable* at two places in 10.2.1 may be subject to some interpretation as stated, contractors should be aware that owners often insist that the words *all possible* be substituted. Whatever the wording, it is clear that the contractor is the key party responsible for construction site safety. Far beyond the scope of this book is the interesting question of the contractor's responsibility for the safety of subcontractor employees, material delivery persons and others; suffice it to say that different courts and arbitration panels have given different rulings!

The contractor's on-site operations are also subject to various safety laws and regulations. In the UK, for example, the Health and Safety Executive (2000a) enforces the *Construction Health, Safety and Welfare Regulations of 1996*. The Executive makes clear the multi-party responsibility for enforcing the regulations, including employees themselves:

> The main duty-holders under these Regulations are employers, the self-employed and those who control the way in which construction work is carried out. Employees too have duties to carry out their own work in a safe way. Also, anyone doing construction work has a duty to cooperate with others on matters of health and safety and report any defects to those in control.

The regulations cover the following topics:

- Safe place of work
- Precautions against falls
- Falling objects
- Work on structures
- Excavations, cofferdams and caissons

Table 6.5 Sample entry from construction health and safety checklist

Roof work
Are there enough barriers and is there other edge protection to stop people or materials falling from roofs?
Do the roof battens provide safe hand and foot holds? If not, are crawling ladders or boards provided and used?
During industrial roofing, are precautions taken to stop people falling from the leading edge of the roof or from fragile or partially fixed sheets which could give way?
Are suitable barriers, guard rails or covers, etc. provided where people pass or work near fragile material such as asbestos cement sheets and rooflights?
Are crawling boards provided where work on fragile materials cannot be avoided?
Are people excluded from the area below the roof work? If this is not possible, have additional precautions been taken to stop debris falling onto them?

From Health and Safety Executive (1999).

- Prevention or avoidance of drowning
- Traffic routes, vehicles, doors and gates
- Prevention and control of emergencies
- Welfare facilities
- Site-wide issues
- Training, inspection and reports.

In the UK, the Health and Safety Executive is invested with enforcement and prosecution powers in the case of violations. A sampling of recent convictions includes an accident in which a worker died from a fall from a ladder (£18 000 fine), a crushed hand when a temporarily suspended wall collapsed (£2000), a scaffolding that lacked sufficient guards and boards (£1000), an improperly protected asbestos removal operation (£120 000) and failure to provide any welfare facilities on a demolition site (£450) (Health and Safety Executive, 2000b). Unfortunately many of these convictions result from investigations following accidents, rather than from site visits that might prevent such occurrences.

Among the Health and Safety Executive's publications is an excellent *Construction Health and Safety Checklist* (Health and Safety Executive, 1999). A sample entry for one type of work is shown in Table 6.5.

In the USA, the *Occupational Safety and Health Act* (OSHA) requires many facets of compliance by employers; every employer involved in interstate commerce must provide a workplace free of recognised safety and health hazards and must comply with various OSHA standards. OSHA provides substantial penalty powers, with fines of up to US$ 10 000 and/or imprisonment for proven wilful violations resulting in the death of an employee (Clough and Sears, 1994).

As in the case of quality management, it is good practice to have an overall company safety programme and policy, as well as a site safety programme for each project. With the objective of identifying specific jobsite hazards and teaching employees to conduct their work in a way that minimises injury risk, the company-wide programme ought to include the following (based on Clough and Sears, 1994):

- safety policy statement
- company safety organisation
- safety training and personal protection
- first aid training
- fire protection
- safety record keeping
- jobsite inspection
- accident and hazard reporting.

The contractor's chief executive must be the organisation's leader in terms of safety, first by issuing a policy statement that (1) conveys the idea that safety is the workplace's highest priority, (2) states who has authority and accountability for providing a safe work environment and (3) requires adherence to all provisions in the company's safety manual (Mincks and Johnston, 1998) and then by acting in ways that demonstrate their commitment to the safety policy and programme.

The site safety programme must include the development of a reasonable plan and the implementation of that plan as soon as field operations are begun. It is founded on indoctrination and training, inspections, briefings and written documentation and reports. A 'baker's dozen plus one' of specific elements, based in part on Clough and Sears (1994) and Hislop (1999), includes the following.

1 Assignment of prime responsibility for jobsite safety to a top-level field supervisor.
2 Provision of a safety indoctrination to all new personnel prior to their starting work, including a set of written work rules similar to those shown in Figure 6.5.
3 Availability of suitable first aid facilities and personnel trained in first aid; large projects will have a staffed clinic, whereas small jobs will provide first aid kits and trained supervisory personnel.
4 Required use of personal protective clothing and equipment, including hard hats, goggles, respirators, and harnesses, depending upon the nature of the task.
5 Installation and use of fall protection measures, including safety netting and barricades, as appropriate (see Figures 6.6 and 6.7).
6 Provision of fire-fighting equipment and materials that are appropriate and easily accessible.
7 Periodic safety presentations and discussions for work crews on such topics as material handling and equipment operation, including feedback and suggestions from the workers.
8 Regular meetings of supervisors and project management personnel to review job safety experience and the programme's effectiveness.
9 Operation of periodic job safety inspections, conducted by project management personnel and safety experts, to identify hazards and make plans to remedy them.
10 Insistence that regular equipment maintenance include safety inspections.
11 Reporting and investigation of all accidents promptly, with recommendations and actions to prevent their reoccurrence.
12 Insistence on excellent jobsite housekeeping, with proper storage areas, immediate cleanup of spilled substances and regular sweeping and trash removal as required.
13 Utilisation of posters, warning signs and notices of the project safety record, to remind workers of the special importance of safety on the job (see Figures 6.8 and 6.9 for examples)
14 Establishment of an incentive programme to encourage safe work.

XYZ Contractors welcomes you to this project. Our company has set jobsite safety as its highest priority. We believe that everyone has a responsibility to work safely and maintain the site as a safe workplace. For that reason, and for the mutual benefit of everyone on the jobsite, certain safety rules have been established. Everyone participating in this project must abide by these rules; there will be no exceptions. Violations will be treated as follows: First violation – verbal warning, with written notice to follow; second violation – dismissal from the project.

1. Hard hats and other personal protective gear appropriate to your task must be worn at all times.
2. Fall protection is required for all work taking place above 2 metres above ground, unless the space is permanently enclosed.
3. All stair openings, floor holes and platform edges must be properly barricaded.
4. A competent person must inspect all scaffolds, ladders and lifting devices daily before they are used.
5. All operators of power machinery and equipment must carry evidence that their certifications are current.
6. No person shall ride in or on the bucket, platform, sling, ball or hook of any hoisting equipment.
7. Flammable liquids shall be kept in approved containers, and transfer systems shall be properly grounded.
8. Work areas, stairways, walkways and change areas shall be kept clean at all times.
9. Regular cleanup shall be performed at the end of each shift.
10. Personal vehicles shall not be operated on the jobsite except on the main access road and in the employee parking area.
11. No drugs or alcohol may be used or in your possession on the jobsite.
12. Any accident resulting in injury, however minor, must be reported to your supervisor, who is required to report, through channels, to XYZ's jobsite safety engineer.

Your understanding and cooperation are appreciated. A.K. Singh, Project Manager

Figure 6.5 Sample jobsite safety rules (based in part on Hislop (1999) and Halpin and Woodhead (1998)).

Two comments on items in the list given above are in order. First, all subcontractors must subscribe to this plan and see that their employees abide by it; some speciality subcontractors such as steel erectors, painters and roofers may have supplementary plans covering their especially hazardous work. Second, incentive programmes must be used with caution. If rewards are given for accident-free activities, workers and their supervisors will likely be reluctant to report 'minor' and apparently inconsequential accidents and injuries; experience statistics may be low in the midst of a rather unsafe workplace. Nonetheless, incentive systems have been effective on most projects. One contractor operates a programme on its large projects in which the names of all employees who have been accident free for several (say, 6) months are included in a drawing; the winner of the drawing receives a new pickup truck. Reduced costs of accident insurance, due to improved jobsite safety, are more than enough to pay for these rewards.

Environmental management

The impact of construction on the environment appears in many forms. These include the selection of environmentally safe materials and products to be incorporated into the project; the

Figure 6.6 Safety netting on scaffolding, Singapore.

Figure 6.7 Thatched roofing project, South Pacific style. Can you find at least four violations of good safety practice?

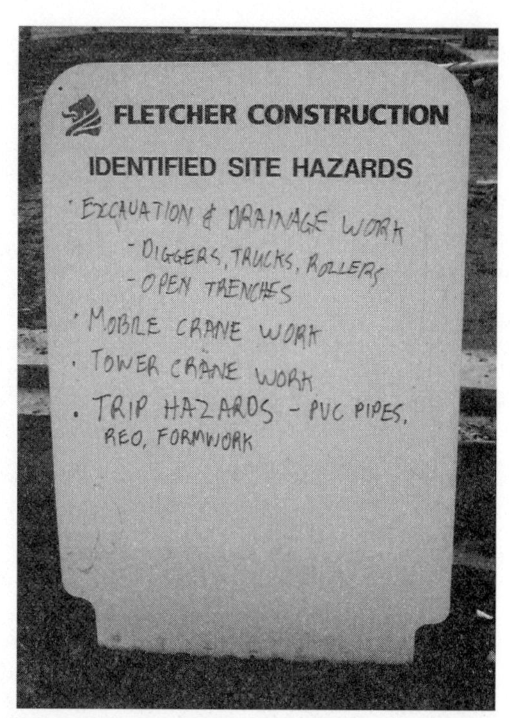

Figure 6.8 Two site safety signs, hospital project, Nelson, New Zealand.

Figure 6.9 Accident free safety sign, Fairbanks, Alaska.

planning and design work that can alter landforms, drainage, vegetation and wildlife; and the long-term impact of projects, such as silting of dams, their effects on fish and other animals and the flooding of upstream regions and the drying of downstream regions. An entire book would be required to begin to cover all of these issues. A balanced treatment here, as we consider the various phases of the project life cycle, might call for more attention to environmental impacts of activities during planning and design and project operation, in addition to on-site construction. However, we limit our brief discussion to impacts of the construction process on the environment: how what the contractor does while assembling the project can adversely impact the environment and measures that can be taken to lessen these impacts.

In line with changing attitudes in all sectors of society, modern construction management is subject to a variety of relatively recent laws and regulations stipulating how various construction operations must be conducted to minimise environmental impacts. Many of the permits that must be acquired by the owner, owner's representative or contractor, as discussed in previous chapters, are related to environmental protection. Often the special conditions or other contract documents will specify that the owner has agreed to certain stipulations as a condition of the issuance of a permit. For example, a land management agency (the Bureau of Land Management for federal land in the USA) may issue a permit for the building of or use of a roadway across federal land, subject to the condition that the road may be used during a limited time period annually, because of dust conditions, the presence of wildlife or the susceptibility to thawing of permanently frozen ground in the summertime. Such limitations would appear in the contract documents. Because of the seasonal spawning of certain species of fish, the extraction of gravel from a riverbed may be restricted to times when the fish are not present. A permit or permits allowing the extraction of the gravel would be predicated upon the existence in the contract documents of those time restrictions (Perkins, personal communication).

Thus, contractors must perform in certain ways because 'it's the law'. In a wider sense, whether an action would be illegal or not, contractors have an ethical obligation to protect the environment while assembling their projects.

Environmental issues subject to monitoring and control by the construction contractor include the following.

Water drainage and runoff

As land is exposed and disturbed, soil erosion can increase markedly. Land patterns can be altered and some of the soil can reach waterways. Measures such as silt fences (Figure 6.10) and dams may be required. Even if increased erosion does not occur, drainage patterns can be altered, with consequent negative impacts on adjacent lands. Temporary drainage ditches may be necessary to direct runoff in specified directions during construction operations.

Compacted soil from equipment operations

On level ground, the movement of trucks, lorries and other equipment can cause soil compaction, increasing its impermeability and causing rainwater to stand on the surface until it evaporates (Lowton, 1997). Temporary drainage works and the shaping of travelled ways may be needed to carry rainwater away from the site.

Mud, dust and slurry from tyres

A further impact of rain and also of the use of water on the site is that the treads of tyres can collect large quantities of mud or slurry, carrying it to roadways and other areas off site (Lowton,

Figure 6.10 Silt Fence, 350–400 mm high, beside highway project clearing, Wasilla, Alaska.

1997). The result can be clogged drainage works, dust and general unsightliness. Minimising such pollution of local roads requires diligent cleaning of tyres prior to their leaving the site or restricting operations to dry periods.

Air pollution

Dust can be produced not only from vehicle tyres but also from any operation that involves soil, such as foundation excavation and the compaction of an embankment. Often the contractor must dedicate one or more water carriers to the task of wetting the soil surface to minimise the production of dust. Other sources of air pollution include mixing plants that produce asphalt; dust is produced from the drying of aggregate in the plant's rotary kiln. Bag houses may be required to collect the particulates and prevent them from being discharged into the atmosphere. Plants that crush aggregates also produce dust; a permit to operate such a plant may disallow its operation in dry or windy conditions.

Contamination from petroleum and other spills

Spills at petroleum storage and transfer sites can make their way to the groundwater or to watercourses. Dykes are used around storage areas to contain spills until they are removed. Liners are placed under transfer locations, with dykes around them, for the same purpose. In case of an accident with a piece of equipment on the worksite, there is potential for petroleum contamination. A petroleum spill contingency plan covers spills on land and in water. Regulations typically require a spill of any amount in water to be reported to authorities, whereas a spill above a certain volume on land must be reported. To contain and clean up spills in water, containment booms may be needed, followed by pumping to remove the petroleum product. Spills on land may require stockpiling the contaminated soil, followed by removal and disposal to a location where it is incinerated or otherwise treated.

Discharge into waters

Often the contractor must discharge water into a watercourse or water body, either from a pipe or from a trench or channel. Laws often are strict regarding the composition of such water. In the USA the *Clean Water Act* requires a contractor to secure a pollution discharge elimination system permit before discharging such pollutants. The act defines *pollutant* very broadly (Erickson, 1998):

> dredged spoil, solid waste, incinerator residue, sewage, garbage, sewage sludge, munitions, chemical wastes, biological materials, radioactive materials, heat, wrecked or discarded equipment, rock, sand, cellar dirt, and industrial, municipal and agricultural waste discharged into water.

It is clear that many, though not all, of these substances may result from construction operations. The permit strictly controls the amounts that may be discharged. Negligent violations can result in fines of up to US$ 25 000 per day plus 1 year of imprisonment, while knowing violations are subject to fines of up US$ 50 000 per day and 3 years of imprisonment.

A special comment is in order regarding dredging spoil. Not only may silts and other soil components be extracted and transported by the dredging process, but also hazardous waste may be contained in the soil sediments. The result may require that the contractor remove and treat materials such as persistent organic pollutants, including (polychlorinated biphenyls PCBs) and heavy metals such as mercury and lead.

Solid waste disposal

Despite all efforts to be efficient in the use of materials, any construction operation produces large quantities of solid waste in such forms as timber, metals, plastics, paper, stone and concrete, gypsum and rubber. In the past, much of this waste might have been burned on the site, but concerns for air pollution have led to requirements that solid waste be transported to landfills. Trucks transporting solid waste must have their beds covered while travelling public roadways and often the contractor must separate the various types of waste for disposal at different locations. In addition, care must be exercised to prevent hazardous waste from being carried to and buried in a municipal landfill.

Products of demolition and renovation

Special care must be taken when the contractor carries out demolition or renovation of existing structures and other facilities, as the process may expose undesirable materials and components, both expected and unanticipated. Asbestos-containing materials such as insulation, flooring, fireproofing, soundproofing and roofing are often encountered from past constructions when asbestos was not considered dangerous. Lead, primarily from paints, mercury, petroleum, PCBs and other chemicals used for cleaning and boiler scaling may be discovered. The contract documents will (or ought to!) require certain steps to be taken. If such conditions are known to exist prior to the start of construction, a specialty contractor will remove the undesirable materials before the general contractor begins work. If the materials are found during the course of regular construction, work will be suspended, the area will be sealed off and an industrial hygiene evaluation will be conducted to determine the appropriate next steps. The contract documents should provide remedies for the contractor due to lost time and extra expense.

Worker sanitation

In our section on worksite layout, we noted the need to provide sufficient, well-placed toilets and hand washing locations for workers. During project operations, these facilities must be emptied, serviced and cleaned regularly. Workers must be required to use these toilets, rather than the behind-the-tree alternatives, as the latter quickly become highly malodorous and attractive to various vermin.

Endangered species

The contract documents are likely to contain instructions to the contractor for actions to be taken when endangered species of flora and fauna are encountered. If these species are known to exist on or near the site – an eagle-nesting area, for example – the documents can contain detailed instructions. In case the contractor encounters such conditions unexpectedly, proper actions will still have to be taken to comply with applicable law and regulations. The mating and birthing seasons may preclude any construction operations at those times. Nesting areas may need to be protected. Employees must be forbidden to molest such species, whether the employees are on or off duty. In the southwestern USA, an electric power transmission company has entered into a conservation agreement with the US Fish and Wildlife Service that provides long-term protection of one of the world's rarest species, the San Xavier talussnail. This rare snail's habitat is confined to a 5000 square foot area on a single hillside in Pima County, Arizona. The agreement guides construction and maintenance activities on a microwave communication site near the talussnail habitat (Southwest Transmission, 2002).

Wildlife protection

Even if wildlife is not classified as endangered, the contractor may have to comply with public regulations and contract provisions that restrict its operations. The restrictions may be for the protection *of* the wildlife or the protection of the project and its personnel *from* the wildlife. In the former category, annual migration patterns of animals may not be interrupted and construction work may be required to be suspended during those times. If fish use a portion of a stream for spawning during a certain season of the year, operations such as gravel extraction, pipeline laying and bridge building may not be allowed in the stream during those times. In many remote regions, bears can be a nuisance at best and dangerous to humans and their facilities at worst. Care must be taken to keep kitchen waste in bear-tight containers, to keep employee's lunches secure and to prevent employees from feeding bears.

Noise

There are many sources of noise on construction projects, including air compressors, pile driving operations, vehicles, demolition work and drilling and blasting. Rather than being subject to permits, noise concerns are usually governed by local ordinances, which specify certain maximum allowed decibel levels, or sound pressure levels, at the project property line. Typically, these allowable levels are higher during regular working hours and lower in the early morning, evening and night time, when nearby residents tend to be at home.

Archaeology

Sometimes the excavation for a building or the earthmoving for a horizontal project will uncover historic artefacts, including buildings or parts thereof, utensils, firesites, perhaps grains and other products of agriculture and even human remains. The contract documents ought to provide for both (1) a process by which the contractor is to deal with this discovery and (2) a remedy for declaring a changed condition and relief for the contractor for the extra time and expense required. In the USA, the proper procedure includes stopping work at the site, preserving temporarily the found items so they will not be further disturbed and notifying designated authorities for instructions. During the reconstruction of a transmission line located in Arizona, USA, it was determined that use of the existing access road could damage a Hohokam culture archaeological site. The work was stopped and the site was excavated to obtain and preserve information about that early culture. The site was the first of its kind to be excavated in that region and proved to contain much information about Hohokam farming practices (Southwest Transmission, 2002).

Resource management

If the construction manager's job during project operation is about monitoring and control, it is also about the management of resources. Halpin and Woodhead (1998), in their introduction to construction management, state that 'construction management addresses how the resources available to the manager can best be applied'. They suggest that the four primary resources to be managed are the 'four Ms' of manpower, machines, materials and money. We have already dealt extensively with the management of money and will do so again later in this chapter. In this section we address the management of jobsite personnel and issues of labour productivity, plus the management of materials and equipment.

Personnel supervision and labour productivity

Sources of craft personnel, their assignment and supervision

In Chapter 5's consideration of the worksite organisation structure, which must be established during the mobilisation phase, we discussed the roles of various supervisory personnel. Now, during project operations, their task is to provide the supervision that will get the job done correctly and efficiently. As shown in Figure 5.26, a typical supervisory structure for craftspeople includes a general foreperson and a foreperson for each trade. Large projects may require more than one foreperson for some trades. Thus, we see a several-tiered structure to the labour effort: general foreperson (sometimes called lead craftsperson), foreperson and craftsperson. Among the craftspeople on union projects, there will be journeymen, apprentices (those in training) and sometimes helpers. Pay scales vary with the tier, with differences of a few dollars between forepersons and journeyman craftspeople, for example. Under terms of a typical collective bargaining agreement, all journeymen of a given craft, such as carpenters, performing the same kind of work, are paid the same hourly rate, regardless of their individual skill level. In the case of non-union contracting, formal apprenticeship programmes may not exist and the contractor has more flexibility in setting individual pay rates in accordance with ability and productivity.

Other important knowledge the construction manager must know about on-site personnel management includes the methods by which craftspeople are hired. On a union project, the source of labour is the union hiring hall. Assuming the contractor is signatory to the collective bargaining agreement that covers the type of craft needed, a request is made to the hall for a certain number workers of that craft. If available, the workers will be assigned to the project with a minimum of procedure or paperwork, because most conditions of their employment are already covered in the agreement. If their work is satisfactory, they will probably stay until their part of the work is completed. In the case of unsatisfactory work, violation of safety rules and other detrimental conduct, the contractor can discharge an employee with relative ease.

On a non-union project, the contractor has more flexibility in setting wage rates for individual workers and there will likely be more steps in the contractor's hiring process. The contractor will either recruit through local and regional media or utilise the services of a hiring agency or open shop contractors' association. More screening for background and ability will generally occur than in the case of union-supplied labour.

Collective bargaining contracts provide for the presence of union stewards on the jobsite, to represent the union and deal with disputes involving the union. Stewards are hourly employees of the contractor who perform their union duties on the contractor's time. Most are expected to perform as regular craftspeople for some or most of the work day, although some collective bargaining agreements provide that on very large jobs some stewards carry out their union duties full time. It is important to understand that stewards are not supervisory employees and thus do not have a role in directing the work. Nor does the union's business agent; that person may visit the project site on occasion but their main function is to manage the union's office, assign workers to projects and maintain contact with local contractors and their projects.

A qualified workforce is essential to successful construction management. Under union contracting, the union supervises apprentice programmes that consist of a combination of classroom instruction and on-the-job training. While on the job, apprentices work beside their journeyperson colleagues, earning somewhat less per hour, often in graduated steps depending on the amount of training already completed. In the case of non-union contracting, the contractor assumes a larger role in training, although contractors' associations often conduct such programmes for the benefit of their members.

If several unions are represented on the jobsite, the contractor's supervisory personnel may have to contend with jurisdictional disputes arising from the assignment of craftspeople to tasks. Such a dispute is the result of a claim of more than one union that its members are entitled to perform the work, based on the various collective bargaining agreements. Such disagreements are really between unions, leaving the contractor caught in the middle. Do carpenters or ironworkers insert anchor bolts into concrete wall formwork? Who places concrete in ground slabs, labourers or cement masons? Are mechanical connections to certain permanent equipment completed by millwrights or plumbers? In many cases, the various collective bargaining agreements clarify the answers sufficiently, but if they do not, a dispute may arise. The ideal situation in the case of a possible dispute is to hold a meeting with the concerned forepersons prior to the start of the work task. If an understanding cannot be reached in this way, the contractor will have to make the assignment and the result may be some sort of disruptive action by the aggrieved union. (In the USA, legislation forbids outright strikes or other coercive action by a union as a means of trying to gain a work assignment.) In countries where a single union represents several construction crafts, such as Australia's Construction Division of the Construction, Forestry, Mining and Energy Union, such disputes are almost non-existent; likewise they do not occur on non-union projects. The following 30-year-old editorial has been widely quoted (such as in Halpin and Woodhead, 1998) as representing an example of the kinds of impacts that jurisdictional disputes can cause:

> The nozzle-dispute on the $1-billion Albany, N.Y., mall project has caused hundreds of stoppages on that job, which employs over 2,000 persons. The argument revolves around whether the teamster driving a fuel truck or the operating engineer running a machine shall hold the nozzle during the fueling operation. Both unions claim the job. Because holding the nozzle involves a certain amount of work, the question is why the union should want it, since regardless of which man does the job, the other still gets paid. The answer undoubtedly is that the union that gets jurisdiction will eventually be able to claim the need for a helper. This particular dispute has been reported as plaguing contractors in many states, including West Virginia, Oklahoma, Missouri, California and Washington. (Low Productivity: the Real Sin of High Wages, 1972)

Part of the job of the labour supervisor is to complete and submit accurate time records of those under their supervision as well as the task(s) to which they have been assigned. Usually a time card is completed daily for each work crew, as described earlier in this chapter. One purpose is simply to assure that only those on the job receive wages, thus preventing theft of the contractor's money. Time card records are used as the basis for the amounts paid to each employee and they are also used for cost accounting purposes, by reporting hours worked in each work item cost category. This latter information allows the contractor to compare actual with budgeted costs, as well as providing an historic record that is useful in preparing estimates for future projects. Some contractors use a system of brass tags for checking workers in and out of the jobsite. A tag remaining on the tag storage board during the workday indicates an absent worker, thus providing a comparison with the time card record.

Motivating employees to do their best work is both a nebulous concept and an essential aspect of personnel management. Because labour on a typical construction project is essential to every task and a large proportion of the project budget, the project's success in terms of schedule, budget and quality depends heavily on how well the contractor and its supervisors can motivate the workforce. At the risk of implying that motivation is simply a matter of following a 'cookbook' or a checklist, which it is not, the following ideas may be helpful in motivating construction employees.

- Information about the company and how this project fits into the company's overall programme, furnished to the employee as soon as they are hired.
- Likewise, information about the project and the role the employee is expected to play.
- Provision of steady work, to the extent the project environment allows.
- Opportunities to air grievances, with clear procedures to be followed.
- A focus on achieving feelings of fair treatment among employees on such matters as work assignments and reactions to grievances.

- Benefit packages, such as insurance, holidays, profit sharing and bonuses, with clear explanations of the conditions under which they will be granted.
- Public recognition in employee meetings and such media as a company newsletter, for especially worthy performance, following the adage 'praise in public and criticise in private'.
- A good safety programme, well planned and well administered.
- Opportunities for growth within the organisation, including clear guidelines for promotions and various formal training programmes.
- Overall good communications to make assignments clear, to provide timely feedback on work performance, both positive and negative and to instil a feeling of involvement as part of the project team.

Another area of concern to the construction personnel supervisor is the provision, use and security of tools. Some collective bargaining agreements specify which tools are to be provided by the craftspeople and which by the contractor. Usually the craftsperson will furnish their own hand tools, the selection being unique to the craft, with the company providing power tools, as well as cords and ladders. In the carpenter's personal tool kit, for example, one might find hand-held hammers, various hand saws, screwdrivers and chisels, wrenches (spanners) and pliers, hand levels and other miscellaneous hand tools. The contractor might provide the carpenter with pneumatic hammers, circular saws, drill motors and drills, powder-actuated tools and electrical cords (Mincks and Johnston, 1998). Training in the use of newly introduced tools is an important supervisory function. Tool security includes minimisation of theft of small tools throughout the workday, as well as loss of large and small tools during periods when the project is inactive. Bar-coded labelling systems have recently been introduced as a means of tracking tool inventory, similar to a book inventory in a library.

Drug and alcohol abuse is a major problem worldwide and construction jobsites are not immune from its negative impacts, including accidents, low productivity, poor quality work, absenteeism, high employee turnover, high insurance costs and even damage to and destruction of property. An effective substance abuse programme must be part of the contractor's jobsite supervision, as well as a companywide priority. Elements may include urine testing, searches, supervisor training, counselling and rehabilitation programmes and discipline and termination for cause. Sometimes the extent to which the contractor can investigate individual cases of abuse is limited by the collective bargaining agreement.

A final issue with respect to the supervision of employees is jobsite crime. We have already discussed the loss of tools through theft. Projects sometimes lose major pieces of plant to their employees and disgruntled workers can also damage work in progress, temporary facilities and project documents. These crimes are not confined to employees, as malicious damage can be inflicted by trespassers on installed work and construction plant. The impacts of these crimes are usually much more costly than only the value of the damaged equipment or goods, because the construction schedule can suffer serious delays and increased insurance rates may result from the incidents. Crime prevention programmes are yet another modern day necessity on the construction jobsite.

Labour productivity

Productivity in the construction industry is a major concern among all segments, especially owners and contractors. In quantitative terms, productivity is often defined as units or value of output, adjusted for inflation, divided by hours of labour or equipment input (Adrian, 1987).

Thus, we might say that the productivity of a carpenter crew installing wall panels was 15 m^2/labour hour or the productivity of a trenching machine is US$ 85/machine hour. A descriptive definition of productivity is the efficiency by which materials are handled or placed by labour and equipment.

In many countries construction productivity has improved during at least the past fifty years, but it has done so at widely varying rates per year (Haas et al., 1999). Some studies place the *decline* in productivity in the US construction industry at an average annual rate of 5.1% between 1967 and 1973 and 1.1% between 1974 and 1985, whereas there was an overall annual *increase* in the industry's productivity of about 0.8% during approximately 25 years ending in the mid 1980s (Adrian, 1987). Recent studies indicate improvements in US construction productivity, with significant variations among different industry segments (Haas et al., 1999). Singapore reports negative construction productivity growth rates ranging between –2.9% and –4.3% annually between 1995 and 1998 (Ministry of Manpower and Ministry of National Development of Singapore, 1999). When construction productivity increases over time, it tends to do so at a much slower rate than many other industries; the 25-year 0.8% US rate cited above compares with annual increases as high as 3.64% in agriculture and 4.60% in transportation. If hourly wage rates increase more rapidly than productivity, the wage cost per unit of installed construction increases, as illustrated in one of the questions at the end of the chapter. This per-unit cost increase is one of the classic ways of defining inflation.

Other studies have suggested that as much as 45% of the time a worker spends on the construction jobsite is non-productive (Adrian, 1987). Clough and Sears (1994) report that at least half of this wasted time is due to poor job management. While factors other than those related to labour can cause wasted time and low productivity, such as inefficient plant and tools, poor design and overly restrictive governmental regulations, we focus now on factors that affect labour productivity on the construction jobsite.

Training

This includes both supervisory and craft personnel. Many contractor organisations provide little or no training for forepersons and other supervisors. While union apprenticeship programmes provide the industry with at least minimally skilled craftspeople, the trend towards non-union contracting has resulted in fewer well-trained employees and contractors are reluctant to contribute greatly toward such programmes if members of their mobile workforce are likely to work for another contractor in the next season.

Working conditions

Many aspects of the conditions under which employees are asked to work can influence productivity, including jobsite layout and accessibility, the level of housekeeping and cleanup, the potential for unsafe conditions and accidents and such inconveniences as congestion and poor lighting, heating and ventilation.

Employee motivation

We have already covered this all-important aspect of personnel supervision. Clough and Sears (1994) report that some of the most common worker complaints that influence motivation and thus disrupt productivity are unavailable material, unsafe working conditions, having to rework already-completed parts of the job, unavailability of tools or plant, lack of communication and disrespectful supervisors.

Tardiness and absenteeism

Both of these factors may be due to lack of motivation. Tardiness can be controlled by establishing and enforcing reasonable and understandable working time rules.

Programming and schedule control

The extent to which the contractor effectively coordinates the timing of the various project activities can have a major impact on non-productive time. Through proper planning and control of the schedule, all required predecessor activities should be completed before a crew is sent to begin its task.

Material management

We deal with this topic in the next section. If materials are delivered late, if they are incomplete or defective, if they are stored a long distance from the place where they will be installed or if they must be double handled, delays or unnecessary work will occur.

New technology

'All but the most basic of tasks on a site have seen changes due to advances in technology over recent years' (Haas et al., 1999). While reluctance to embrace these changes is understandable in a fragmented industry with diverse standards and whose contractors may enjoy only a brief strategic advantage from such adoption, the extent to which the contractor is willing to implement these changes can impact productivity markedly.

Length of the working week

There is some evidence (Clough and Sears, 1994) that scheduled overtime can be counterproductive. A schedule under which hours worked extend beyond the normal 37.5 or 40 per week, if continued for several weeks, tends to result in decreased productivity.

Changes in the work

Changes can come from the discovery of a latent condition, errors in construction documents or requests by the owner. Whatever the source, the workflow can be interrupted, crew and supervisor time can be consumed and confusion can result.

Project uniqueness

The fact that each project has at least some unique aspects means that some of the work will be unfamiliar and thus will have to be learnt anew, resulting in inefficiencies not present in more repetitive 'assembly line' types of operations. Efficiencies can be realised for those work packages that can be planned and carried out in a cyclical, repetitive manner.

Environmental conditions

Precipitation and extremes of temperature and humidity have negative effects on construction productivity. Dramatic decreases have been documented in extreme hot and cold regions, where the contractor must provide extra liquids and extra rest periods, in the case of hot weather for example, or where employees are burdened with bulky warm clothing and extra time to warm up in cold regions (Koehn and Brown, 1985; McFadden and Bennett, 1991).

Contractual arrangements

The collective bargaining contract may provide for a quota of non-working forepersons, mandated time off with pay for holidays and various union activities and crew sizes and work assignments that otherwise could be more efficient.

Note that the last four of these 12 factors are essentially 'givens' within the project context; the construction manager has little influence over them. That means that management *does* have considerable control over the first eight factors; recall the report from Clough and Sears (1994) that at least half of all time wasted on the construction jobsite is caused by poor management. Enough said!

Materials management

The management of a construction project's materials begins well before operations start at the jobsite, as we have discussed in previous chapters. Technical specifications prepared by the design professional designate the required material qualities and sometimes specific manufacturers as well. In the contractor selection phase, the contractor estimates the costs of materials, based on historical experience and prices from suppliers; the contractor also develops a preliminary plan that includes methods of installing the various materials. An especially important part of project mobilisation is buyout, in which orders for materials and equipment are placed.

It is helpful to categorise materials into three types: bulk materials, such as lumber, reinforcing mesh, water piping and electrical rough-in materials, that require little or no off-site fabrication; manufacturers' standard items needing some fabrication, including metal decking, wall coverings, pumps and lighting fixtures; and items that are fabricated or customised for the particular project. Examples of the last are reinforcing and structural steel, custom cabinetry and special doors and windows (Halpin and Woodhead, 1998). The first category requires relatively short times for delivery after an order is placed and requirements for submittals to the owner or its representative for approval prior to delivery (to be discussed in more detail later in this chapter) are usually confined to standard literature that proves the material meets specifications. The second and third categories require increasingly complex submittals, usually including detailed drawings and samples; the time from placement of the order to delivery on site includes manufacture as well as submittals and approvals and can take several months. We review the steps in the procurement process and then describe the handling and storage of materials on the jobsite.

In Chapter 5, we listed the following steps in procuring materials for a construction project: receipt and evaluation of offers, purchasing or placement of the purchase order, approval by the owner or owner's representative, expediting or contact with the supplier to assure timely delivery, fabrication, shipping, delivery and inspection. Note that the complexities involved in approvals and manufacture will depend on the type of material, as explained in the previous paragraph. After the order for a certain material is placed, the owner will want assurances that the material conforms to its requirements. In the case of floor tiles or paint, the contractor will obtain from the supplier samples and product data sheets and furnish them to the owner's representative. For reinforcing steel, special drawings, called *shop drawings*, will be prepared by the supplier for approval by the owner. In the case of soil aggregate, certified copies of gradation and hardness tests may be required. Complicated equipment, such as the steam supply system for a power-generating station, will require many sets of drawings, as well as material test results, manufacturer qualification statements and other data. In planning its programme, the contractor must provide sufficient time for the preparation and approval of this documentation, which is a pre-condition of the start of any required manufacture; several months may be required for this step for the power station equipment noted above. Several cycles of submittal–review–resubmittal may be expected in some cases and the programme must anticipate such time requirements.

Because the contractor cannot assume that delivery dates established in the various purchase orders will be strictly adhered to, a series of follow-up actions, called *expediting*, is usually required. This function may be conducted in the company's home office, but on large projects expediting is often based at the jobsite. The expeditor contacts the supplier periodically, by e-mail, letter, telephone or personal visits, to assure that document preparation and manufacture is proceeding according to schedule. Detailed records are essential to the expeditor's role, to track all the steps from placement of order to on-site delivery of every material order. Frequent contact between the expeditor and the project scheduler is essential, because unavoidable delays in deliveries must be predicted as far in advance as possible so that their impact on the overall programme can be assessed and any necessary changes in plans can be made in a timely manner.

After the required documentation is approved, any needed fabrication can proceed. Of course, this process varies widely depending upon the product. The required time might range from a few hours of pipe cutting to a year or more in the case of our steam supply system. The material is then prepared for shipping by the manufacturer and transported to the jobsite. Management of the delivery schedule has important impacts on the project. If the materials are delivered too early, the contractor's money is tied up in material inventory, space on the jobsite must be allocated for storage and the materials could be subject to pilferage, breakage or spoilage. If they are delivered too late, negative impacts on the programme can occur. Achieving a balance between too early and too late is always a challenge. Some projects rely on *just in time* (JIT) deliveries, in which the material arrives on the jobsite at the moment its installation is to begin, thus avoiding the extra expense and space required if storage is needed.

Upon delivery at the project site, the contractor's personnel will unload the material and it will be inspected. (Unloading carries the potential for interesting jurisdictional disputes; who is assigned to do the unloading, the truck driver, their helper, the labourers who work in the material storage yard or the plumbers who will install it into the project?) The inspection is carried out to assure that the material complies with the drawings and specifications and has been supplied in the proper quantity. Although tedious, the inspection process is crucial to a smoothly functioning installation task and it must be carried out immediately after delivery to allow any identified corrections to be made promptly. Inspection may include material testing for proof of quality; the engineering staff or an outside testing agency can be involved in obtaining concrete samples, analysing sand and gravel and overseeing the laboratory testing of reinforcing steel.

All of the preceding activities related to materials have been preliminary to the main activity, of course, which is the actual installation of the gravel, reinforcing steel, electrical cable, steam supply system and other materials and equipment into the project. We spent considerable time in Chapter 5 describing the importance of good jobsite layout and its importance in efficient material flow. An effective installation programme depends heavily on how well the storage and handling of materials is managed. After a shipment of materials arrives at the site, they may all be stored for later installation, some may be installed and some stored or all may be installed immediately. If they are stored, the location must be out of the way of active movement of personnel and equipment but close enough to allow short travel distances to the facility under construction. Items must be sorted upon their arrival at the storage location, for ease of later identification. The fewer times the materials are handled, the more efficient the installation is likely to be.

As one example, a high-rise building project provides special challenges and opportunities with regard to material storage and handling. Each floor, as its deck is completed, can provide space for storage of materials. In organising the storage, the manager must assure that only materials for that floor are hoisted to it. The quantity and variety of materials, including wall framing timber and drywall, plumbing and electrical supplies, ductwork, door and window framing, the doors and windows themselves, cabinetry, paints and floor coverings and the fact that work, as well as storage, must be conducted on the floor, requires careful layout of the materials (Halpin and Woodhead, 1998).

Adequate physical protection of stored materials is an important consideration. In establishing the site layout, the contractor must provide for protection against rain, snow, very hot or very cold temperatures, sunlight and wind, depending on the location and the type of material. In addition, protection against fire and theft is also required. And once these arrangements are created, it is incumbent upon the manager to assure that they are used (Mincks and Johnston, 1998).

The case study at the end of this chapter provides some insights into the multiple facets of managing the structural steel procurement, manufacture (in Korea), transportation and storage for a project in Fairbanks, Alaska.

Equipment management

In this short section we consider the last of the four Ms, the primary resources that must be managed on the construction site. Machinery in this context means the machines that carry, hoist or otherwise move the various materials and components to and around the site, plus other machines that remove, process or convert materials. They are moved to the site for use during the construction operations phase and are not incorporated into the finished project. Some regions of the world use the term *plant* to refer to this machinery, while others call it *equipment*, sometimes leading to confusion between these items and the permanent equipment that becomes part of the finished project. Sometimes the term *rolling equipment* is used to refer to the equipment that travels on tyres or tracks, to distinguish these types from such fixed equipment as concrete batching plants. Equipment plays a larger role and has a larger proportion of the budget in a highway or heavy construction project than it does in a project to construct a building. Hendrickson and Au (1989) provide a helpful breakdown of common types of construction equipment.

- Excavation and loading. Crawler, truck or wheel mounted cranes, shovels, draglines, tractors with blades (bulldozers) and scrapers.
- Compaction and grading. Various types of rollers, some vibratory, used to provide mechanical means to increase soil density. Motor graders and grade trimmers to shape soil and bring it to proper elevation.

- Drilling and blasting. Percussion, rotary and rotary-percussion drills to provide holes in rock for the placement of explosives. Tractor-mounted rippers for penetrating and prying rock. Automated tunnelling machines with multiple cutter heads capable of cutting full diameter tunnels in rock.
- Lifting and erecting. Derricks and tower cranes, both of which are mounted in a stationary position, plus truck- or tractor-mounted mobile cranes. Material and personnel hoists.
- Mixing and paving. Portland cement concrete mixers mounted on trucks, self-propelled Portland cement concrete paving mixers, truck-mounted bituminous distributors and bituminous paving machines.
- Construction tools and other equipment. Air compressors, pumps and such pneumatic or electric tools as drills, hammers, concrete vibrators and saws.

In managing the cost aspects of jobsite equipment operations, the contractor records and works with two types of data, the *time* the equipment is operated and the *rate* to be charged for each hour of operation. The actual charge against the project budget depends on whether the equipment is leased or rented for the job or is owned by the company. If it is leased for the project duration or some part thereof, the lease cost will be directly charged to the project. Time cards, similar to labour time cards, will be used to record the actual time spent by the equipment and operating personnel; if the lease payments are based on a fixed monthly charge plus an hourly rate, the recorded hours will become part of the payment basis. In the case of an hourly rental arrangement, the actual hours will be multiplied by the hourly rate to determine the payment.

If the jobsite equipment is owned by the company, it is common practice for the company to establish 'rental rates' that are charged to its projects by the hour, day or week. Often, a separate division of the company will oversee all equipment management and occasionally, a legally separate company will be formed. A fair rental rate will be crucial to both the project, so that it is not overcharged, and to the company, so that it recoups all of its costs. The definition of 'costs' is a matter of company policy; surely costs include the capital recovery cost of the purchase price (sometimes denoted by depreciation plus interest), licensing and insurance. Other costs incurred as a consequence of ownership and operation include fuel, oil and lubricants, routine servicing and maintenance, major overhauls and labour costs of operating personnel. The company must decide whether to embed these costs into the rental rates or have the projects charge actual costs against their budgets. Except for labour costs and possibly fuel, oil and lubricants, it is probably in the best interest of company and its equipment fleet to include the other costs in the rental rates. Otherwise, a 'penny wise and pound foolish' project manager may neglect upkeep and maintenance on equipment assigned to the project, in the interests of showing a higher job 'profit' than if such costs were incurred and charged to the job.

No matter how the costs are charged, the maintenance of construction equipment is an important jobsite function. Note that our jobsite organisation chart in Figure 5.26 provides for a master mechanic, with shops and equipment, reporting directly to the project superintendent. In some cases, the reality is that maintenance and repairs are performed in less than ideal conditions, such as a broken scraper axle being welded in place adjacent to a mudhole beside the roadway embankment where it failed, late at night in the presence of traffic, mosquitoes and artificial light. Records of maintenance efforts and costs are essential both to make proper charges and to help decide when equipment should be replaced.

Two equipment-related decisions usually not directly within the purview of the jobsite manager are (1) whether to lease, rent or purchase and (2) when a piece of equipment should be replaced rather than repaired and kept in service. For most contractors, these questions are studied and resolved by central office staff, often with substantial amounts of forecasting and economic analysis.

Documentation and communication

The on-site management of a construction project involves great amounts of paperwork, even for relatively small projects. The purpose, of course, is to communicate directions, questions, answers, approvals, general information and other material to appropriate members of the project team, so that the project can proceed in a timely, cost-effective and quality manner. We discuss in this section some of the documents necessary for the execution of the contract, primarily from the contractor's point of view. Included are considerations of shop drawings and other submittals, variations or change orders, the documents required for the contractor to receive payment and, once again briefly, the value engineering process. Many pieces of 'paperwork' are in electronic form and thus we include a description of many types of such documents, as well as computer-based approaches to the organisation and management of all project documents. Finally, we consider the use of collaborative web-based means for facilitating communication of all relevant information among the project team.

Submittals

The drawings and specifications prepared by the design professional are the basis for determining the contract price and for overall job planning, but in many cases they are not sufficient for the fabrication and installation of the various job elements. Further details are provided by supplemental documents furnished by subcontractors, material suppliers and in some cases the contractor itself. The term *submittals* refers to the totality of the shop drawings, product data and samples; all of these documents are submitted to the owner or owner's representative for approval prior to fabrication and manufacture of the items they represent. A portion of a *Contract Administration Plan* (US Army Corps of Engineers, 2000) contains the following:

> The primary responsibility for scheduling, management adequacy, accuracy, and control of submittals lies with the contractor doing the work. The contractor is responsible for incorporating all submittals for government approval into the NAS [network analysis system]. Approved submittals become contract documents. Therefore, the contractor must understand that the handling of submittals is as important as the actual construction and installation activities.

Note well that the contractor is responsible for managing the submittal process, including the incorporation of all submittals into the network-based project program (network analysis system).

Shop drawings are the supplier's or fabricator's version of information shown on the drawings already prepared by the design professional. These special drawings contain more detail, including all information needed to fabricate and erect the materials. In the case of structural steel, for example, the shop drawings would include details about welding and connections that are typically not part of the structural engineer's drawings.

Product data can consist of manufacturer's catalogue sheets, brochures, diagrams, performance charts and illustrations. In the case of an electric motor, for example, the data might include dimensions, weight and required clearances, plus such performance data as revs per minute, power characteristics and consumption and noise level curves. Often these data sheets must be modified to delete information that is not applicable.

Samples are especially important for demonstrating that the contractor will furnish the various required architectural features to the satisfaction of the design professional. Examples include wall and floor covering samples, colour chips for all paint to be used and furniture finishings. Sometimes the specifications require that a construction sample, such as a small sample of a brick wall, be erected at the jobsite for approval prior to the beginning of that material's installation.

Technical specifications usually stipulate in detail the quantity, size and contents of shop drawings and other submittals. In the case of a single prime general contractor, all of this information flows through the general contractor. Consider our example of the structural steel shop drawings. It is likely that a steel erection subcontractor will be contracted by the general contractor. This subcontractor will likely purchase the steel from a supplier/fabricator. Thus, the fabricator's shop drawings will be forwarded to the subcontractor, who will, in turn, forward them to the general contractor. The general contractor will review them to assure compliance with contract requirements, after which the drawings will be sent to the design professional, sometimes via the owner's project representative. After review, the design professional will return them to the contractor, with an indication of the level of 'approval'. The law is quite clear in most jurisdictions that, regardless of actions taken on the submittals by the design professional, the contractor is responsible for performing in accordance with the contract documents. To minimise their potential implied liability, many design professionals use such wording as 'no exception taken', 'make corrections noted', 'revise and resubmit' or 'rejected' to indicate their response, rather than words like 'accepted' (Mincks and Johnston, 1998). The contractor will return the drawings to the steel fabricator via the subcontractor. Revisions, if needed, will then follow the same sequence. As it is involved in steel erection, the subcontractor will be interested in reviewing the shop drawings each time they pass through its office (both coming and going!).

Scheduling the approval process for each material must contain sufficient contingencies to permit revision and resubmittal if needed. If done properly, the project programme will identify as critical, or nearly so, those approval sequences whose timing must be managed most closely. Often the contract documents will specify the amount of time allocated for review.

When one contemplates the hundreds of submittals required on a typical construction project, it is clear that a system must be installed and used to keep track of the status of each. A *submittal log* is the document that is the backbone of this process. Used in conjunction with the overall procurement schedule, it records the scheduled and actual dates and actions taken by each party to the process, for each submittal. In a later section, we shall consider the overall document management challenge, of which submittal management is one part and the role of information technology in this management responsibility.

Variations

During the execution of the construction contract, one matter that requires careful documentation and thorough communication is the case of changes in the work. In any but the smallest and simplest of projects, it is common that some changes will be made. *Changes*, also known as *variations*, arise for many reasons: the owner may decide to add some new item to the project, delete some portion of it or add to, reduce or modify an already defined part of the job. Unexpected site conditions discovered during the course of construction, including soil conditions, archaeological findings, endangered species or hazardous materials, may require a change. Discrepancies may be discovered in the contract documents. Codes or regulations may change after the contract is signed, requiring a change in the contract. Products originally specified may not be available. Hollands (2000) has identified 28 circumstances in the New Zealand *Conditions of Contract for Building and Civil Engineering Construction* (Standards New Zealand Paerewa Aotearoa, 1998), in addition to owner-desired modifications, which can give rise to changes.

A typical contract provision used for Alaskan highway construction that relates to variations begins as follows:

> 104–1.02 CHANGES. The Department reserves the right to make, at any time during the progress of the work, without notice to the sureties and within the general scope of the Contract, such changes, deviations, additions to or deletions from the Plans or Specifications, including the right to alter the quantity of any item or portion of the work, as may be deemed by the Engineer to be necessary or desirable and to require such extra work as may be determined by the Engineer to be required for the proper completion or construction of the whole work contemplated. Such changes will be set forth in writing as a Change Order and shall neither invalidate the Contract nor release the surety. The Contractor agrees to perform the work, as changed, the same as if it had been a part of the original Contract. (Alaska Department of Transportation and Public Facilities, 2001)

Two features of the paragraph quoted above are of special interest. First, it is the owner (in this case, the Department) who decides whether changes will be made in the contract. Second, the changes must be 'within the general scope of the Contract'. A change may be as simple as a different paint colour, for which it is likely no change in the contract price or schedule would be made. If the owner wishes to add a second building to a manufacturing facility, or to reduce the project by half, the contractor might argue that such a change was not within the scope of the original contract and therefore it was not bound to perform the change. Interesting legal cases have arisen over the interpretation of the term *project scope* in the context of contract variations and a body of legal theory around the matter of *cardinal changes* has arisen. Simon (1989) suggests that a cardinal change exists 'when the identity and character of the thing contracted for is significantly altered, or when the method or manner of anticipated performance is so drastically and unforeseeably changed that essentially a new agreement and undertaking is created.' Some courts have held that the owner requiring the contractor to perform a cardinal change commits a breach of contract, although one court decided that a general contractor could properly reduce a subcontractor's already contracted-for work by 80% (Simon, 1989).

An advantage of the measure-and-value (unit-price) type of contract is that additions or reductions to contract quantities can usually be handled easily. If an earthwork contract provides for 8000 m^3 of fill material at a unit price of US$ 6.50/m^3, the owner could, with a minimum of paperwork, direct that the quantity be increased to 8800 m^3. This additional 10% would be compensated at the same US$ 6.50/m^3. A similar approach would be taken to reductions in quantities. A typical measure-and-value contract usually provides for an exception to this straightforward method when the final quantity varies from the originally estimated quantity by more than a specified percentage. The Alaska Highway Construction specifications quoted above provide that, 'Where the final quantity of a major contract item varies more than 25% above or below the bid quantity, either party to the Contract may request an equitable adjustment in the unit price of that item' (Alaska Department of Transportation and Public Facilities, 2001).

When the owner desires to initiate a variation in a lump-sum construction contract, the process is somewhat more complicated. The contract contains detailed language outlining the process to be followed and the various types of documentation required. The process begins with a decision by the owner or design professional to consider a change in the work, for any of the reasons listed above. The design professional then prepares a *variation (change order) request*, outlining the work to be added, deleted or modified and forwards the request to the contractor. The contractor estimates the cost impact of the change (either an addition or reduction) as well as the effect on the schedule, completes a *variation (change order) proposal* and forwards the proposal to the design professional. Some contracts restrict the contractor's proposed price to its estimated costs plus a stipulated percentage for overhead and profit. If the design professional and owner agree that the proposal should be accepted, a *variation (change order)* is written and forwarded to the contractor. If there is disagreement, the parties may negotiate in an attempt to resolve differences. In the end, the owner may decide not to pursue the change. The contract may provide that, if agreement cannot be reached at this stage and if the change is required, the owner or its representative can direct the contractor to carry out the work, with the final terms of payment and schedule impact to be decided later. In some cases, the contractor may be directed to do the work, with payment to be based on the actual costs plus a fee or on a time and materials basis.

It is important that the contractor does not proceed with a change until the written variation is issued. To do so may make it impossible to obtain payment for any extra work. Different courts have ruled differently on the validity of oral variation directives and thus the contractor is well advised to wait until directed in writing. The term *constructive change*, as opposed to a *directed change*, refers to an informal action by the owner or its representative that results in a de facto change in the contract requirements. If the on-site owner's representative provides an interpretation of defective drawings or specifications or conducts a faulty inspection, these actions may result in increased costs to the contractor. Even in the absence of a written variation, the contractor may be entitled to extra compensation. Claims for such constructive changes often lead to time-consuming legal disputes.

Measurement, progress payments and retainage

All projects except those with very limited duration have contract provisions that allow the contractor to apply for partial payments as the project progresses. Otherwise, the contractor would have to fund the entire project with borrowed funds or from its own equity; with either of these methods, the owner would ultimately pay the finance costs through their inclusion in the lump-sum price, the various unit prices or the fee charged in a cost-plus contract. This section discusses the means by which the contractor seeks its periodic partial payments.

The basis for payment depends upon the type of contract. In every case, some sort of measurement is carried out to assess the amount of work completed. It is common practice to prepare payment requests every month. In the case of a cost-plus contract, the 'measurement' consists of the various documents that demonstrate that the contractor has incurred the costs during the period. With its requests, the contractor will submit audited material and equipment invoices, certificates of material taken from inventory, audited payroll information, billings for any construction plant whose expenses are allowed as 'costs' and approved subcontractor billings. As we emphasised in our earlier discussion of this type of contract, the contract documents ought to clarify which expenses are allowable as reimbursable and which must be covered by the fixed or percentage fee. A summary page is used to total the several costs, after which the appropriate fee is added to determine the total amount earned in that period.

If the project is being constructed under a measure-and-value (unit-price) contract, careful measures of each pay item must be made as the work is installed. If the tender is accepted for the embankment and roadway project whose unit price schedule is shown in Table 4.4, the various areas, volumes, lengths and weights will be measured and recorded as the basis for payment. The quantity surveyor is often involved in these measurements. Student engineer interns often find themselves assigned the task of counting the trucks that deliver gravel or other roadway materials to a construction site; these counts, on behalf of the owner, lead to the volume or weight measure upon which the contractor's payment will be based. After the hectares of clearing, cubic metres of unclassified excavation, tonnes of asphalt concrete and other measures for the month are completed and agreed upon (agreement between contractor and owner may require considerable discussion and negotiation!), the quantities are multiplied by their respective unit prices and the results are totalled to determine the amount earned for the period.

To illustrate the preparation of a periodic payment request for a lump-sum contract, we use the concrete wall project whose cost estimate was developed in Chapter 4 and whose cost summary and forecast at the end of a certain period was shown in Table 6.1. The first three columns of Table 6.6 contain a schedule of values prepared by the contractor prior to the start of field construction. The scheduled values are the basis for payment and have been derived from the direct costs for the respective items in Tables 4.1 and 6.1, with the addition of a proportionate share of the various add-ons for overheads, mark-up, sales tax and bonds. Note that the total scheduled value equals the tender price of US$ 25 013; if the tender price had been modified by one or more variations, the schedule would be kept up to date and the total would equal the revised contract price.

The remaining parts of Table 6.6 are completed by the contractor at the end of each period and submitted as its payment request. It is important to distinguish between the two documents shown in Tables 6.1 and 6.6. The relevant differences can be summarised as follows.

- The summary and forecast in Table 6.1 is an internal document, used by the contractor to monitor and control its costs, whereas the progress payment request in Table 6.6 is prepared for use by the owner to allow payment to the contractor.
- Table 6.1 focuses on the control of direct costs by comparing estimated direct costs with actual costs. Except for site overheads (which could be monitored with a similar table), direct costs are all that the field supervisors have control over. Table 6.6 includes the entire project value, because that amount is ultimately to be earned by the contractor's organisation.
- Although we show the same number of items (13) in both Tables 6.1 and 6.6, the number of pay items in an actual 'real-life' project may be considerably less than the contractor's cost items that are used for cost-control purposes. Adequate control of costs may require much more detail.
- Completion percentages may not be identical in the two documents, for two reasons. First, the owner may not allow as high a percentage as the contractor realistically believes has been completed. Second, the percentages in the contractor's cost summary may include materials that have been delivered to the site but not incorporated into the work, whereas the progress payment request will probably not include such material in the percentages.

Table 6.6 Schedule of values and progress payment request, concrete wall project

Item number	Item	Scheduled value (US$)	% Complete to date	Completed value to date (US$)
1	Mobilisation	904	100	904
2	Clearing and grubbing	292	100	292
3	Silt excavation	686	100	686
4	Rock excavation	955	100	955
5	Backfill	441	0	0
6	Footing forms – place and remove	786	100	786
7	Footing reinforcement	209	100	209
8	Footing concrete	625	100	625
9	Wall forms – place and remove	13 633	40	5 453
10	Wall reinforcement	2 026	30	608
11	Anchor bolts	455	20	91
12	Wall concrete	3 407	0	0
13	Clean up and move out	594	0	0
	Total	25 013		10 609
	Materials stored on site			375
	Total completed and stored on site			10 984
	Less 10% retainage			1 098
	Net earnings to date			9 886
	Less previous certificates for payment			2 325
	Current payment due			**7 561**
	Balance to finish, including retainage			15 127

- Most progress payment requests provide for recognition of materials stored on site as a separate item in the request.

Payment requests for lump-sum contracts usually display the total value completed through the end of the period for which payment is requested; in Table 6.6, that value is US$ 10 609. Note that some of the percentages used to determine this value are different from those in Table 6.1, as explained above. To this total is added, first, the value of materials stored on site, US$ 375; the contractor must provide invoices to support this part of the request. Sometimes the form provides for the total value of all delivered materials minus the value of materials already incorporated into the project. The total of the completed value and stored materials is US$ 10 984, from which the owner will retain 10%, or US$ 1098, as an incentive for the contractor to maintain work progress and produce a quality project. We shall discuss retainage further below. From the net earnings of US$ 9886, the amount already certified for payment in previous periods, US$ 2325, is subtracted, giving a total current payment due of US$ 7561. Finally, the net earnings to date are subtracted from the total scheduled value to find the balance that the contractor can expect in future periods, assuming no changes in the contract price.

Our simplified example is such a short project that we have assumed it would be monitored weekly and its payment requests would be prepared weekly as well. As indicated above, the most common practice is a monthly progress payment frequency. This concrete wall project also includes no subcontracted work. If it did, the payment request would incorporate any subcontractor requests for payment, with supporting documentation.

On cost-plus contracts, it is normal practice for the owner to pay all of the contractor's proven costs, without retaining any percentage of the cost plus fee; occasionally such a contract will allow the owner to withhold a specified amount after most of the project is completed, with this amount to be paid at the end of the job (Clough and Sears, 1994). The somewhat controversial practice of *retainage* (also called *retention*) is a common feature of most measure-and-value and lump-sum contracts. Obviously, if 10% is retained until the end of a large project, a considerable sum of money that the contractor has earned is involved. Some contracts provide for reduced percentages as the project nears completion. The avowed purpose of retaining some of the contractor's earnings is to provide an inducement to complete the work on time, because the retainage will be released upon completion, and to produce a quality project. The counter argument is that the owner is already well protected by surety bonds and various contractual remedies. However, bonds protect the owner only if the contractor breaches the contract, so owners feel a need to retain moneys for such cases as contractor failure to remedy defective work or contractor-caused claims against the work that the owner may have to settle (Clough and Sears, 1994). Although changes in retainage practice have been made, such as reducing the withholding amounts, it is likely this practice will remain a topic of discussion in the industry. Some contract conditions (such as Standards New Zealand Paerewa Aotearoa, 1998) allow the contractor to furnish a bond in addition to any other required bonds, in lieu of retentions.

Value engineering

In Chapters 3 and 4, we have covered the essential role of value engineering, or value management, in the planning and design and contractor selection phases of the project life cycle. Recall that, during planning and design, a value engineering consultant may be engaged to study alternative designs and their cost impacts. Also, as part of the tendering process, the contractor may develop and propose different designs; the tender would then include a proposed price for the original design as well as a tender for the new proposals. Now, during the operation phase, value engineering plays an equally important part in striving to achieve lowest overall cost consistent with required performance, reliability, quality and safety. Indeed, a key concept to the management of the construction life cycle is that value engineering is a continuous, not a static, process. Whether the contractor practises value engineering during the mobilisation phase or the operation phase or both is unimportant. Once the construction contract has been signed, the contractor can play a vital role in increasing the project's value for money and be rewarded with additional income. The process described below is thus an important aspect of the project's documentation and communication.

If the contract provides for value engineering, detailed instructions will be included in the general and/or special conditions. Usually they will stipulate that the owner will consider value engineering change proposals (VECPs) that have the potential to result in cost savings without damaging such essential functions and characteristics as service life, economy of operation, ease of maintenance, desired appearance and safety. In the submission of such a proposal, the American Association of State Highway and Transportation Officials (1998) suggests that the contractor be required to include the following:

1 A statement that the submission is a VECP.
2 A description of the existing work and the proposed changes for performing the work, including a discussion of the comparative advantages and disadvantages of each.
3 A complete set of plans and specifications showing proposed revisions to the original contract.
4 A detailed cost estimate for performing the work under the existing contract and under the proposed change.
5 A time frame within which the owner or its representative must make a decision.
6 A statement of the probable effect the proposal would have on the contract completion time.
7 A description of any previous use or tests of the proposal, conditions, results and dates, project numbers, and the owner's action on the proposal if previously submitted.

Often a quick response by the owner is important in maintaining the schedule (see item 5 above). If the owner is interested in considering the proposal but the proposed timeframe cannot be met, it is expected that the contractor will be so notified in a timely manner and a mutually agreed timeframe established. If the owner decides to accept the proposal, a change order will be issued. A typical contract provision for the sharing of cost savings gives half of the net savings to the contractor, although the US Public Buildings Service grants a 55% share to the contractor for fixed-price contracts (US General Services Administration Architecture and Construction, 2000).

What kinds of design changes might result from the value engineering process? In one case, the contractor on a large highway project proposed a redesign for several 10 m high retaining walls that would replace very heavy drilled piles and tie-backs with cantilevered lightweight tubular pipe piles. The accepted proposal resulted in cost savings to the project, extra income for the contractor and an installation method that was virtually noiseless, vibration free and accurate to within a few millimetres (Bedian, 2002). Other examples include the use of pre-cast rather than cast-in-place concrete barriers for highway median and sidewalk parapets, a pre-cast bridge deck in lieu of a cast-in-place deck and the installation of empty conduits in a highway median for future utility usage (Connecticut USA Department of Transportation, 2000). In this last case, the project cost was increased, but the owner was convinced, after considering the likely savings in future expenditures, that the overall costs during the entire life cycle would be lower than they would have been under the original design.

Other project documentation and document management

In the previous parts of this section, we have presented four especially important aspects of project documentation – submittals, variations, the progress payment process and value engineering proposals. Now we consider briefly a large number of other documents that the contractor must utilise during project operations and then we discuss the use of electronic means for the management of all of the project's documents. It has been estimated that 1–2% of a construction project's cost is for paperwork (Project Center, 2001). If that is so, then an efficient management system that can reduce those costs could result in significant increases in the contractor's profit. Mincks and Johnston (1998) list several features of effective construction jobsite documentation and communication.

- Objectivity and truthfulness. Fair, honest, without bias.
- Timeliness. Distribution as quickly as possible.
- Retrievability. Readily accessible and closely tied to specific issues.
- Appropriate distribution. Sufficient detail and frequency for individual recipients, leaving out irrelevant information.
- Standard, uniform information. Consistency in content and format, for ease in evaluation and comparison with other reports for this project and other projects.
- Completeness and comprehensiveness. All facets of the event or issue, without gaps in the information; continuous throughout the project.

The list of documents given below must be considered incomplete, as each project will have its own specific requirements, but it should convey an understanding of the wide variety of documents that must be managed. For ease of identification, the documents are organised into a series of categories. Some of the classifications are somewhat arbitrary; in several cases a type of document could be placed in more than one category. Note that, in nearly every category, individual documents will be added or modified as the project proceeds. Thus, a system for managing them must provide for this dynamic character. If the system were completely manual, each type of document would need (1) a 'folder' large enough to receive and store all documents and (2) a log that would list the current status of every document in the folder. The same concept is the basis for any construction document management system.

General

- Contacts. A list of all persons and organisations with whom the contractor corresponds during the life of the project, with relevant contact information.
- Programme/schedule. The master project schedule, including the original baseline and all updates, together with any supplementary related information such as detailed subnetworks, resource studies and analyses of schedule trends.
- Accident reports. A report for each accident, plus summary data and related analyses.
- Punch lists and other project closeout documents. Lists of deficiencies identified during inspections as the project nears completion; certificates, warranties and record drawings.

Contract documentation

- The contract itself, as amended by variations (change orders) during the project.
- Drawings. Design drawings from the design professional, as revised throughout the project, with any supplementary sketches that may be issued.
- Specifications. Technical specifications, plus general and special conditions and other parts of the project documents manual.
- Subcontracts. Copies of agreements with all subcontractors.
- Insurance. Certificates of all insurance carried by the contractor, as well as certificates for insurance required to be carried by all subcontractors.
- Bonds. Similar to insurance; contractor's copies of its performance and payment bonds, plus proof that subcontractors have furnished surety bonds, if so required.

Communication records

- Meeting minutes. Records of all meetings held at the jobsite or elsewhere if pertaining to the project; regularly scheduled general and safety meetings; other special meetings of any kind.
- Telephone records. Brief records of all telephone calls placed or received at the jobsite.
- Conversation records. Often simply a memorandum to the file to record an understanding resulting from a conversation that was less formal than a meeting.

Project status documentation

- Daily reports. A standard report that includes weather conditions, work in progress, number of direct employees and subcontractor personnel on site, visitors, equipment on site, material deliveries and special issues.
- Weekly and/or monthly reports. Summary of accomplishments for the period, comparison of actual with planned schedule progress, cost status, change orders and special issues.
- Progress photography. Periodic still and video photography that records project progress.

Correspondence

- Letters. All letters written by the contractor, as well as those received at the jobsite office; should include a record of correspondence conducted at the home office if related to the project.
- Field memoranda. Various types of correspondence, less formal than letters, issued at the jobsite to subcontractors, forepersons and other individuals; job directives, safety issues and disciplinary matters.
- Transmittals. Accompany submittals to owner's representative, subcontractors and material suppliers, accompany requests for payment and accompany certificates and other documentation at project closeout.
- Requests for information (RFIs) issued from contractor to design professional or owner for clarification of design information or to present any other questions; includes summary log containing the status of each request.

Materials management

- Purchase orders. Issued to material suppliers, as explained in an earlier section.
- Submittals. Shop drawings, product information and samples, plus status information on their review and approval.
- Expediting and delivery information. Status of manufacturing, shipping and delivery of each material item, with comparisons against project schedule. Receiving information for delivered items.
- Material inventory status. Inventory of all stored materials, with quantities and dates when items were added and withdrawn.
- Quality control reports. Concrete test reports, soil laboratory results, off-site testing of steel components and so on.
- Invoices. Also part of cost system. All requests for payment of materials, with summary log of status of each.

Financial management

- Requests for payment. Prepared by contractor to request periodic payments as the project progresses, as explained earlier; based on measurements and progress to date.
- Cost and budget tracking reports. Comparisons of actual costs to date against planned costs, for individual work items and the project as a whole; analysis of cost trends.
- Variation requests. Requests from the owner for proposals for changes in the work; summary log of status of each.
- Variation proposals. Prepared by the contractor for the owner, in response to variation requests; summary log of status of each.
- Variations. Orders from the owner authorising changes in the work; summary log of status of each. Note that one log may suffice for variation requests, variation proposals and variations.

This rather overwhelming series of lists demonstrates that the contractor must develop and maintain an effective system for coordinating all documents and knowing the status of each. Computer-based systems are available to assist contractors with this vital document management function. Such a management system ought to have the following characteristics.

- It must provide a means for recording and tracking the status of every document. Because the status of any document usually changes over the course of the project, the system must provide for this dynamic nature.
- It must be able to generate documents. For example, it is desirable that it prepare transmittal documents, meeting minutes, daily reports and similar paperwork as well as track their status. Integration of a word processor is required.
- It must have the capability to analyse the project's document management performance. As an example, it is helpful to know how much time the various parties take in reviewing each submittal, what their average time is and how this performance compares with other projects.
- Graphical presentation of the document system's status is helpful. Some software provides pie charts of the project's cost status, as well as bar charts that represent the time for each step in the submittal review process.
- Search features allow the easy retrieval of any desired document.
- The system must make the information easily available to those who need to use it. Most such systems allow for the transmittal of information via electronic mail and some provide Internet or intranet access. The use of such web-based collaborative communication techniques is the subject of the section that follows our discussion of an example computerised document management system.

A large number of document management systems have been developed (Construction Industry Computing Association, 1998, 2002); a comparative review is well beyond the scope of this book. In general, the software is organised into the kinds of categories we have listed above. For most categories, a log of the status of each document in the category is provided. In many cases, the documents themselves can be created electronically within the program; as a document is generated and manipulated, the program records the actions, such as the date it was prepared. As a simple example, consider the preparation of minutes for the weekly project progress meeting. A template containing fields for date, time, attendees, discussion, actions, assignments,

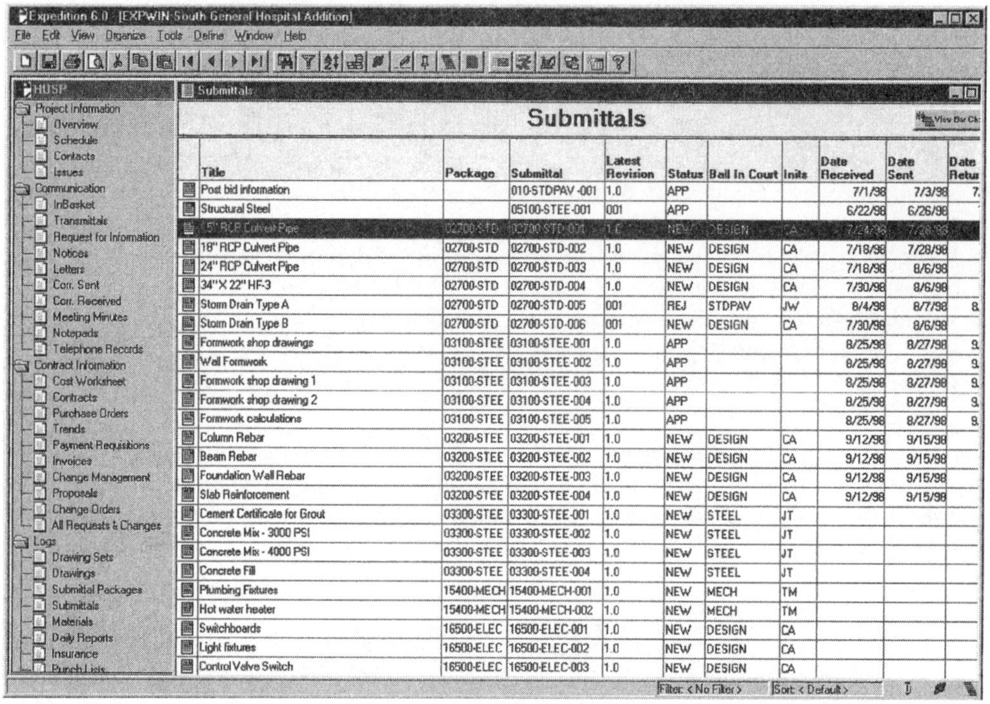

Expedition 6.0 - [EXPWIN South General Hospital Addition]

File Edit View Organize Tools Define Window Help

HUSP
- Project Information
  - Overview
  - Schedule
  - Contacts
  - Issues
- Communication
  - InBasket
  - Transmittals
  - Request for Information
  - Notices
  - Letters
  - Corr. Sent
  - Corr. Received
  - Meeting Minutes
  - Notepads
  - Telephone Records
- Contract Information
  - Cost Worksheet
  - Contracts
  - Purchase Orders
  - Trends
  - Payment Requisitions
  - Invoices
  - Change Management
  - Proposals
  - Change Orders
  - All Requests & Changes
- Logs
  - Drawing Sets
  - Drawings
  - Submittal Packages
  - Submittals
  - Materials
  - Daily Reports
  - Insurance
  - Punch Lists

Submittals

Submittals

Title	Package	Submittal	Latest Revision	Status	Ball In Court	Inits	Date Received	Date Sent	Date Retur
Post bid information		010-STDPAV-001	1.0	APP			7/1/98	7/3/98	7.
Structural Steel		05100-STEE-001	001	APP			6/22/98	6/26/98	
15" RCP Culvert Pipe	02700-STD	02700-STD-001	1.0	NEW	DESIGN	CA	7/24/98	7/28/98	
18" RCP Culvert Pipe	02700-STD	02700-STD-002	1.0	NEW	DESIGN	CA	7/18/98	7/28/98	
24" RCP Culvert Pipe	02700-STD	02700-STD-003	1.0	NEW	DESIGN	CA	7/18/98	8/6/98	
34"X 22" HF-3	02700-STD	02700-STD-004	1.0	NEW	DESIGN	CA	7/30/98	8/6/98	
Storm Drain Type A	02700-STD	02700-STD-005	001	REJ	STDPAV	JW	8/4/98	8/7/98	8.
Storm Drain Type B	02700-STD	02700-STD-006	001	NEW	DESIGN	CA	7/30/98	8/6/98	
Formwork shop drawings	03100-STEE	03100-STEE-001	1.0	APP			8/25/98	8/27/98	9.
Wall Formwork	03100-STEE	03100-STEE-002	1.0	APP			8/25/98	8/27/98	9.
Formwork shop drawing 1	03100-STEE	03100-STEE-003	1.0	APP			8/25/98	8/27/98	9.
Formwork shop drawing 2	03100-STEE	03100-STEE-004	1.0	APP			8/25/98	8/27/98	9.
Formwork calculations	03100-STEE	03100-STEE-005	1.0	APP			8/25/98	8/27/98	9.
Column Rebar	03200-STEE	03200-STEE-001	1.0	NEW	DESIGN	CA	9/12/98	9/15/98	
Beam Rebar	03200-STEE	03200-STEE-002	1.0	NEW	DESIGN	CA	9/12/98	9/15/98	
Foundation Wall Rebar	03200-STEE	03200-STEE-003	1.0	NEW	DESIGN	CA	9/12/98	9/15/98	
Slab Reinforcement	03200-STEE	03200-STEE-004	1.0	NEW	DESIGN	CA	9/12/98	9/15/98	
Cement Certificate for Grout	03300-STEE	03300-STEE-001	1.0	NEW	STEEL	JT			
Concrete Mix - 3000 PSI	03300-STEE	03300-STEE-002	1.0	NEW	STEEL	JT			
Concrete Mix - 4000 PSI	03300-STEE	03300-STEE-003	1.0	NEW	STEEL	JT			
Concrete Fill	03300-STEE	03300-STEE-004	1.0	NEW	STEEL	JT			
Plumbing Fixtures	15400-MECH	15400-MECH-001	1.0	NEW	MECH	TM			
Hot water heater	15400-MECH	15400-MECH-002	1.0	NEW	MECH	TM			
Switchboards	16500-ELEC	16500-ELEC-001	1.0	NEW	DESIGN	CA			
Light fixtures	16500-ELEC	16500-ELEC-002	1.0	NEW	DESIGN	CA			
Control Valve Switch	16500-ELEC	16500-ELEC-003	1.0	NEW	DESIGN	CA			

Filter: <No Filter> Sort: <Default>

Figure 6.11 Sample submittal log from Expedition® (Primavera Systems, Inc., 2002).

drawing references and other information that must be recorded is part of the program; the user completes the appropriate blanks and files the document. In the process, the status log of meeting minutes is updated. On the other hand, a submittal record for product information simply records and tracks the status of these documents; the documents themselves are not likely to be in electronic form but will come to the project office in 'hard copy' form. A variation request may come to the contractor in electronic form; if so, it can be filed electronically and then tracked by the variations log in the document management system as it is responded to by the preparation of a variation proposal. Similarly, contract drawings and shop drawings may be in electronic form; if so, they can be made part of the system and then logged as they are used in various ways. If they are not in electronic form, their status will still be tracked using the appropriate logs in the system.

As an example, we show in Figure 6.11 a sample submittal log from the Expedition® document management system (Primavera Systems, Inc., 2002). For each submittal, the screen shows its title, the submittal number, the document package of which it is a part, its status (new, approved and so on), the organisation and person responsible for the next action ('ball in court') and the dates it was received by the contractor, sent to the next party and returned from them. Not shown on this screen print, because they are beyond the right-hand edge of the screen, are the date the contractor forwarded the submittal after it was returned, plus information about involved organisations and people, the number of days the document was held, whether it is overdue and by how many days and the date its approval is required. This log is only a summary and additional information for any individual document can be displayed by clicking on the leftmost column; Figure 6.12 shows details about the 15" culvert pipe. Also of interest in Figure

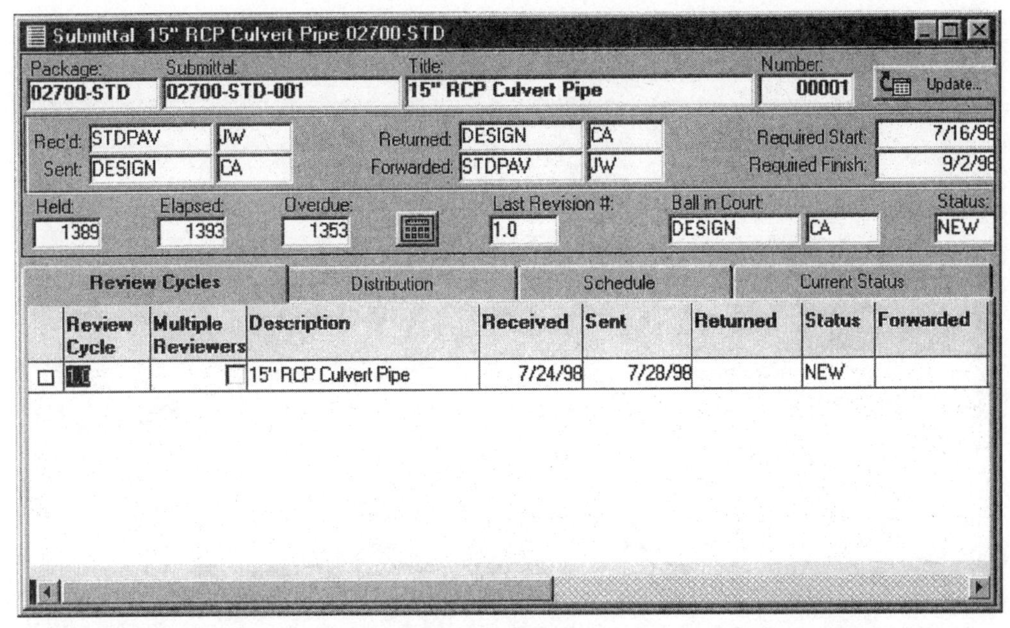

Figure 6.12 Sample submittal detail from Expedition® (Primavera Systems, Inc., 2002).

6.11 is the left-hand side of the screen, which shows all document folders, organised into four groupings: project information, communication, contact information and logs.

Electronically enhanced project communications

The 2001 report on *Modernising Construction* in the UK (Comptroller and Auditor General, 2001) identified 'effective communication and coordination of all those involved in the construction supply chain' as a key characteristic of good project management. The widespread availability of both the Internet and intranets within organisations provides significant opportunities for contractors, owners and all other members of the construction project team to enhance communications among the entire team. We have already considered many computer-based systems that assist the contractor prior to and during the operations phase, including programming and schedule control, cost estimating, accounting and cost control and document management. Now we discuss the use of project-specific websites to allow communication of this and other information rapidly, conveniently, accurately and inexpensively throughout the project organisation.

Construction organisations use web-based databases of project information that can be accessed or 'pulled' by members of the project team as a means of speeding information flow by circumventing the traditional chains of command that 'push' information to team members. Thus, websites can 'eliminate information push and enhance information pull' (Thorpe and Mead, 2001). In some cases, the contractor may be required by the contract to establish such a site or participate in a site that has already been established by the owner or its representative. If the contractor uses websites for its projects, it must decide whether to (1) host the sites itself by providing the server (the computer where the website is stored and

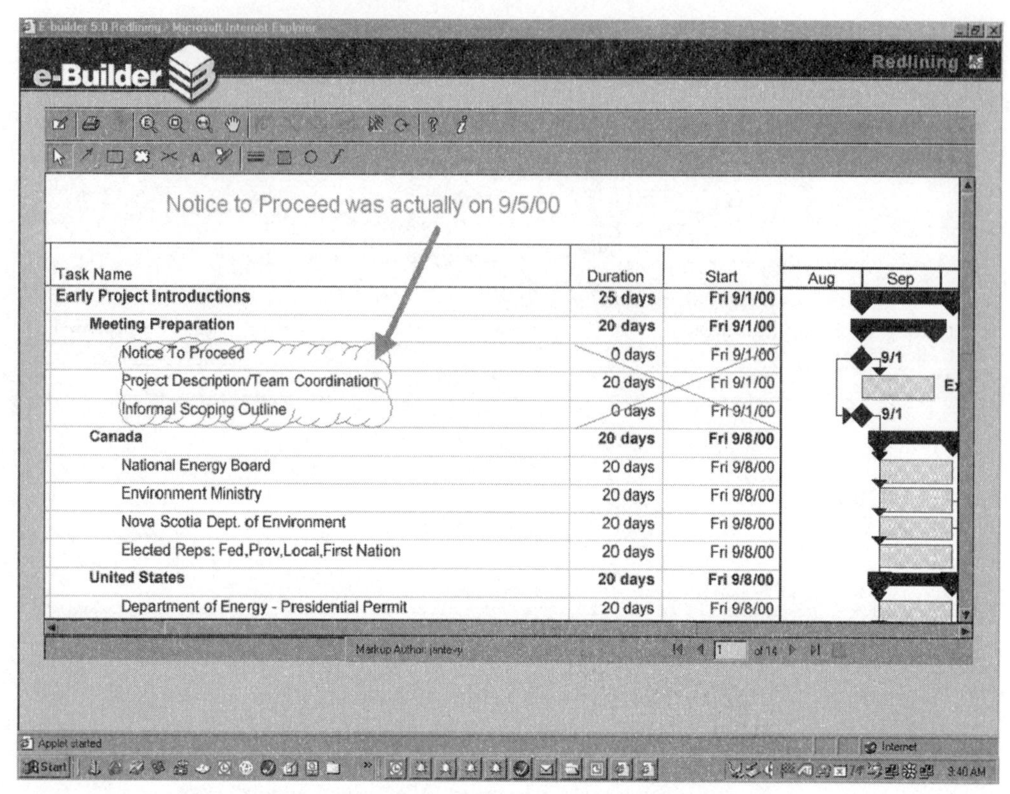

Figure 6.13 e-Builder™ website sample. Red-lined project schedule, wherein suggested modifications are shown in red (courtesy of e-Builder, Inc.).

from which the other computers in the network access the site) and maintaining the website software on that computer or (2) invoke the services of an application service provider (ASP), whose server and program are available to any member of the project team with a personal computer and web browser. Those who choose the first option often purchase website software designed expressly for construction projects, although several large companies have designed their own packages.

Whether the website is self-hosted or maintained by an ASP, the contractor is still responsible for deciding what types of information will be included on the site and then providing the databases of current information. Many systems provide security over confidential information through the use of passwords. Once a project team member is granted permission to access the website or a portion thereof, the information becomes available for viewing and use. The various capabilities included in these web-based systems include collaboration, workflow, work process management and conventional communication (Kraker, 2000). Collaboration capabilities can include review of drawings, in some cases permitting commenting and red-lining; review of computer-aided design models and photos, both regular progress photos and photos taken to discuss and resolve particular issues or problems; message boards and discussions; and scheduling or hosting on-line meetings. Enhanced workflow can be made possible through transmission of meeting minutes, RFIs, transmittals, submittals, approvals and change notifications and the management of correspondence; note that this is the document management function considered in the previous section.

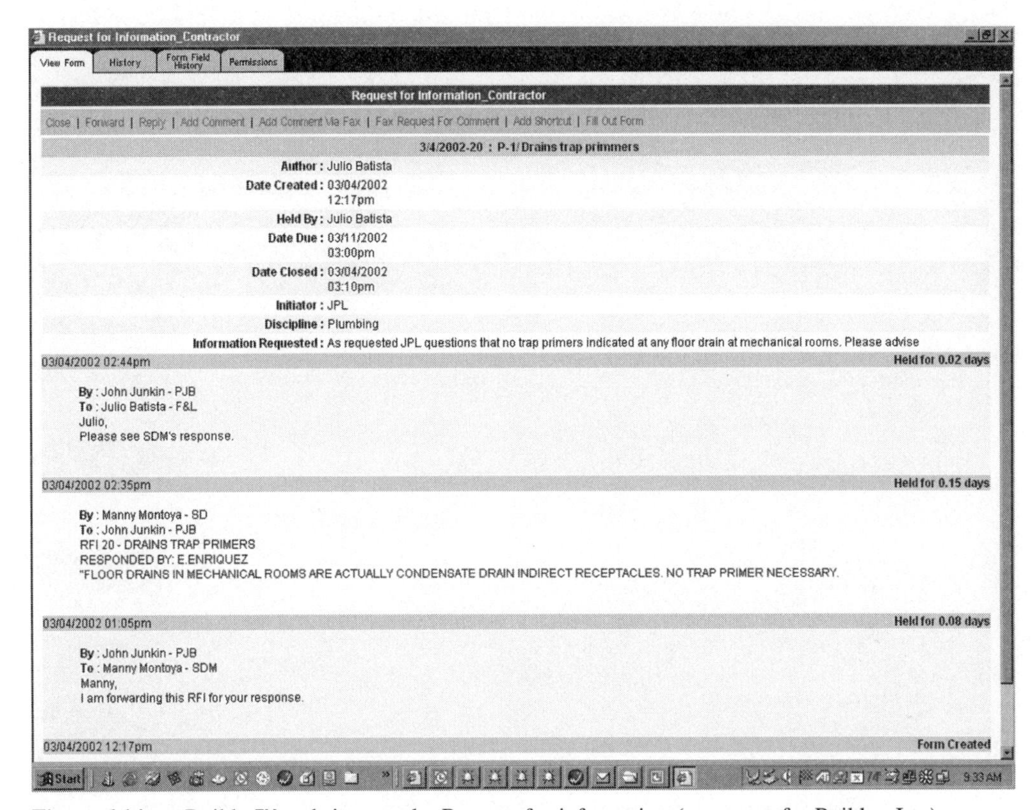
Request for Information_Contractor

View Form | History | Form Field History | Permissions

Request for Information_Contractor

Close | Forward | Reply | Add Comment | Add Comment Via Fax | Fax Request For Comment | Add Shortcut | Fill Out Form

3/4/2002-20 : P-1/ Drains trap primmers

Author : Julio Batista
Date Created : 03/04/2002 12:17pm
Held By : Julio Batista
Date Due : 03/11/2002 03:00pm
Date Closed : 03/04/2002 03:10pm
Initiator : JPL
Discipline : Plumbing
Information Requested : As requested JPL questions that no trap primers indicated at any floor drain at mechanical rooms. Please advise

03/04/2002 02:44pm — Held for 0.02 days

By : John Junkin - PJB
To : Julio Batista - F&L
Julio,
Please see SDM's response.

03/04/2002 02:35pm — Held for 0.15 days

By : Manny Montoya - SD
To : John Junkin - PJB
RFI 20 - DRAINS TRAP PRIMERS
RESPONDED BY: E.ENRIQUEZ
"FLOOR DRAINS IN MECHANICAL ROOMS ARE ACTUALLY CONDENSATE DRAIN INDIRECT RECEPTACLES. NO TRAP PRIMER NECESSARY.

03/04/2002 01:05pm — Held for 0.08 days

By : John Junkin - PJB
To : Manny Montoya - SDM
Manny,
I am forwarding this RFI for your response.

03/04/2002 12:17pm — Form Created

Figure 6.14 e-Builder™ website sample. Request for information (courtesy of e-Builder, Inc.).

In the realm of work process management, the project website can be used for compiling and displaying the project programme and the associated job progress reporting; for compiling the cost estimate, including the solicitation of material prices and subcontractor tenders; for displaying the project budget and updating the cost status; and for other accounting functions. Typical project website software exchanges work process management data with other applications, such as project cost accounting software. Finally, project websites often allow team members to send and receive fax messages, to communicate by e-mail within and outside the project and to interface with the various functions of personal digital assistants.

Figures 6.13–6.15 show sample screens from e-Builder™, one project website product developed for construction project use by e-Builder, Inc. (2002). In Figure 6.13, a schedule document has been red-lined and made available to the team. Figure 6.14 shows details of a RFI generated by a contractor regarding floor drains, while Figure 6.15 is a brief construction status summary.

The management of large amounts of information with rapid response times is the feature of project websites that makes them well suited to construction project management. A sports arena project in California, USA involved more than 32 000 RFIs among its other paperwork. It was estimated that response times were reduced 60% through the use of a project website (Antevy, 2002). Websites can provide all team members with the same information in a reliable and easily retrievable manner (O'Brien, 2000). Some cautions are in order, however. Team members need both tools for accessing the information and a collaborative commitment to the project, its success

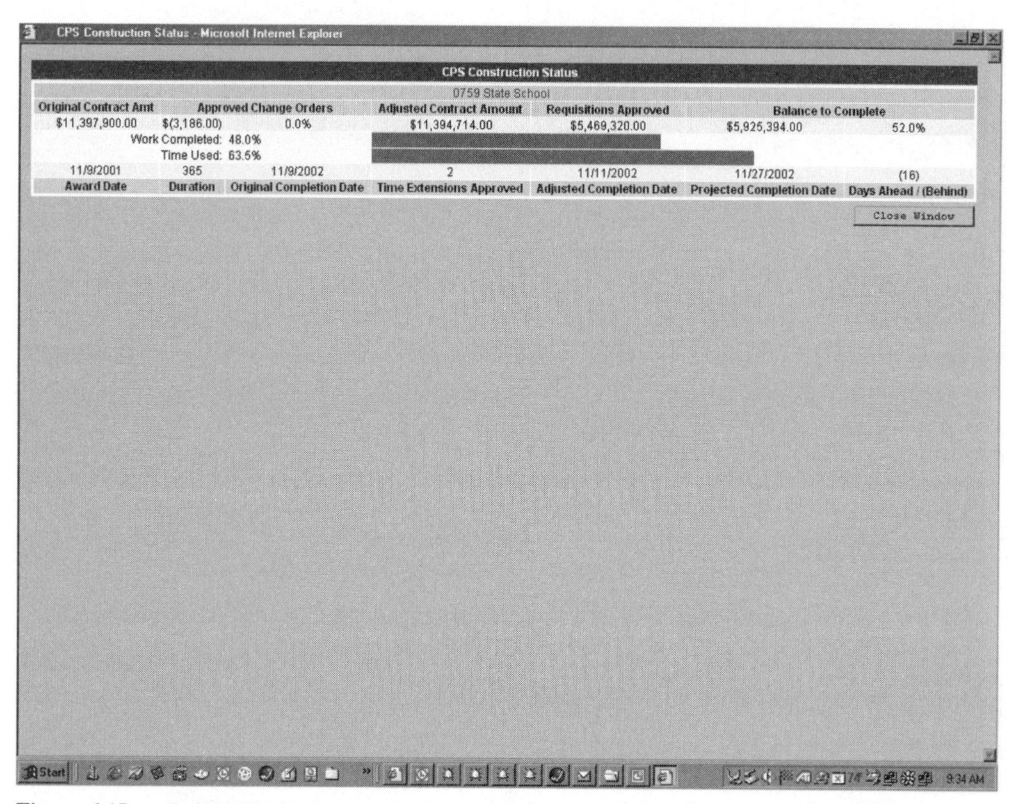

Figure 6.15 e-Builder™ website sample. Construction status summary (courtesy of e-Builder, Inc.).

and its use of a website. There is a tendency towards resistance to change with the introduction of any new technology such as this form of communication. The issuance of passwords to selected team members for certain information, while necessary, tends to violate the spirit of a holistic approach and the claim that all information is available to everyone. Finally, project websites provide yet another communication channel; they may add confusion in an atmosphere where communication is already occurring by many other means (Bennett, 2000). O'Brien (2000) provides some excellent recommendations for effective implementation of project websites.

The challenge for the modern construction manager interested in utilising project websites is to keep up to date in a very rapidly changing field. Of all the information technology products available to the contractor, it is likely that web-based collaborative communication tools, as well as on-line purchasing, or e-commerce, will evolve most dramatically during the few years after the publication of this book. Thus, a survey of particular products and their features would not be useful here.

The providers of such software and services tend to enter and leave the market quickly, and their products' features are constantly changing. As of this writing (late 2002), the *Directory of Project Management Software* (Project Management Center, 2002) at http://www.infogoal.com lists more than 125 providers of web-based project management software. A careful reading of the list indicates that only a small proportion offer project website software or services. A directory such as this will be helpful to the manager wishing to determine the latest available products suitable for their particular needs.

Doherty (2001) suggests that during the decade of the 2000s 'it is inevitable that almost everything will be linked and connected to the Internet, with project-specific websites (known as extranets) and e-marketplaces acting as integrative hubs'. If that is so, the technology will be available for vast improvements in construction project communication worldwide. The manager's task will be to select those features that will be most useful and provide leadership for their effective implementation.

Selected legal issues

In this final section of our consideration of the contractor's activities and responsibilities during the project operations phase, we must confront the fact that legal matters often arise as the fieldwork progresses. We begin by describing a typical process for filing a claim, usually against the owner. Because the claims process sometimes escalates into a contract dispute, we next discuss means for preventing disputes and then present various approaches to their resolution. Lastly, we consider a selection of specific issues that frequently lead to disputes.

Claims process

Although the owner may file claims against the contractor, this discussion is confined to claims filed by the contractor. The issues that are the subjects of claims tend to depend upon the type of contract. For example, in a fixed-price contract, a claim may be filed by the contractor if it finds conditions at the site to be different from those represented by the contract documents, whereas the contractor in a cost-plus contract is not likely to file a claim in that case. A more common claim under a cost-plus contract arises out of the interpretation of the term *reimbursable costs*; the contractor may claim that the costs were incurred but were not allowed as reimbursable, while the owner may counter that such costs are to be covered by the fee.

The kinds of conditions giving rise to claims by the contractor include late payments, changes in the work, constructive changes, changed conditions, delay or interference by the owner, acceleration by the owner resulting in loss of productivity, errors or omissions in design, suspension of the work, variations in bid-item quantities, failure to agree on variation pricing and rejection of requested substitutions (Clough and Sears, 1994). The contract documents ought to provide guidance for the contractor wishing to file a claim. The latest version of the American Institute of Architects' (1997) *General Conditions of the Contract for Construction* specifies a typical procedure. The contractor must refer the claim to the architect within 21 days of the occurrence of the event that leads to the claim. Until the claim is resolved, the contractor must continue to work on the project and the owner must continue payments to the contractor. Within 10 days after it has received the contractor's claim, the architect must take one of five actions: (1) request more information about the claim from the contractor, (2) notify the contractor when it will take action on the claim, (3) reject the claim, (4) recommend approval of the claim or (5) propose a compromise. Depending upon the architect's response, the contractor is then given another 10 days to submit more information, modify the claim to make it more acceptable or respond that the original claim remains unchanged. If this response does not lead to resolution, the architect must then state in writing that it will render a final and binding decision within 7 days (Collier, 2001).

It is in the best interests of the parties involved that claims be resolved as quickly as possible. Most claims are settled using a process such as that outlined above. Many are resolved quickly, to the satisfaction of both parties. If an unforeseen condition is encountered, say an unexpectedly difficult soil layer, with proper documentation the contractor may convince the owner or its representative rather easily that additional compensation and/or time should be granted. When a claim cannot be resolved through such a process, the issue can be said to have become a *dispute*, which is a legal disagreement involving two or more parties. Rubin et al. (1999) make an important point about the term *claim*, in their discussion of variations, or change orders:

> . . . in the change order definition the word 'claim' is used to mean a legitimate request for that adjustment in the case of a change. In usage, however, the word often connotes a dispute or demand. 'Claims' should be a neutral term; . . . a claim doesn't have to lead to a lawsuit. In the true sense of the word, a claim is a request for an equitable adjustment due to a change under the contract. However, those requests too often end up in less than neutral situations.

In the following section we consider the disputes that sometimes arise from unresolved claims, including ways to prevent them and methods for resolving them.

Dispute prevention and resolution

The American Institute of Architects' (1997) *General Conditions of the Contract for Construction* discussed above specifies that the architect's 'final and binding' decision at the end of the process, if the process extends that far, is subject to arbitration. Thus, under that set of contract documents, a procedure for attempting to resolve disputes is stipulated. We describe in this section a number of dispute-resolution methods in use in the construction industry, after considering ways to prevent such disputes from reaching that stage.

A construction contract dispute arises when one of the parties is unsatisfied with the result of a claim. In the example of the discovery of an unexpected soil layer cited previously, the owner may deny the contractor's claim for additional compensation or time or may grant only part of what the contractor sought. If so, the contractor may choose to pursue the matter further, using a process outlined in the contract documents or some other process. When the decision is made to pursue the matter further, the issue becomes a *dispute*. Indeed, Murdoch and Hughes (2000) suggest that '*disputes* arise when a conflict becomes an altercation'.

The contractor is well advised to remember that it must have exhausted the claims process before it pursues the dispute resolution process. Any authority in charge of conducting a formal dispute resolution will inquire whether all the required steps have already been followed to resolve the claim. If they have not, it is likely the request for consideration of the dispute will be denied. The contractor who 'files a lawsuit' immediately when a potential claim arises will be told to follow the rules and then come to the court if still unsatisfied. An application for arbitration is also likely to be denied under the same circumstances.

It is helpful to divide such construction contract disputes into two aspects, legal and technical. The legal aspect has to do with whether the party seeking redress has any legal entitlement to a remedy, based on interpretation of the contract and other documents. If so, the technical data is used to ascertain the magnitude of the recompense. Most disputes contain both of these aspects. If a large television transmitter tower collapses in the middle of the night in an isolated area with no one watching, as actually happened at Emley Moor, the investigation of the reasons for the collapse is a technical matter. The legal documents indicate which party or parties may be entitled to compensation or another remedy and the technical investigation determines the extent of the damage, the allocation of blame and the magnitude of the compensation (Murdoch and Hughes, 2000).

Dispute prevention

The construction project will be a better project if disputes among the parties can be prevented altogether. Doing so involves the entire project team and it requires measures that can be taken throughout the project life cycle. The following list suggests ways by which construction contract disputes can be prevented or at least minimised.

- Accurate, complete design documents. The more clearly the project is defined through the documents, the less likely there will be conflicts. Good design (sound engineering and mutually consistent drawings and specifications) can be enhanced through design reviews: internally, by the owner and by independent peers.
- Contract provisions that allocate risk equitably among the parties. Contracts that make the contractor entirely responsible for such risks as weather, unforeseen conditions, delay by the owner, labour unrest, material shortages and inflation are likely to be more dispute prone than those that spread such risks among the parties.
- Constructability reviews. We have encouraged the use of such reviews during the design process, utilising experts on practical aspects of construction (who often can be found among senior field supervisors) to advise on the plausibility of the design from the contractor's point of view, including the cost and schedule impacts of various alternatives.
- Partnering. This is a familiar topic to the reader of earlier portions of this book. It is a means of fostering cooperation throughout the building process. The parties anticipate problems and design an approach that commits them to resolving issues before disputes arise. Partnering can be applied from the initial stages of contract development, or it can be introduced in conjunction with an experienced facilitator, as a means of creating a positive contract management environment in a project that has become soured by disputes.
- A reasonable, sufficiently detailed project programme and adherence to the programme by all parties. Whichever party has overall responsibility for the programme, all must agree it to and each must endeavour to fulfil its schedule obligations. Adequate flexibility must be part of the management of the programme, to enable the parties to deal with unforeseen circumstances that may delay execution.
- Maintenance of thorough project records. Project record keeping is not just the contractor's responsibility. Each party must maintain appropriate documentation, from minutes of discussions to variations; distribution of such records to those authorised and needing to know is also a wise means of minimising misunderstanding and conflicts. Site diaries should be kept by the responsible supervisors appointed by each of the parties and should be factual and brief.
- Minimising the number of contract changes. This suggestion can be realised through the careful and complete design in item 1. Because changes to the contract can result in misunderstandings by the owner, design professional and contractor, including conflicts about their costs, fewer changes ought to result in fewer disputes.
- Dispute review boards or single experts for smaller projects. We shall shortly consider several methods of alternative dispute resolution, including dispute review boards. Often the presence of such a board, which maintains a presence throughout the construction operations phase and is ready to evaluate problems and suggest resolutions, will tend to deter disputes or solve them in their early stages.

Most of the foregoing has to do simply with good management, which is what this book is about!

Resolution through court systems

Over many years, parties involved in disputes, including those involved in construction contract disputes, have relied upon their country's court systems for the resolution of their disagreements. Although construction contracts and their participants now often seek to avoid the legal process, or *litigation*, because of its perceived costly and time-consuming nature, in favour of various alternative resolution methods, the court system still provides the forum for resolving many such conflicts. The court system in each country has its own unique aspects, but certain characteristics of the legal process are common throughout most of the world. An excellent discussion of the resolution of construction disputes via the court system and by various other means in the UK is found in Murdoch and Hughes (2000). Here we outline briefly the process usually followed in seeking to resolve a dispute by litigation.

The process applies not just to construction disputes but also to other private conflicts wherein one party seeks redress against another. This is the part of the law known as *civil law*, as opposed to *criminal law*; in the latter, parties are brought to trial for alleged crimes, which are violations of public laws. The kinds of conflicts that arise between parties to a contract are dealt with in civil law. Suppose the contractor decides to file a lawsuit against the owner for allegedly delaying the contractor by not having the land available, not approving submittals in a timely fashion or ordering the contractor to stop work; the contractor would be referred to as the *plaintiff* or the party bringing the complaint ('filing the suit'). The owner, the party being sued and who must defend itself, would be called the *defendant*. The various activities that take place as a lawsuit proceeds can be conveniently divided in to the *pre-trial* activities and the *trial* activities (Bennett, 1996).

After engaging an attorney, the first step for the plaintiff in the pre-trial phase is to *file a complaint*, which designates the defendant and describes the relief being sought. This complaint, together with the *summons*, is then delivered to the defendant; the summons indicates that a suit has been filed, specifies the court which has jurisdiction and directs that the defendant respond within a stipulated timeframe. The defendant may respond in one of several ways: provide no answer at all, in which case the court decides in favour of the plaintiff with no further proceedings; issue a statement claiming there are insufficient facts to support the case; deny the claims altogether; state that an affirmative defence will be provided, including new information sufficient to preclude recovery by the plaintiff; and/or file a counterclaim. Either side may file a request for a *summary judgement*; in this case the position taken is that the facts are not disputed and therefore a trial is not required. The requesting party seeks in such a request to have the judge enter a judgment on its behalf, after conducting a hearing and with no further proceedings.

The process of *discovery* is an important aspect of trial preparation in many legal systems. The notion is that each side is entitled to learn as much as possible about its adversary's position, prior to the beginning of the trial. In the discovery phase, the parties obtain sworn testimony of witnesses in the form of depositions at hearings, provide written questions for the other side to answer, ask for copies of documents or other physical evidence and conduct physical and mental examinations of people involved in the case. Before the trial begins, the judge convenes one or more pre-trial conferences, at which procedures and schedules are decided, the issues are clarified and attempts are made to settle the case out of court.

In some cases, a judge will hear the case and render a decision; such a trial, without a jury, is called a *bench trial*. In other cases, depending upon the legal system, the type of case and whether or not one of the parties so requests, a jury of lay persons will listen to the trial testimony and decide, based upon its understanding of the facts, in favour of one of the parties. Thus, in the case of a *jury trial*, the first step in the trial phase is to select the jury. After that, the trial proceeds as follows, whether or not a jury is involved.

Counsel for each side presents opening statements, after which the plaintiff's side presents testimony to support its case. This *direct examination* of witnesses and the presentation of their testimony may be followed by *cross examination* of those witnesses by the defendant's side. After that, the plaintiff's side may conduct a *redirect examination* of its witnesses and the defendant's side may follow with a *re-cross-examination*. The plaintiff 'rests', having completed its presentation. The defendant may ask the judge to dismiss the case on the basis that there is not enough evidence to support the case. If this motion is unsuccessful, the defendant presents its side of the case, beginning with its direct examination and continuing as outlined above. After closing arguments, the judge issues instructions to the jury, if there is one; the jury then conducts its deliberations and reaches a *verdict*. The rules vary in different jurisdictions as to whether the jury's decision must be unanimous, three-quarters majority, simple majority or some other proportion. If they decide in favour of the plaintiff, the jury must also decide on the appropriate amount of damages. Various post-trial motions may be introduced, such as a motion for a new trial, if one side believes the judge committed a prejudicial error, the damages were inappropriate or other reasons. If they deny those post-trial motions, the judge then enters a *final judgment*. Although it is 'final', it can be appealed to a higher court. The appeal process does not re-try the case, but instead is confined to deciding whether any errors of law were committed during the trial. No jury is involved, with attorneys presenting *briefs* stating their positions to the appellate court. The trial court's decision may be affirmed, reversed or modified, or if a decision cannot be made for one party or the other, the case may be remanded back to the lower court for a new trial.

Courts of law have strict rules relating to *evidence*, which is the means by which facts are proven in a trial. Evidence consists of *real evidence*, which is something physical, such as a document or a piece of building material and *testimony*, or statements by witnesses. The types of evidence allowed are defined in each jurisdiction; for example, usually *hearsay* evidence, or second-hand information, is not allowed. In general, witnesses may express only the facts as they observed them, such as the colour of an object or the motion of a person or vehicle and not opinions based on their observations. However, an *expert witness*, such as a professional engineer, is allowed to offer *opinion testimony* if the topic is within their area of expertise.

It is evident from this brief introduction to resolution of disputes through the court system that the process is very formal. It tends to be time consuming and is often expensive. It is also well defined, with relatively few surprises (despite various popular television portrayals!) and the parties have a fair idea of the steps to be followed. Records of trials are usually kept and many systems allow the use of prior decisions as a means of establishing arguments for a party's position through *judicial precedent*. (The opposing side will no doubt seek other precedents to strengthen its case!) Murdoch and Hughes (2000) state that 'the use of courts for settling disputes is expensive, uncertain and time consuming. It is also very public.' The US 9th Circuit Court of Appeals used stronger words in claiming that lawsuits are 'clumsy, noisy, unwieldy and notoriously inefficient. Fuelled by bad feelings, they generate much heat and friction, yet produce little that is of any use' (Dispute Prevention and Resolution, Inc., 2000). Worldwide, the construction industry has been at the forefront of the use of various alternatives to litigation for the resolution of construction disputes. We now describe some of those *alternative dispute resolution* methods.

Alternative resolution methods

This discussion of alternative means for resolving construction disputes is limited, by necessity, to only a few facts about each. Further information can be found in a number of references, including those upon which this discussion is based (Bennett, 1996; Bockrath, 2000; Harris, 2001; Hollands, 1998; Murdoch and Hughes, 2000; Rubin *et al.*, 1999; Sweet, 2000). We begin with less formal methods and move toward those that begin to resemble the law court-based process described in the previous section.

Negotiation is a voluntary process by which the parties come together to attempt to resolve their differences without the aid of an outside third party. Certainly negotiation is used to find agreement on many types of matters not related to construction or to disputes, such as the negotiation of the contract itself. Hollands (1998) suggests that negotiation is 'interest-based bargaining', in which at least some of each party's interests or needs are satisfied in some way. Careful preparation for the negotiation is essential, as is the selection of a skilled negotiator. A good way to begin attempting to resolve a construction dispute ought to involve some sort of negotiation, with or without the assistance of attorneys, but always without a third-party facilitator. If an agreement is reached, it is important to incorporate the terms into a written document as soon as possible.

If the parties decide to involve a third party, a first step might be through *conciliation*. The conciliator is independent of the parties involved in the dispute and must maintain strict confidentiality. This person talks with the parties in private, seeking to ascertain the facts of the case and to establish some common ground for its resolution. After the private meetings, they may bring the disputants together for an open discussion. Beyond that, it is up to the parties themselves to come to an agreement if they can and to decide upon its terms.

The outside party has a larger role in *mediation*. Like the conciliator, the mediator is an independent person who meets with the parties in private as a sort of 'shuttle diplomat' and then brings them together to attempt to resolve the dispute. While the mediator does not make a decision in the case, the role includes making recommendations for its resolution. In this sense, mediation can be called 'an extended version of conciliation' (Murdoch and Hughes, 2000). Harris (2001) labels mediation as 'supervised negotiation', a term that could apply to conciliation as well. Like negotiation and conciliation, mediation is simple, quick, inexpensive, private and confidential. Like conciliation, it also relies upon an expert third party.

Dispute review boards (DRBs) are an increasingly popular method of resolving construction disputes. A distinguishing characteristic is that a DRB is chosen at the outset of the project and is available throughout its duration to hear and attempt to resolve disputes. A board usually consists of three members, one chosen by the contractor and approved by the owner, one chosen by the owner and approved by the contractor and the third chosen by the other two and approved by both contractor and owner. Members have technical expertise and construction experience. They are provided with construction documents, visit the site regularly and are made privy to progress and status information. If a dispute arises, the board convenes quickly, gathers required information and makes it recommendation.

The relatively relaxed culture of DRBs often results in considerably less formality than some other dispute resolution mechanisms. Recommendations are not binding, but the parties generally accept the board's recommendations, reasoning that construction experts have considered their concerns impartially and it is unlikely that a more favourable outcome would result from taking the matter further. If the case is continued through litigation, many jurisdictions allow the DRB deliberations to be introduced as evidence in the trial. In the international arena, the World Bank's recommended construction contract documents provide for the assignment of disputes to a DRB; either party may reject the board's recommendation, after which the dispute moves on to arbitration. The presence of a dispute review board throughout the operations phase of the project often leads to fewer disputes. The parties tend to resolve minor misunderstandings by themselves, not wishing to appear foolish or petty in front of a panel of experts who may be called upon later to resolve a major issue. In this sense, DRBs may be regarded as dispute avoidance, as well as dispute resolution, mechanisms.

A *mini-trial* is basically a structured form of mediation, involving top management personnel from each side taking the part of a 'jury'. A neutral advisor coordinates the proceedings, which follow somewhat the routine in a court trial, except that the discovery period is compressed and statements by both sides are limited. A mini-trial is a private, non-binding procedure; unlike mediation, 'it gives the parties the psychological satisfaction of having their "day in court", without the attendant cost, delay and aggravation' (Rubin et al., 1999). Preparation for a typical mini-trial consists of developing short 10–15-page position papers to exchange with the other side. During the hearing, attorneys or others from each side make brief presentations to a panel consisting of (1) key executives from both sides who have the full power to settle the dispute and (2) the neutral advisor. Management representatives can interact and can ask clarifying questions. A short rebuttal period is permitted, after which the panel meets in private to attempt to resolve the dispute. Because its results are non-binding, a mini-trial that leaves one party believing it has not been treated fairly may be unsuccessful, leading to other attempts at reconciliation, such as arbitration or litigation. The active participation by top executives of each side tends to lead to satisfactory results; one report noted that 90% of cases utilising mini-trials have been resolved successfully (Bennett, 1996).

The previous five alternative dispute resolution methods might be considered non-adversarial; collectively, they are sometimes referred to as *reconciliation*. They are private, they are voluntary, their results are non-binding; at any time in the proceedings, either party may terminate its participation and resort to litigation. The remaining two methods, together with resorting to the court system, are decidedly more adversarial.

In some jurisdictions, especially the UK since 1996, parties to construction contracts may refer any dispute arising from their contracts to an independent third party under *adjudication* proceedings (Murdoch and Hughes, 2000). Under one form of adjudication, the contract must contain provisions for its use, including the right of either party to give notice of a filing at any time, a specific timetable for handling disputes and certain rights and duties of the adjudicator. Under such a contract, a party seeking relief begins the process by notifying every other party of the reasons for its filing and the type and amount of relief being sought. If the contract includes the name of the adjudicator, the notice is sent to that person as well; if not, the notice goes to a body charged with selecting such a person and in accordance with a specified timeframe and qualifications, an appointment is made. After the aggrieved party provides relevant documents to the adjudicator and other parties, the adjudicator is given considerable latitude in deciding a procedure, requiring further documentation, calling witnesses, making site visits, holding hearings to question the parties and witnesses and setting timetables. The intent of the process is that the adjudicator is clearly in charge and must take the initiative. After considering all of the evidence, the adjudicator must reach a decision on all matters referred within a specified timeframe, such as 28 days. The decision includes the prevailing party, the amount of damages and any appropriate interest. The intent of English law is that an adjudicator's decision is binding and enforceable, but in some instances it may be overcome by subsequent arbitration or litigation. The process is clearly intended to be more efficient than litigation, while providing a binding result.

The most formal dispute resolution technique short of litigation is *arbitration*. We noted earlier in this chapter that the American Institute of Architects' (1997) *General Conditions of the Contract for Construction* provide for arbitration as a last resort, as an appeal from the architect's 'final and binding' decision on a claim. The New Zealand *Conditions of Contract for Building and Civil Engineering Construction* (Standards New Zealand Paerewa Aotearoa, 1998) provide a simple, easy-to-understand approach to construction claims and disputes. In less than three pages, the conditions set forth three levels for resolving these controversies: first, the engineer's review, second, mediation and third, arbitration. Many other contacts specify similar procedures. Arbitration involves the submission of a dispute to an impartial person or people for a final decision. Even in the absence of a contractual provision specifying arbitration, the parties can agree to use the procedure after a dispute has arisen.

The typical arbitration process begins with the filing of a demand for arbitration by one of the parties. Following the other party's response and, optionally, its counterclaim, the arbitrator or panel of arbitrators is chosen. It is common practice to utilise a panel of three, one chosen by one party, one by the other and one by the other two panel members. In arbitration, discovery (as defined earlier in our section on litigation) is considered a privilege rather than a right (Harris, 2001). Thus, the rules may require a showing as to why discovery should be allowed and the arbitrator may have to decide whether to allow it. Rules regarding evidence tend to be less formal than for litigation, with hearsay evidence often allowed and testimony by affidavit rather than in person also often permitted. After hearing the evidence, calling witnesses if they so desire and deliberating among themselves, the arbitrators reach a decision and make the award. In some cases, only the decision is announced and not the reasons behind it. In such a case, the decision cannot be used as a precedence in a subsequent litigation. It is well established in most jurisdictions that the results of a binding arbitration will generally be upheld by a court of law, unless there was some flaw in the process, such as fraud, partiality on the part of the arbitrator(s), lack of a full and fair hearing, a decision outside the scope of the case as submitted and/or miscalculation in figures or mistaken description of people or objects described in the award (Bennett, 1996).

The construction industry has led many others in the introduction and use of arbitration for dispute resolution. Not surprisingly, standard sets of rules have arisen, as a result of considerable experience. In the UK, for example, disputes are usually arbitrated under either the Construction Industry Model Arbitration Rules or the Institution of Civil Engineers' Arbitration Procedure. Construction disputes in the USA, when carried to arbitration, usually make use of the Construction Industry Arbitration Rules of the American Arbitration Association. The last, for example, now provide for three different 'tracks' – fast track, regular track and large/complex track – for disputes of different monetary value; each track has its own rules governing the number of arbitrators and the timing of the various parts of the process (Sweet, 2000). Table 6.7 lists some claimed advantages and disadvantages of arbitration, as compared with litigation.

Special characteristics of international construction projects, in which a contractor from one country undertakes a project in another country, often lead to the contractor's insistence on the use of international arbitration for resolving disputes. Sweet (2000) says the following about these kinds of projects:

> The contracts themselves may be expressed in more than one language, often generating problems that result from imprecise translation. They may also involve

Table 6.7 Advantages and disadvantages of arbitration as a dispute resolution mechanism

Advantages
The method provides for resolving disputes as they arise, rather than resorting to a lawsuit at the end of the project.
An arbitrated case can be processed more quickly than a court trial, since the trial has to be fitted into crowded court schedules and typically involves more evidence and procedural sparring than are permitted in arbitration.
Arbitration tends to be less costly than a court trial.
A panel of technical experts, or a single expert, may be more capable of sorting out complex technical arguments than a judge, whose primary training is in the law, or a jury of lay persons.
Court proceedings may preclude the admission of certain facts or evidence that are relevant to the case.
While negotiating their contract, the parties have the opportunity to specify the procedure for dispute resolution to meet their exact needs.
Disadvantages
If discovery is not allowed, inadequate preparation of presentations prior to the hearing may result.
If hearsay evidence is allowed, unjust decisions may result.
Arbitration decisions can be reached without concern for statute and common law.
The right of appeal from an award is limited.
Some important parties may not be subject to arbitration, such as subcontractors in a construction industry case.
Decisions are not part of the public record and thus cannot be used as precedent.

From Bennett (1996).

> payment in currency that varies greatly in value. Contractors in such transactions may often have to deal with tight and often changing laws relating to repatriation of profits and import of personnel and materials. Perhaps most important, neither party may trust the other's legal system, and the contractor may believe it will not obtain an impartial hearing if forced to bring disputes to courts in the foreign country, particularly if the owner is an instrument . . . of the government.

The model contract produced by FIDIC provides for international arbitration but not 'unless an attempt has been first made by the parties to settle such disputes amicably' (Sweet, 2000).

Other interesting variants of some of the procedures considered above could be described. For example, sometimes a combination of mediation and arbitration is used, with the mediator taking the role of arbitrator for a certain issue. Under *mediation and final offer arbitration*, the process begins as an ordinary mediation; if the parties agree upon a finding regarding the prevailing party but cannot agree on the amount of money damages, each side makes one final offer in writing. The mediator, now acting as arbitrator, chooses one of the offers (no other choice is allowed) and issues the award (Hollands, 1998).

Example common issues

We end this chapter with three types of construction contract issues that often lead to disputes. While the laws of various countries may deal with the issues differently, we can at least point out the nature of the disputes and possible outcomes. The first has to do with the discovery of conditions different from those anticipated by the parties. The second involves delays in prosecuting the work, while the third arises if the contract is terminated prior to its completion.

Differing site conditions

An especially perplexing problem in the drafting and interpretation of construction contracts is the handling of unexpected conditions at the building site. The terms *differing site conditions* and *changed conditions* are both used to refer to this situation. Rubin et al. (1999) remind us that

> . . . in fact, the site does not change. The terms refer to situations in which construction conditions turn out to be different from those represented in the contract documents or from what the parties to the contract could reasonably have expected from the information available.

While the majority of unforeseen and differing conditions are found in subsurface work such as tunnelling and foundation construction, such situations can also arise in rehabilitation and restoration work, retrofitting to meet revised seismic codes and projects that improve energy efficiency. These latter cases often involve a lack of as-built drawings or any plans and specifications from the original work.

Whether the contractor is entitled to any sort of relief in these situations is often dependent on whether the owner or owner's representative was given proper notice of the discovery of the condition. For many of its projects, the US government uses the following changed conditions clause (quoted in Rubin et al., 1999):

> (a) The Contractor shall promptly, and before the conditions are disturbed, give a written notice to the Contracting Officer of (1) subsurface or latent physical conditions at the site which differ materially from those indicated in this contract, or (2) unknown physical conditions at the site, of an unusual nature, which differ materially from those ordinarily encountered and generally recognized as inhering in work of the character provided for in the contract.
> (b) The Contracting Officer shall investigate the site conditions promptly after receiving the notice. If the conditions do materially so differ and cause an increase or decrease in the Contractor's cost of, or the time required for, performing any part of the work under this contract, whether or not changed as a result of the conditions, an equitable adjustment shall be made under this clause and the contract modified in writing accordingly.
> (c) No request by the Contractor for an equitable adjustment to the contract under this clause shall be allowed, unless the Contractor has given the written notice required: provided, that the time prescribed in (a) above for giving written notice may be extended by the Contracting Officer.
> (d) No request by the Contractor for an equitable adjustment to the contract for differing site conditions shall be allowed if made after final payment under this contract.

The importance of timely notice and the availability of an 'equitable adjustment' to the contract are important aspects of the clause cited above. In addition, at paragraph (a), the clause distinguishes between two types of conditions for which relief may be granted. These types have come to be known as *Categories I and II*, in which Category I is 'a condition encountered at the work site that is different from that represented in the contract documents' (Bennett, 1996). Simon (1989) refers to this category as 'the administrative equivalent to a breach of contract', because the owner essentially failed to provide what it promised when entering into the contract. Category II is 'a condition that is materially different from what a contractor would reasonably expect to encounter on the project. In this case, relief is predicated on having a clause in the contract that spells out how such relief is to be provided' (Bennett, 1996).

The literature contains numerous interesting case examples. In one Category I type case, the plans indicated a buried, solid pipe used for carrying water to a trench. The contractor began its excavation for structural foundations and encountered water flowing from the pipe; in fact, the pipe was perforated to allow water to be absorbed by the soil. Additional payment was granted to the contractor because the investigation discovered 'an implicit representation of a changed pipe condition because the pipe was depicted as solid-walled pipe rather than a perforated pipe, . . . a condition materially different from that which was indicated in the contract' (Loulakis and Cregger, 1991).

The contractor is expected to conduct a reasonable site inspection prior to submitting its tender. In one case, the contractor's claim for an adjustment in the contract price was denied when the court found that a thorough site inspection would have allowed the contractor to ascertain with reasonable certainty the quantity of rock to be removed (Simon, 1989).

Unusual hydrostatic pressure or subterranean water conditions may be designated Category II changed conditions; they often result in the need for more costly and time-consuming construction methods. A contractor encountered an underground water quantity and flow rate substantially greater than anticipated. Although the contract documents advised of the need for dewatering, they were silent as to the potential magnitude of the problem. Simon (1989) describes more details as follows:

> More importantly, at a nearby site a prior contractor incurred a similar problem which was litigated and eventually ruled against that contractor. The problem facing the board [of contract appeals] was whether, under a changed conditions clause, the government had a responsibility to advise the adjacent contractor of the experience the first contractor had. In ruling in favor of the contractor, the board held that the government did have that responsibility. A key element of the board's reasoning was that, although the adjacent contractor was on notice of the need to dewater, it would have had to hire an expert to determine the extent of the problem. The board held that, under these circumstances, the contractor's duty to inquire was overcome by the government's failure to alert it to the first contractor's experience.

This author represented the owner in the investigation and defence of a Category II type claim by a plaintiff contractor who cited 'unexpected' permanently frozen ground ('permafrost') underneath a rock and gravel material source as the basis for a request for additional money and time. The case outcome favoured the owner, because the owner argued successfully that the presence of permafrost in the area of interior Alaska where the project was located is common knowledge to reasonably experienced contractors working on those types of projects in that locale.

Among other cases, a contractor was allowed recompense when it encountered unexpected difficulties in removing existing windows that were embedded in mortar; it was shown that this condition was not normal, the contract drawings did not reveal the condition and a 'reasonable visual examination of the windows could not have revealed the condition'. In another case, a dry lake bed was no longer dry when the contractor arrived to commence work, but it was ruled not to be a changed condition, while a contractor required to work in a 'quagmire condition caused by an inadequate site drainage system' provided by the owner was allowed relief under the changed condition theory, because conditions were materially different from what could be reasonably expected (examples from Simon, 1989).

Some leniency was granted a contractor who seemingly did not abide by the timely notice requirement. After the site had been backfilled and the structure built, the contractor submitted a request for additional payment because of a sizeable increase in the volume of rock excavated. The quantity of rock could be estimated from an independent survey that was introduced as evidence and it was further shown that the owner's representative was on site and was aware of the changed condition (Rubin et al., 1999).

Finally, always read the contract documents carefully! One set of such documents contained the disclaimer that the owner takes on no responsibility 'for the accuracy or completeness of information on the drawings concerning existing conditions and the work required' (Contractor Cannot Recover Costs for Changed Condition in the Face of Disclaimers, 1989). The citation makes clear the outcome of that case!

Delays

Two types of delays are of special interest to the contractor during the project operations phase. The first is that which has been, or may have been, caused by the contractor and for which the owner expects some sort of compensation. The second type is caused by the owner; in this case the contractor may expect both additional time to complete the project and some amount of monetary payment. Each of these types of delays can result in complicated legal entanglements. We consider a few of the highlights.

Simon (1989) states it well when he says that this area

> is probably the single most devastating area of claims and potential losses that any party to the construction process may face. The expression that time is money is of great significance to each party, be it the owner who is financing a project, a contractor who is building at a fixed or guaranteed price, an architect or engineer administrating the contract at a percentage of a fixed cost, or any other related entity. Although time plays such a significant part in the determination of the entity's survival, the basic proposition is that encountering a delay or suspension does not automatically entitle one to damages or equitable adjustment of the contract price. Not all delays are compensable.

Delays on construction projects can be complicated and difficult to analyse. A far from exhaustive list of causes includes failure to make the site available as planned, weather, labour strife, fabrication problems, shipping problems, lack of timely review and approval of documents, lack of other timely decisions, poor overall planning, poor schedule monitoring and control, ineffective jobsite coordination of personnel and materials and low labour and equipment productivity due to many reasons discussed earlier. There may be overlapping or concurrent delays. Their origins, their impacts and the respective responsible parties may be difficult to identify. Any party to a construction contract where delays may occur is well advised to (1) read and be thoroughly familiar with the contract provisions and (2) maintain complete, accurate and timely records (Rubin et al., 1999). (These two pieces of advice apply equally to all aspects of construction contract administration!)

If the contractor fails to complete the project within the time stipulated in the contract or possible additional time allowed due to variations or contract-specified excusable delays, the contractor may be subject to the assessment of *liquidated damages*. (Read the contract!) These damages are an amount stated in the contract by which the contractor's payment will be reduced for each calendar day the project is late; they are assessed in lieu of determining the actual damages the owner has suffered. If the contract states that liquidated damages will be US$ 1500 per day and if the project is completed 8 days late, US$ 12 000 will be assessed against the contractor.

Three points regarding liquidated damages are especially important. First, such contract provisions are held to be enforceable only if *time is of the essence*. In US law, courts have consistently decided that time is not considered to be 'of the essence' unless there is a statement in the contract to that effect; an owner wishing to be able to enforce a liquidated damages provision must be sure the contract is complete in this regard. Second, liquidated damages are not intended to penalise the contractor but to compensate the owner. Thus, a liquidated damages clause that is in reality a penalty clause will not be enforceable unless the contract includes a companion clause providing a bonus for early completion.

The third point about liquidated damages is that they must be a reasonable approximation of the actual damages the owner would suffer due to late completion. If the owner is required to pay US$ 200 per day in rental charges because its new building is unavailable and those are the only damages suffered by the owner due to late completion, a clause setting liquidated damages at US$ 1500 per day is not likely to be enforceable. In one case, a court allowed an owner its actual costs due to a delay, to cover expenses of its employees and overhead charges. The owner had sought to enforce a liquidated damages provision of the contract that would have resulted in approximately US$ 125 000 additional moneys assessed against the contractor. The court said the owner had not adequately explained the nature of the additional damages it claimed and the daily rate of US$ 250 (the basis for the US$ 125 000 assessment) was not related to any possibly foreseeable actual damages (Simon, 1989).

The law distinguishes between excusable delays, for which the contractor is not held responsible and non-excusable delays, which are those caused by the contractor and for which liquidated damages tend to be assessed. A typical contract section is the following (Engineers Joint Contract Documents Committee, 1996):

> Where Contractor is prevented from completing any part of the work within the Contract Times (or Milestones) due to delay beyond the control of Contractor, the Contract Times (or Milestones) will be extended in an amount equal to the time lost due to such delay if a Claim is made therefore as provided in paragraph 12.02.A. Delays beyond the control of Contractor shall include, but not be limited to, acts or neglect by Owner, acts or neglect of utility owners or other contractors performing other work as contemplated by Article 7, fires, floods, epidemics, abnormal weather conditions, or acts of God.

The model general conditions from which the above quotation is taken also contain provisions for the cases when delays are within the contractor's control and when delays are beyond the control of both the owner and the contractor.

If the contractor is delayed due to some act of the owner, such as lack of availability of the building site or late arrival of owner-furnished materials, what remedies are available to the contractor? Certainly a time extension is likely to be granted, as provided in a contract section such as that quoted just above and thus no liquidated damages assessed, if it can be shown that the contractor was not at fault. But what about some extra money to cover lower productivity while waiting, extra overhead costs and possibly other expenses? Some contracts contain explicit *no damages for delay* clauses, such as the following (Jervis and Levin, 1988):

> In the event the contractor is delayed by factors that are beyond the control of and without the fault of the contractor, the contractor may be entitled to an extension of the contractual performance period. This extension of time shall be the contractor's sole remedy in the event of a delay. The contractor shall not be entitled to an increase in the contract price as a result of any delay of any nature whatsoever, including delay caused by the acts or omissions of the owner.

The intent is clear: a time extension, but nothing else, may be granted, even if the owner is at fault! Such is the stuff that makes for interesting lawsuits. The no damages for delay clause essentially means that the owner or owner's representative is not in default if it causes delay. Once again, we strongly urge the reading and understanding of all contract sections prior to signing the contract! It would not be pleasant for the contractor to discover, after an owner-caused delay had caused a significant impact on project progress, that the contract contained such a provision. In a case that denied a contractor's claim for delay damages after the architect was slow in correcting errors and omissions in the drawings, the Supreme Court of the State of Washington, USA stated that the decision to allow or deny money damages is dependent upon the language of the contract, the nature of the default and various other circumstances. 'The specific provision here in question states positively that the contractor shall not be entitled to any claim for damages on account of hindrance or delays from any cause whatsoever' (Simon, 1989).

Rubin et al. (1999) suggest that even with a no-damages-for-delay clause in the contract, the contractor may, in the US court system at least, be allowed money damages if the owner withheld information such as that about non-availability of a building site and the contractor was led to believe the site would be available on time. A helpful distinction in considering excusable delays is to label them *compensable*, if they are caused by the owner (a no-damages-for-delay clause absent) and *non-compensable*, such as an act of God or an epidemic, for which only a time extension is usually granted.

Is there such a thing as a 'reasonable delay'? Suppose the contractor fails to have the project ready for a subcontractor's activities on the date specified in the subcontract; is the subcontractor entitled to some sort of relief? The answer probably turns on the specific language in that subcontract, such as whether it contains a no-damages-for-delay clause in favour of the prime contractor. Different courts have produced different outcomes. In one case, a subcontractor sought money damages after it was denied access to the site on the date stated in its contract with the general contractor. While it allowed an extension of time to complete the subcontractor's portion of the work, the court did not allow money damages, stating that the delay must be 'unnecessary, unreasonable or due to defendant's fault' in order to permit such recovery and the delay in this case did not meet those criteria (Simon, 1989).

Finally, consider the interesting situation in which the contractor completes the project on time but had planned to finish ahead of schedule. Perhaps it had discovered an innovative installation method that would save several weeks compared to the programme contemplated by the design engineer; its low tender price was based on saving considerable amounts of project overhead due to the shorter schedule. Is the contractor entitled to money damages if it can prove that the delay in finishing early was caused by the owner or the engineer? In general, such compensation *may* be awarded if the contractor can prove that (1) it had the ability to finish early, (2) the other party was aware of its plan to finish early and (3) but for unreasonable delay by the other party, the project would have finished early as planned by the contractor. A case illustrating one possible outcome involved a project whose contract completion date was 1 October; but for various delays and changes by the owner, the project could and would have been completed by 6 April of that year. Actual completion was 24 April of the following year. The contractor was granted extra overhead costs for 314 days, a substantial portion of the 384 days between 6 April and the next year's 24 April; the other 70 days were covered in a separate dispute.

Contract termination

Despite the frequent use of terms like breach of contract, suspension and early termination, the fact is that most contracts run their full course and are terminated when the project is completed, with both sides at least reasonably satisfied with the results. But sometimes a contract is terminated before the project is complete, 'an expensive, time-consuming and difficult undertaking for all concerned' (Bennett, 1996). Cashion (2001) calls termination for default 'one of the most serious actions which can be taken on a construction project', while Sweet (2000) entitles his chapter on the subject 'Terminating a construction contract: sometimes necessary but always costly'. One of the reasons that parties to construction contracts do not more frequently exercise their rights to terminate is the serious consequences that can occur to the party initiating the termination if it is later found that the termination was without proper cause.

Usually a construction contract gives both the owner and the contractor the right to terminate the contract before completion under certain circumstances. However, it is much more common for the owner to initiate the action. The grounds for contractor-initiated termination are generally limited to stoppage of work by the owner for at least a specified number of days, through no fault of the contractor, stoppage of work due to non-issuance of a certificate of payment and failure of the owner to provide evidence that it will fulfil its financial obligations. The second of these reasons is most common; 'rarely does anything but non-payment justify termination by the contractor' (Cashion, 2001). It is also important to understand that termination can occur through mutual agreement of the parties; there may be good reason for both sides to want to 'call the deal off' part way through. The two sides can agree to 'unmake' their bargain, just as they have agreed to make it. Because this type of termination is also rare, the balance of this discussion is devoted to termination actions initiated by the owner against the contractor.

A well-made construction contract will include language that provides the grounds and the process for termination of the contract by the owner and by the contractor. Under a typical contract, the owner is entitled to terminate the contract for *default* and for *convenience*. The usual *default* acts or omissions by the contractor, which entitle the owner, at its option, to terminate, include the following (as discussed in Cashion (2001), based on the American Institute of Architects (1997)):

- substandard, defective or non-conforming work;
- failure to pay subcontractors and suppliers;
- failure to pursue the work diligently, including failing to supply enough workers with proper skills and enough materials and failing to comply reasonably with the project programme;

- violation of laws, ordinances and regulations, including non-payment of various taxes, failure to secure required permits and licences and non-compliance with safety laws;
- other substantial breaches of the contract.

The owner or its representative must give proper notice of the termination to the contractor and to its surety, if there is one, allowing a specified amount of time for the contractor to 'cure' the default. If, after this time period, the contractor is still in default, the contract can be terminated. The owner can then take possession of the site, any unused materials, plant and tools, retain some or all of the subcontractors and complete the project in any manner it wishes. If a surety is involved, the performance bond will provide for the surety to have responsibility for project completion, either with its own forces or agents; with a contractor selected by the surety through a normal selection process, to be paid by the owner with reimbursement to the owner by the surety up to the amount of the bond; or by simply paying the owner the amount of its liability.

A public utility issued a contract for the installation of pollution control equipment for an energy plant that used coal-fired generators. The low tender was for approximately US$ 7 million. Variations added another US$ 1 million. After about 80% of the US$ 8 million contract price had been paid, the utility terminated the contract due to the contractor's alleged failure to maintain the schedule; evidence indicated the project was about 60% complete at that point. The utility spent another US$ 5 million to complete the work. The ensuing trial, brought by the contractor claiming unfair termination, lasted 81 days and produced 36 000 pages of testimony and other documentation. The trial court jury found in favour of the contractor and awarded it US$ 17 million, including US$ 12 million in punitive damages. The utility appealed and the award was reduced to 'only' US$ 2 million (Sweet, 2000). This case illustrates the risks involved in contract terminations and the monetary amounts that can be involved, as well as the time and expense of the litigative process.

The contract may provide the owner the right to terminate the contract for its convenience. A manufacturer may decide to build a new factory and then, while the project is underway, decide the project is no longer financially feasible. Perhaps a laboratory facility becomes outmoded even while under construction. Or funding for a school construction project may be withdrawn unexpectedly. If a *termination-for-convenience* clause is absent from the contract, the owner would be in default if it simply told the contractor to stop work and ceased making payments. If the owner invokes such a clause, the contractor is normally required to stop work, cease placing orders, cancel orders currently outstanding and do all else needed to conclude performance of the contract. Certainly the contractor will be dealt with less harshly than if the contract is terminated for default. Payment to the contractor includes reimbursement for work performed to date, unavoidable losses it may have suffered, expenses required to protect the property and profits on work performed (and, in some contracts, profits on work *not* performed!).

A state Department of Transportation was required to provide a right of way for a highway project, prior to the contractor's commencing work. The agency failed to do so and the contractor filed a lawsuit, claiming the department had breached the contract. The department then terminated the contract, using the termination-for-convenience clause in the contract as the basis for its action. In the court case that followed, it was decided that the department had violated the intent of the termination-for-convenience provision and the contract had been improperly terminated (Simon, 1989).

Finally, we note that the law may operate to allow termination of the contract even if the contract is silent regarding termination rights. An examination of the facts, circumstances and intent of the parties will try to determine whether there has been a 'material breach' that would lead fairly to a termination. An owner of some land entered into a contract with a builder who agreed to construct a house and dig a water well. A dispute arose over additional costs for drilling and casing the well to a depth greater than anticipated. The owner refused to pay these extra billings and the builder terminated the contract. Was the work stoppage by the builder a breach of the contract? If so, was it 'material'? The court noted that the owner was deprived of the benefits of receiving a completed home and the builder had no intention of curing its failure to perform. It concluded that the builder's breach 'constituted a material failure of performance' and thus absolved the owner from all liability under the contract (Sweet, 2000).

Contract termination is a time-consuming and expensive process, with likely delays in project completion, increases in project price and risks that the action may be overturned through the dispute process. All who study and advise on construction contract terminations urge that the parties exhaust all other options before resorting to termination. As we terminate this section and this chapter, comments by Sweet (2000) seem especially appropriate:

> It may sometimes be necessary to terminate a construction contract. Performance may be going so badly and relations may be so strained that continued performance would be a disaster. But . . . the drastic step of termination should not be taken precipitously.

Discussion questions

1 (This question reviews some of the materials in Chapter 5.) The programme and budget for a wastewater system installation project is based on the following activities, precedences, durations and cost estimates.

Activity	Immediate predecessor	Duration (days)	Budgeted cost (US$)
Order and deliver septic tank		5	2000
Order and deliver piping		4	750
Deliver construction plant		2	200
Obtain permits		6	400
Layout and prepare site		2	600
Excavate for septic tank	Obtain permits Layout and prepare site Deliver plant	1	300
Excavate trenches	Obtain permits Layout and prepare site Deliver plant	1	300
Excavate for leach field	Obtain permits Layout and prepare site Deliver plant	3	800

Activity	Immediate predecessor	Duration (days)	Budgeted cost (US$)
Install septic tank	Excavate for septic tank Order and deliver septic tank	1	400
Install trench piping	Excavate trenches Order and deliver piping	2	600
Install leach field	Excavate for leach field Order and deliver piping	4	1200
Make connections	Install septic tank Install trench piping Install leach field	2	300
Prepare as-built drawings	Make connections	1	250
Inspection	Make connections	1	100
Backfill and compact	Inspection	2	400
Test system	Backfill and compact	1	200
Remove plant and cleanup	Backfill and compact	2	500

(a) Prepare an activity-on-node network schedule diagram for this project. In each node, show the activity's description, duration and budgeted cost.
(b) Find the number of working days the project is expected to require.
(c) Label the critical path.

2 Assume the project in Question 1 has progressed to the end of the 10th working day. The following status information is available for each activity.

Activity	Percentage complete	Actual cost to date (US$)	Estimated cost to complete (US$)	Estimated days remaining
Order and deliver septic tank	100	1850	0	0
Order and deliver piping	100	750	0	0
Deliver construction plant	100	200	0	0
Obtain permits	100	550	0	0
Layout and prepare site	100	500	0	0
Excavate for septic tank	100	250	0	0
Excavate trenches	0	0	300	1
Excavate for leach field	50	400	300	1
Install septic tank	50	250	300	1
Install trench piping	0	0	600	2
Install leach field	0	0	1100	3

Activity	Percentage complete	Actual cost to date (US$)	Estimated cost to complete (US$)	Estimated days remaining
Make connections	0	0	300	2
Prepare as-built drawings	0	0	250	1
Inspection	0	0	200	2
Backfill and compact	0	0	400	2
Test system	0	0	200	1
Remove plant and cleanup	0	0	500	2

Update the network schedule. Identify the critical path. What is the remaining project duration, based on this analysis? Is the project behind, ahead of, or on schedule? By how many days? If the project is behind schedule, what actions can be taken to bring the schedule into compliance with the original completion date?

3 Based on the information in Question 1, find the budgeted cost of work scheduled as of the end of the 10th working day, assuming that all activities are scheduled to start and finish at their early times. Assume that the cost of each activity is spread evenly over its duration.
4 Based on information in Question 2, find the budgeted cost of work performed as of the end of the 10th working day.
5 Based on information in Question 2, find the actual cost of work performed.
6 Calculate the cost variance and the schedule variance for the project analysed in Questions 3, 4 and 5. What conclusions do you draw from these variances?
7 Based on information in Question 2, find the estimated total cost of the project at completion and the estimated cost variance at completion.
8 Write a paragraph that summarises the cost and schedule status at the end of the 10th working day for the wastewater system installation project you studied in Questions 1–7, including any recommendations you wish to offer.
9 We noted in our discussion of Table 6.1 that a quick study of the quantity and cost information could yield important facts about the general status of the project. List those work items that were begun and completed in week one, those that were begun in week one and completed in week two, those that were begun and completed in week two, those that were begun in week two and are not yet completed and those that have not yet been started as of the end of week two. Are any work items missing from these lists? If so, describe their progress since project inception.
10 Design a progress payment form that could be used with the unit-price contract whose unit price schedule is shown in Table 4.4. Assume the following quantities have been completed as of the end of the month for which payment is sought: clearing and grubbing, 0.75; unclassified excavation, 13 050; borrow, 24 670; aggregate base course, 2500; aggregate surface course, 665; 300 mm corrugated steel pipe, 85; 500 mm corrugated steel pipe, 172. The retainage percentage is 10%, and the amount paid through the end of the previous month is US$ 553 275. Complete the form, and show the amount the contractor will be paid, assuming the request for payment is approved.

11 Distinguish between a contracting organisation's company-wide quality management programme and a plan that would be used a specific project. Who prepares each? Who is charge of implementing each?
12 Obtain construction accident statistics for your country or region similar to those in Table 6.3 and Figures 6.3 and 6.4. Compare your statistics with those in the chapter and comment on similarities and differences and the likely reasons for these similarities and differences.
13 For each item in the list of common construction site hazards in Table 6.4, describe one type of accident that might happen, a possible injury that could result and one action the contractor might take to protect the worker or the public against the hazard.
14 According to the text, prime responsibility for jobsite safety should be assigned to a top-level field supervisor. Write a brief job description for this person and tell where the position should be placed in the project organisation chart.
15 For a construction project with which you are familiar, select four of the environmental issues subject to monitoring and control described in the text (or others that are of special importance on this project) and tell how the contractor might monitor and control them.
16 Discuss with a contractor in your locality the environmental laws and regulations which affect its work. What penalties can be levied for violations?
17 One way of defining inflation in the construction industry is to measure the difference between the increase in the cost of a worker and the increase in the productivity of that worker. If a worker is able to increase their productivity at the same rate as their wage increases, there is no inflation.

    Consider the following hypothetical example. Assume that in 1997 a concrete block layer's labour rate was US$ 14.00 per hour and they received a 9% annual increase each of the following 5 years. Thus, their wages in 1998 through 2002 increased at a rate of 9% per year. Assume further that, in 1997, they placed 120 blocks per 8-hour day, or 15 blocks per hour and that their annual increase in productivity for the following 5 years was 2% per year.

    (a) What might have been the reasons for this increase in productivity?
    (b) Based on the above information, calculate the labour cost per block placed for each of the 6 years.
    (c) Calculate the inflation rate for each of the last 5 years; that is, find the percentage increase in the labour cost per block for each year from the previous year.

18 Discuss specific ways by which the contractor's supervisory personnel can positively influence four of the factors affecting jobsite labour productivity identified in the text.
19 Draw a flowchart that tracks the management of a certain material or piece of permanent equipment from the time the procurement process begins until the item(s) is (are) installed and inspected. Pick a specific item such as structural steel or motor control centres or gravel fill. Make assumptions as necessary and state those assumptions in your answer.
20 Design a submittal log, indicating the required column headings; show how this fits into the procurement schedule.
21 In a measure-and-value (unit-price) contract, a common provision stipulates that the unit prices for items whose actual quantities are more than some stated percentage (such as 25%) greater than or less than the estimated quantities will be considered for renegotiation. This means, for example, that if a contract includes a unit price of US$ 95 m^3 for aggregate base course, based on an estimated quantity of 4000 m^3, and the actual installed quantity is 5200 m^3, the owner may seek to renegotiate a lower unit price, whereas, if the actual

installed quantity is 2500 m^3, the contractor might believe it is entitled to a higher unit price. Explain the rationale for each of these positions.

22 Talk with a contractor in your area about the concept of value engineering. Have them describe a project in which they developed a value engineering proposal. If they have not participated formally in a value engineering process, have them tell you about a project which they believe could have been designed differently with resulting cost savings for both owner and contractor.

23 If you were to design a document management system for a construction project, you might begin by developing a tree diagram, similar to a work breakdown structure, in which major categories of documents were subdivided into a series of hierarchies. Sketch the beginning of such a diagram and show details for at least one of its branches.

24 Visit the Project Management Center website. Select three providers of project website software. Identify and compare the features of each.

25 As the newly hired Information Technology Manager for XYZ Contractors, you have been tasked with the job of implementing project-specific websites on your company's projects. Prepare a list of 10 action items you must accomplish during your first three months on the job.

26 Review the claims section of a construction contract in your area. Write a brief report on the steps the contractor must take in presenting a claim.

27 Review the dispute resolution section of a construction contract in your area. Write a brief report on the steps the contractor must take in pursuing a dispute.

28 Distinguish between a dispute review board and an arbitration panel. Could both be used on the same project? If so, under what circumstances?

29 One item on a unit-price bridge construction project was a continuous concrete protection wall, which the successful tenderer priced at US$ 18 per linear metre. The contractor used a slip-forming method for the wall, rather than a set of fixed forms as anticipated by the owner. Because the slip-forming method saved the contractor considerable money, the owner sought a price reduction. What was the result of the ensuing litigation? Why?

30 A contract required a construction project to be completed by 15 September. The owner's representative directed the contractor to suspend operations for the months of June and July and the completion date was revised to 15 November. Is that change appropriate? Discuss and defend your answer.

References

Adrian, J.J. 1987. *Construction Productivity Improvement*. Prentice-Hall.

Ahmed, S. M., L. H. Forbes and L. C. Leung. 2001. Improving construction risk management through safety best practices – a Hong Kong study. *Proceedings Third International Conference on Construction Project Management*, Singapore 20–30 March, 220–228.

Alaska Department of Transportation and Public Facilities. 2001. *Standard Specifications for Highway Construction*. Statewide Design and Engineering Services Division.

American Association of State Highway and Transportation Officials. 1998. *Guide Specifications for Highway Construction. AASHTO Value Engineering Task Force*. http://www.wsdot-.wa.gov/eesc/design/aashtove/Constve.htm.

American Institute of Architects. 1997. *General Conditions of the Contract for Construction*. AIA Document A201.

American Society of Civil Engineers. 2000. *Quality in the Constructed Project: A Guide for Owners, Designers and Constructors*, 2nd ed. ASCE Manuals and Reports on Engineering Practice No. 73.

Antevy, J. 2002. *Internet Solutions: Project Specific Web Sites*. http://www.e-builder.net/corporatehistroy.html.

Asian Development Bank. 2000. *Guide to Contractors and Agencies Involved in Urban Sector Projects*. http://www.adbindia.org/guide.htm.

Bedian, M.P. 2002. Value engineering and its rewards. *Leadership and Management in Engineering*, **2**, 36–37.

Bennett, F.L. 1996. *The Management of Engineering: Human, Quality, Organizational, Legal, and Ethical Aspects of Professional Practice*. John Wiley.

Bennett, F.L. 2000. Information technology applications in the management of construction: an overview. *Asian Journal of Civil Engineering (Building and Housing)*, **1**, 27–43.

Bockrath, J.T. 2000. *Contracts and the Legal Environment for Engineers and Architects*, 6th edn. McGraw-Hill.

Building and Construction Authority. 2002. *About CONQUAS (Construction Quality Assessment System)*. http://conquas21.bca.gov.sg/standards.htm.

Building Research Establishment. 2002. *Performance Improvement in Construction: CONQUAS UK*. http://www.bre.co.uk/CPIC/service3.html.

Cashion, G.L. 2001. Default termination. In Enhada, C.Y, C.T. Gatlin and F.D. Wilshusen, eds. *Fundamentals of Construction Law*. American Bar Association Forum on the Construction Industry.

Chung, H.W. (ed.). 1999. *Understanding Quality Assurance in Construction: A Practical Guide to ISO 9000 for Contractors*. Spon Press.

Clough, R.H. and G.A. Sears. 1994. *Construction Contracting*, 6th edn. John Wiley.

Clough, R.H., G.A. Sears and S.K. Sears. 2000. *Construction Project Management*, 4th edn. John Wiley.

Collier, K. 2001. *Construction Contracts*, 3rd edn. Prentice-Hall.

Comptroller and Auditor General. 2001. *Modernising Construction. HC 87. Session 2000–2001*. The Stationery Office. http://www.nao.gov.uk/publications/nao_reports/00–01/000187.pdf.

Connecticut USA Department of Transportation. 2000. *Value Engineering*. http://www.dot-.state.ct.us/BUREAU/eh/ehcn/road/ve.htm.

Construction Industry Computing Association. 1998. *Document Management for Construction*. http://www.cica.org.uk/PublicationsList.doc.

Construction Industry Computing Association. 2002. *Document Management Software List*. http://www.cica.org.uk/PublicationsList.doc.

Contractor Cannot Recover Costs for Changed Conditions in the Face of Disclaimers. 1989. *PEC Reporter* **12**, p. 2.

Dispute Prevention and Resolution, Inc. 2000. *New Developments: Construction Industry Turns to DARTS*. http://www.dpr4adr.com/articles.html.

Doherty, P. 2001. *Emerging AEC Technologies: Beyond Dot.com*. http://www.constructech.com/online_news/columns/010902.asp.

e-Builder, Inc. 2002. *e-Builder*. http://www.e-builder.net/main.html.

Engineers Joint Contract Documents Committee. 1996. *Standard General Conditions of the Construction Contract*. EJCDC No. 1910–8.

Erickson, R.L. 1998. *Environmental Law Considerations in Construction Projects*. John Wiley.

Fédération Internationale des Ingénieurs-Conseils. 2001. *Quality of Construction*. A FIDIC Position Paper. Federation Internationale des Ingenieurs-Conseils.

Haas, C.T., J.D. Borcherding, E. Allmon and P.M. Goodrum. 1999. *US Construction Labor Productivity Trends, 1970–1998*. Report No. 7. Center for Construction Industry Studies.

Halpin, D.W. and R.W. Woodhead. 1998. *Construction Management*, 2nd edn. John Wiley.

Harris, A.E. 2001. Alternate dispute resolution in the 21st century. In Enhada, C.Y, C.T. Gatlin and F.D. Wilshusen, eds. *Fundamentals of Construction Law.* American Bar Association Forum on the Construction Industry.

Health and Safety Commission (UK). 2001. *Health and Safety Statistics 2000/01*. Health and Safety Commission (UK).

Health and Safety Executive. 1999. *Construction Health and Safety Checklist. Construction Sheet No. 17 (Revised).* http://www.hse.gov.uk/pubns/cis17.pdf.

Health and Safety Executive. 2000a. *A Guide to the Construction Health, Safety and Welfare Regulations 1996*. http://www.hse.gov.uk/pubns/indg220.htm.

Health and Safety Executive. 2000b. *HSE Public Register of Convictions*. http://www.hse-databases.co.uk/prosecutions/.

Hendrickson, C. and T. Au. 1989. *Project Management for Construction: Fundamental Concepts for Owners, Engineers, Architects, and Builders*. Prentice-Hall.

Hinze, J. 2000. Incurring the costs of injuries versus investing in safety. In Coble, R.J., J. Hinze and T.C. Haupt, eds. *Construction Safety and Health Management.* Prentice-Hall, pp 23–42.

Hirotani, A. 2001. *FIDIC 2001 Quality of Construction Survey.* http://www.fidic.org/conference/2001/talks/tuesday/bp/hirotani.html.

Hislop, R.D. 1999. *Construction Site Safety: A Guide for Managing Contractors*. Lewis.

Hollands, D.E. 1998. *Dispute – Avoidance and Resolution*. Construction Contract Seminar Notes. http://homepages.ihug.co.nz/~deh/disputes.htm.

Hollands, D.E. 2000. *Variations (Change Orders)*. Construction Contract Seminar Notes. http://homepages.ihug.co.nz/~deh/variations.htm.

Hoyle, D. 2001. *ISO 9000 Quality Systems Handbook*, 4th edn. Butterworth-Heinemann.

Institute of Quality Assurance. 2002. *Introduction to Quality.* http://www.iqa.org/htdocs/quality_centre/d2–1.htm.

Jervis, B.M. and P. Levin. 1988. *Construction Law: Principles and Practice*. McGraw-Hill.

Koehn, E. and G. Brown. 1985. Climatic effects on construction. *Journal of Construction Engineering and Management*, **111**, 129–137.

Kraker, J.M. 2000. Firms jockey for the lead in the race to go on line. *Engineering News Record*, **245**, pp. 50–1, 53, 63, 65–6.

Loulakis, M.C. and W.L. Cregger. 1991. Industry standards of practice may affect contractor claims. *Civil Engineering*, **61**, p. 36.

Low Productivity: The Real Sin of High Wages. 1972. *Engineering News Record*, **188**, pp. 18–23.

Lowton, R.M. 1997. *Construction and the Natural Environment.* Butterworth-Heinemann.

McFadden, T.T. and F.L. Bennett. 1991. *Construction in Cold Regions: A Guide for Planners, Engineers, Contractors, and Managers*. John Wiley.

Mincks, W.R. and H. Johnston. 1998. *Construction Jobsite Management.* Delmar.

Ministry of Manpower and Ministry of National Development of Singapore.1999. *Construction 21*.

Murdoch, J. and W. Hughes. 2000. *Construction Contracts: Law and Management*, 3rd ed. Spon Press.

O'Brien, W.J. 2000. Implementation issues in project web sites: a practitioner's viewpoint. *Journal of Management in Engineering*, **16**, 4–39.

Palmer, W.J., W.E. Coombs and M.A. Smith. 1995. *Construction Accounting and Financial Management*, 5th ed. McGraw-Hill.
Peachtree Software Inc. 2002. *Peachtree Complete Accounting 2002 Release 9.0.01*. http://www.peachtree.com.
Pilcher, R. 1992. *Principles of Construction Management*, 3rd edn. McGraw-Hill.
Primavera Systems, Inc. 2002. *Primavera Expedition®: Complete Project Control*. http://www.primavera.com/products/expedition.html.
Project Center. 2001. *Emerging Construction Technologies*. http://www.new-technologies.org/ECT/Internet/projectcenter.htm.
Project Management Center. 2002. *Directory of Project Management Software*. http://www.infogoal.com/pmc/pmcswr.htm.
Rowlinson, S. 2000. Human factors in construction safety – management issues. In Coble, R.J., J. Hinze and T.C. Haupt, ed. *Construction Safety and Health Management*. Prentice-Hall, pp. 59–85.
Rubin, R.A., V. Fairweather and S.D.Guy. 1999. *Construction Claims: Prevention and Resolution*, 3rd ed. John Wiley.
Simon, M.S. 1989. *Construction Claims and Liability*. John Wiley.
Southwest Transmission. 2002. *Preservation*. http://www.southwesttransmission.org/html/preserve.html.
Standards New Zealand Paerewa Aotearoa. 1998. *Conditions of Contract for Building and Civil Engineering Construction*. New Zealand Standard NZS: 3910: 1998.
Sweet, J. 2000. *Legal Aspects of Architecture, Engineering and the Construction Process*, 6th ed. Brooks/Cole.
Thorpe, T. and S. Mead. 2001. Project-specific web sites: friend or foe? *Journal of Construction Engineering and Management*, **127**, 406–413.
Transit New Zealand. 1995. *Quality Standard TQS1. Quality System for Road Construction, Roads Maintenance and Structures Physical Works Contracts having a High QA Level.*
US Army Corps of Engineers. 2000.*Contract Administration Plan*. http://www.hnd.usace.army-.mil/chemde/cap/chp9.pdf.
US General Services Administration Architecture and Construction. 2000. *Value Engineering for Design and Construction*. http://gsa.gov/pbs/pc/gd_files/value.htm.

Case study: Long-distance construction material management – the Korea–Alaska connection

The author is indebted to Albert E. Bell, PE, President, Ghemm Company Contractors of Fairbanks Alaska, for much valuable assistance in the preparation of this case study.

When the joint venture partnership of Dick Pacific, of Honolulu, Hawaii and Ghemm Company, of Fairbanks, Alaska, sought a steel supplier for its Bassett Hospital Replacement project at Alaska's Fort Wainwright Army base near Fairbanks, they conducted the normal tendering process, with priced proposal competition among several firms around the North Pacific Rim. Because approximately 7.5% of the project's US$ 178.2 million construction contract price is devoted to steel supply and erection, the management of this vital aspect of the project was critical to its success.

Because of its significantly lower price (more than 10% lower than its nearest competitor), the Korean firm of Poong Lim Industrial Company, Ltd. was selected. Thus began a saga of international cooperation, long-distance communication, cultural challenges and multimodal intercontinental shipping that is a rather extreme example of the construction executive's material management responsibilities.

The owner/operator of the hospital project is the US Army Health Facilities, represented on site by the US Army Corps of Engineers, with whom the joint venture contractors executed their contract. The total area supported by steel is approximately 51 000 m^2, including 30 000 m^2 of public access space, plus roof and interstitial mechanical space. Whereas a normal steel usage for structures of this types is of the order of 50 kg m^2, this hospital's 4550 metric tonnes represents nearly twice that figure, at 90 kg m^2. The project's essential status, as a medical facility on a military establishment, resulted in a design that includes provision for a 1000-year earthquake.

As is often the case with steel structures in remote regions where the construction season for outside work is limited to warmer months (temperatures below –40°C are common in the winter in Interior Alaska), the production, shipping and erection of structural steel for this project are on the critical path. After tenders were submitted, negotiations with the Korean supplier involved several trips to Korea and included such terms as scope, bonding, insurance, penalties and letters of credit, as well as price. The production process began with shop drawings. To save time, the Corps of Engineers, responsible for shop drawing approval, placed representatives with approval authority at the fabricator's location in Korea; the joint venture was represented at the fabricator's shop as well, as was the subcontractor responsible for steel erection. The contract required the supplier to provide in-house quality control. In addition, the joint venture contractors engaged an outside quality control agency to monitor the production process and to conduct such testing as ultrasound tests of welding.

Fabrication and delivery were programmed in a manner that would provide about 1100 tonnes to the hospital site in 2002, with the balance, 3450 tonnes, in 2003. Of course, the early deliveries were to support the early stages of construction, so that a portion of the project could be enclosed prior to the coming of winter in October 2002. After fabrication, the steel was trucked from the point of fabrication near Seoul City to the port city of Inch-on, where it was loaded onto sea-going vessels for the journey to Anchorage, Alaska. From Anchorage, the materials were trucked to Fairbanks. Total travel distance was about 7900 km, including 7250 km for the sea route from Inch-on to Anchorage. The map in Figure 6.16 shows the extreme distance over which the fabricated steel travelled on its way from Korea to Interior Alaska. The supply contract specified FOB Fairbanks jobsite (see Chapter 5 for an explanation of a 'free on board' location), but the joint Dick Pacific–Ghemm joint venture assisted with facilitating the shipping process. For example, US shipping law requires the use of US flag vessels for transport of commodities to US ports; exceptions are allowed in cases where such a method is not competitive with foreign vessels. An exception was obtained in this case, due in part to the efforts of the joint venture partners.

As this report is being written (August 2002), the first 35 trailers, carrying 600 tonnes, have arrived at the jobsite. Upon arrival, the components are sorted and placed in designated locations in the lay-down yard, as shown in Figure 6.17. Each piece arrives from the fabricator with an attached label or mark, corresponding to both its associated shop drawing and its place in the erection plan. Figure 6.18 shows two

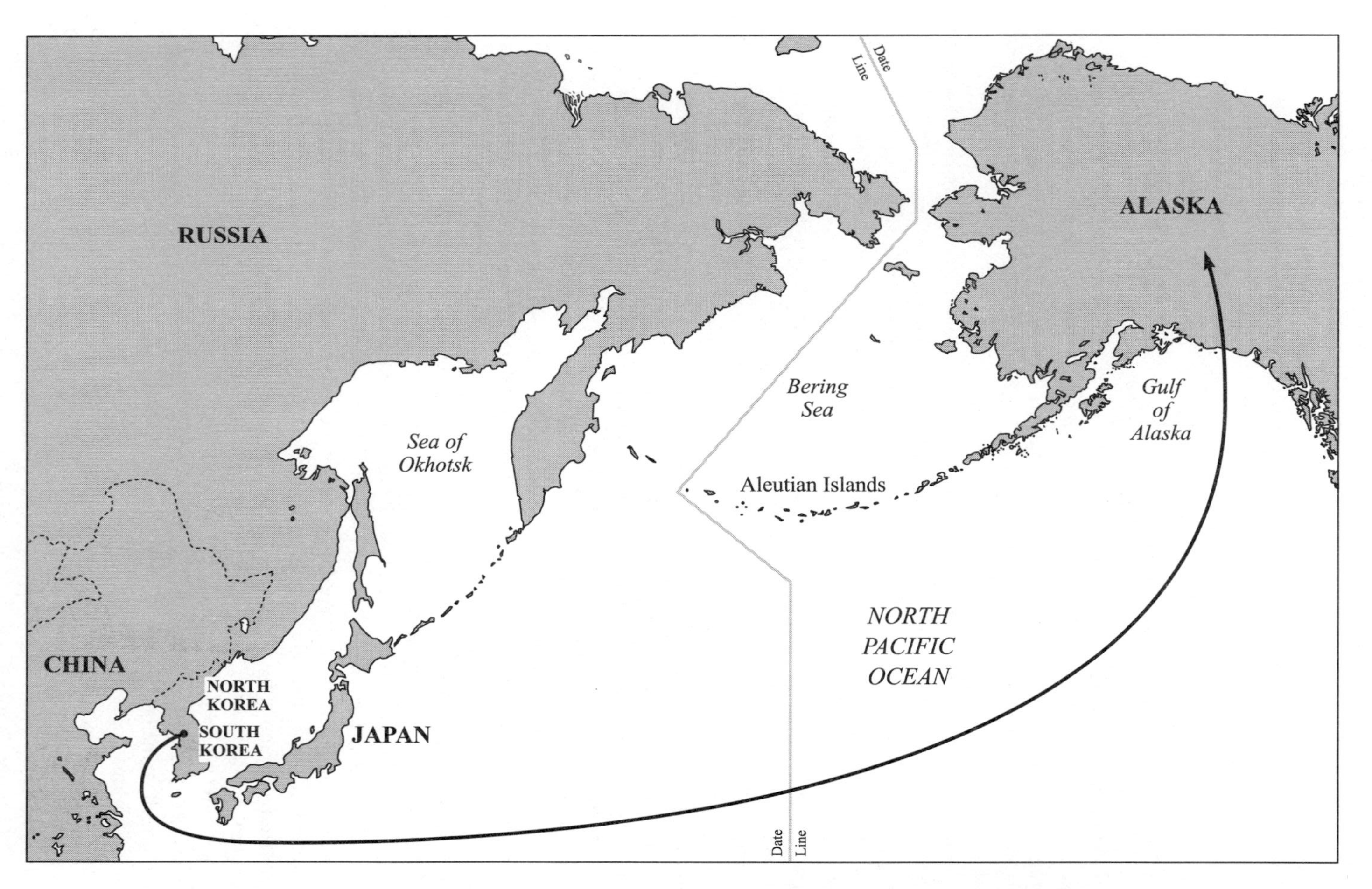

Figure 6.16 Map of supply route, Seoul Korea to Fairbanks, Alaska. Based on Mercator projection map published by US National Imagery and Mapping Agency (1998).

Figure 6.17 Portion of structural steel laydown yard, Bassett Hospital Replacement Project.

examples of the simple yet effective marking system. The management plan includes a map of the yard, with a specific place for each piece, based on its location in the building.

The challenges brought about by the selection of a foreign steel supplier for this project include differences in language, culture and business customs and rules and the long distance between Korea and Alaska. One reason the contractor chose the Korean supplier was that the sponsoring joint venture partner, Dick Pacific, has several native Koreans on its staff, thus easing many of the potential cross-cultural problems. The price was highly competitive, the first shipment was delivered on time and electronic communication is making possible the instantaneous transmission of drawings and other documents. Based on experience to date with this long-distance, time-critical supply contract, it is likely that the structural frame for the Bassett Hospital Replacement project will be completed on time and on budget, thus boding well for the successful completion of the entire project in early 2006.

PROJECT : BASSETT HOSPITAL REPLACEMENT
CONSIGNEE : DICK PACIFIC/GHEMM JOINT VENTURE
4077 NEELY ROAD
FT. WAINWRIGHT, FAIRBANKS, AK 99703,USA
TEL : 907-356-5319
NOTIFY : SAME AS CONSIGNEE
CONTRACT NO. : DACA-02-C-0004
PACKAGE NO. : 62 OF (1)
COMMODITY : STRUCTURAL STEEL
N/W : 5180 KGS. G/W : 5330 KGS
MEASUREMENT : 1190 L× 130 W× 100 H (CM)
POONG LIM INDUSTRIAL CO.,LTD.

Figure 6.18 Marks and labels are essential elements of material management.

7

Project closeout and termination phase

Introduction

The project is almost complete! Before it can be declared finished, however, a number of activities must take place and several responsibilities must be fulfilled. Many contractors are guilty of putting too little emphasis on this final phase in the construction project life cycle, choosing instead to look beyond the end of the current project and on to the next project. A sage observer of project management has said 'Projects proceed smoothly until 95% complete, and then they remain at 95% forever!' Another observed '90% of the effort is expended on the first 90% of the project, and the other 90% is expended on the other 10% of the project.' While many explanations of the difficulties in terminating projects effectively are plausible, we can name at least two. (1) We in the design and construction professions tend to try for perfection; we want the job to be perfect before moving on, when a standard less than perfection is sufficient. (2) The project environment is familiar and comfortable, so there is a reluctance to 'let go' and move to the next project, especially if there is no immediate prospect for a next project. Thus, the construction manager must deal with both the tendency to look beyond the project termination activities and on to the next project, which leads to neglect of this phase and the reluctance to terminate the project, resulting in inefficiencies in executing the many activities required at this phase.

The project closeout and termination phase can be thought of as a project unto itself. Often termed *commissioning*, this phase must be planned and programmed, tasks must be assigned, the phase must be executed effectively and its costs, schedule and quality must be controlled. On large projects, a specialist team is often engaged to assure that project closeout is carried out in the best manner possible. We have chosen to divide the tasks into two categories: (1) completing the work, which includes the physical activities that must be accomplished on the site and (2) closing out the project, involving the multitude of required documents and other paperwork issues, some related obviously to finances but others to certificates, project records and provision to the owner of the required training, operational information, spare parts and the like. The two categories overlap and interact, as we shall see. The presentation in each category is generally in the order in which the activities occur on the project, although many activities take place concurrently.

Completing the work

Testing and startup

Depending on the nature of the project, there may be substantial amounts of testing and initial startup of various systems, especially mechanical and electrical systems. The installer of each system will have made them operational, but the specifications will usually require that they be started and tested under operational conditions at the end of the project. This latter work may be performed by the prime contractor, subcontractor, manufacturer's representative, special testing consultant or owner's personnel. In any case, the owner will be represented on site during the test. Proper records of the test will include the date and location, the persons involved, the methods and the results (Mincks and Johnston, 1998). An example is the testing and balancing of a building's air handling systems, including ventilation and air conditioning. Fans, chillers, automated louvres, controls and other components must work individually and they must also perform satisfactorily as a system. The conditions under which the test is performed should mimic expected actual conditions as closely as possible, although the parties cannot expect to encounter outdoor environmental conditions typical of year-round conditions during a short-term test. Among other systems to be tested are fire protection, pumps, heating, security, elevators (lifts) and communications. Modern facilities often have computer-controlled building services and these systems must be checked out. Systems that fail their tests must be repaired and retested until they are satisfactory.

Cleanup

In earlier sections of the book we emphasised the importance of good site organisation in achieving efficient and safe operations; part of this effort involves routine cleanup at the end of each day and each operation. At the end of the project, however, much more thorough cleaning is required. This cleanup includes, for a building project, the scouring of walls and ceilings, shampooing of carpets, cleaning of exterior walls, window washing and any other cleaning required to remove dust, grease and other accumulations and render the facility usable and attractive to the user and the general public. Specialist cleaning contractors may be employed to provide these services. Cleanup also includes the removal of various temporary facilities such as haul roads, security fencing, utilities, storage sheds and offices, as well as surplus materials, scrap and construction plant. Disturbed landscaping, walkways and drainage facilities may need to be put back to their original condition and any off-site streets used by the those working on the project may need to be repaired and cleaned. While subcontractors may be legally obligated to the contractor to provide their share of the cleanup effort, the owner looks to the contractor for the satisfactory completion of this work, just as with other work on the project.

Preliminary punch lists

The strange term *punch list*, also known as a *snagging list* or an *inspection list*, refers to a list of work items yet to be completed, including repairs and discrepancies, in order to fulfil the contract's requirements. Punch lists are prepared in various forms, but all contain the location of the needed work, its description, and the party responsible for the correction, with some means to indicate when the item has been remedied. On a building project, a list might be prepared for each room and attached to the respective door. Typical items range from the requirement to touch up some defective paint to replacing a malfunctioning motor controller. If a preliminary punch list is used, it is typically prepared by the contractor for the use of its own personnel and its subcontractors, based on a preliminary inspection by the contractor. Following the issuance of this preliminary punch list, the parties are expected to correct any indicated deficiencies prior to the pre-final inspection. It should be noted that the design professional and owner are usually not involved in identifying needed work at this point.

Pre-final inspection

When the preliminary punch list items have been completed to the contractor's satisfaction, the design professional and the owner are notified that the project is ready for its *pre-final inspection*. This inspection is a major event in the project's life! Representatives from all the design disciplines, from the owner's organisation and from all major subcontractors as well as the prime contractor participate in this inspection, which is in the charge of the owner and its representatives. For a large project, several days may be required. A thorough tour and inspection are made of all of the project's features and systems. Some of the testing and startup activities may take place at this time, although there must be reasonable assurance that the components being tested have already been subjected to some checking. Although the owner will have had a representative on the project site throughout the project, this inspection is the first time the owner will have seen the completed project. Obviously, the contractor has great interest in the outcome of this event.

Final punch list

It is likely that the pre-final inspection will identify some remaining deficiencies. The result is a *final punch list* containing those items still requiring effort in order for the project to be accepted. Whereas there may be several preliminary punch lists, perhaps one for each subcontractor, the usual practice is to prepare a single final punch list for the contractor, who will then coordinate with the various parties to assure compliance. The owner's on-site representative may be involved in inspecting and signing off on individual items on the final punch list. If some items are still defective, the punch list must be revised. Numbering and dating each item and dating each version of the list are essential to maintaining an accurate record of the process of correcting deficiencies.

Final inspection

When the contractor believes that all punch list items have been taken care of, the owner is notified that the project is ready for its *final inspection*. This inspection is less time consuming than the pre-final inspection, as it involves ascertaining only whether those items on the final punch list have been corrected. Fewer people are likely to be involved, although surely the contractor, a member of the design professional team and an owner's representative will participate. Depending on the outcome of this inspection, the owner and its design professional will declare the project to be either complete or substantially complete and will issue an appropriate certificate.

Beneficial occupancy

Despite the several possible cycles of inspections and punch lists described above, the project may still not be fully ready for the owner's use. For example, the delivery of some item of equipment may have been delayed or some deficiency correction may require an extended time. In any case, the project comes to the point where it can be used for its intended purpose, whether or not it is fully complete. At this point, *beneficial occupancy* is said to have begun. The issuance of a certificate of completion or substantial completion marks the beginning of this period. The beginning of beneficial occupancy has at least two important legal implications. First, from that point forward, the project is considered contractually 'complete' and liquidated damages, if any, cannot be assessed for the time after that date. Second, the time period for any warranties and guarantees provided by the contractor begins at that time. The owner of the completed project can, at this point, begin to move in and occupy the property, so as to 'benefit from its occupancy'.

It should be noted that some contract documents provide for *partial occupancy*, by which portions of the project are made available for the owner's use at different points in time. The production floor of a manufacturing facility may be turned over to the owner before the offices or vice versa; or an apartment building may be completed several floors at a time, so that the owner can begin renting some units while others are being completed. Thus, at any one time during project termination, various punch lists, inspections and certificates of completion will be at various stages when partial occupancy is allowed.

Keys

Unlike highway projects and other heavy construction work, buildings usually have doors that can be locked. A special concern on building projects is the handling of *keys*. As the project is being built and doors are installed, the contractor needs some means of locking most doors. Upon completion, the owner will insist that new locks and keys be installed, just prior to beneficial occupancy. Under one common practice, the hardware supplier provides construction cores for locks on all doors, giving the contractor easy means of locking and accessing the rooms, with a single key or a small set for the entire project. When it is time for the owner to occupy the building, these locks are replaced with the permanent keying system. The construction documents ought to specify how this system is designed, but in most cases it is still cumbersome, with a series of master keys, sub-master keys and individual keys. The contractor is responsible for being thorough, judicious and organised in arranging this important part of the security system for the owner.

Personnel actions

One of the most emotionally difficult responsibilities of the construction project's upper management team is to terminate the employment of the field and office personnel. Although the project will have seen many workers come and go, as their skills are required and then no longer needed, a core of personnel will have been together since early in the project operations phase, working toward a common purpose and building friendships and professional associations. By definition, a project is a temporary endeavour and thus it is no surprise that the project team must be dissembled, but such partings nonetheless are often difficult. For the contractor's

management personnel, there is likely to be employment on other
immediately in need of a person's services, they may be assigned a te
in the home office, depending on the company's employment agreeme
personnel may seek employment with other organisations. People hi
community are likely to stay there, although the contractor may ch
positions to particularly well-qualified and effective personnel it has hi
In the case of those people whose employment is being terminated,
must be completed, with sufficient notice of termination (even though termination should not be a surprise!). Some employers attempt to assist terminated employees in finding other employment. For employees who are being transferred, arrangements for salary, travel pay, starting date, job description and other matters must be coordinated both with the employee and the new project. It is unlikely that exactly the same project team will ever be assembled again, but that is the nature of project management.

Closing the construction office

Just as the project team must be taken apart, so must the temporary facilities where the team was based. We have already mentioned some of the closure activities in our short section on cleanup, such as fence removal and road repair. The task of closing the office may be as simple as arranging to have utilities terminated and then moving the office trailer from the project site to another site or the contractor's yard. If rental space has been used, that agreement must be brought to an end and the contractor's office equipment and other tangible property removed and taken elsewhere. For some projects, the owner will assume possession and use of the contractor's office; for others, a structure built for the project office may be taken apart. Along with the office, any construction plant remaining on the site must be removed. It is quite like a family who moves from an old neighbourhood; they must say goodbye to their friends and they must also arrange for utility disconnects, the packing and moving of their personal possessions and vehicles and perhaps transfer of ownership of their dwelling.

Closing out the project

In closing out the project, the contractor pursues several activities concurrently. We describe, first, the sequence of efforts related to project financing, followed by a number of other matters involving documentation required before the project is finally complete.

Subcontractor payment

One of the things the contractor must do before it can receive its final payment, including any retainage that has been withheld, is to complete paying its subcontractors. Thus, as the project nears completion, the contractor must work with each subcontractor to determine the amount of work remaining and its value, the approximate completion dates for the remaining work, the amount of payment due, including any retainage amounts withheld by the contractor and any disputed claims or payments (Mincks and Johnston, 1998). As noted above, remaining work is identified formally on the subcontractor's punch list. The contractor may have performed services for a subcontractor, such as providing short-term use of lifting equipment or performing r hauling services for the subcontractor. The usual practice is for the contractor to *arge* the subcontractor by reducing its payment by the value of the services. Likewise, contractors may backcharge the general contractor for miscellaneous services not covered by heir subcontracts. Misunderstandings related to these backcharge amounts, if any, must be resolved prior to final payment. When all issues have been resolved, final payment can be made to subcontractors, allowing the contractor to certify that such payments have been made. If some issues remain in dispute, payment can still be made, but there must be a clear record that notes these exceptions.

Final release or waiver of liens

A *lien* is a claim against property that can be filed by material suppliers, subcontractors, individuals and, in some cases, design professionals, that helps to assure that the filer of the lien will be paid. Lien laws vary widely in different jurisdictions as to when they may be filed, for how long they are in effect and what rights the various parties have in the lien process and in the USA at least, liens cannot be filed against public property. If a material supplier is having trouble being paid by the contractor, for example, the supplier may file a lien against the facility being constructed. Because the facility has some value, the owner of the property may be required to make the payment if the contractor fails to pay, especially if there is no payment bond to protect the owner. The astute owner does not wish to be burdened with the possibility of such a filing after the project is completed and therefore will insist on a *lien release* or *lien waiver* as a condition of final payment to the contractor. Such release indicates that potential lien holders have been paid in full and thus the owner will not be liable for payments the contractor should have made. The owner may require lien releases from subcontractors as well or it may require that the release provided by the contractor cover subcontractor liens.

Consent of surety

On projects that require a payment bond from the contractor, the owner normally requires a *consent of surety for final payment* prior to making final payment to the contractor. This completed form is issued by the bonding company and assures the owner that the surety company approves of the payment to the contractor. The surety can audit the contractor's records to verify that no unexpected financial obligations exist prior to issuing the release. Since the payment bond guarantees that the contractor will pay its obligations, an owner, making final payment before the surety has given consent, could be placed in a vulnerable position if the contractor fails to pay its workers, material suppliers or subcontractors (Fisk, 2003). Protection provided by the bond ceases upon final payment and the contractor could abandon its payment obligations if such a release were not provided by the surety.

Final quantities

The type of contract determines whether the contractor and owner must monitor and agree upon quantities placed into the project. Under a lump-sum/fixed-price contract, the contractor will be paid its contract price adjusted for variations (change orders) regardless of the actual quantities used. If the contract is of the cost-plus type, the contractor must provide documentation for its actual reimbursable costs, as provided in the contract. In the case of a unit-price/measure-and-value contract, the parties must measure the quantities actually put in place, for periodic payments throughout the project and for the final payment. Means for measuring these quantities depend on the item being measured. If payment for a roadway embankment is based on in-place volumes of fill material, for example, a field survey will be conducted to ascertain final profiles and lengths of the various sections and a comparison with the profiles before the material was placed will lead to the determination of the embankment volume.

Request for final payment

No matter what kind of contract, the contractor's final payment is equal to the total contract price minus the total of all previous periodic payments made by the owner. The form of the request is, therefore, similar to that used for all the previous partial payment requests. However, as discussed in previous sections, for final payment the contractor must comply with a number of special stipulations and provide evidence with its payment request. In addition to the testing and startup work, final cleanup, completion of punch list items, locks and keys, closure of the construction office, subcontractor payments, lien release and consent of surety already described, various other certificates and documents will be needed. Other evidence that may be required includes affidavits stating that all payments related to the project have been made and a certificate indicating that any required insurance will be in effect for the stipulated time period after project completion. The work covered by the request for final payment will include any work resulting from variations (change orders). Claims and disputes ought to be resolved by the time of this request; if they are not, the request must contain a statement that it is presented subject to the resolution of these matters.

Liquidated damages

Once the beneficial occupancy date has been established, the parties can determine whether *liquidated damages* are to be assessed against the contractor, by comparing that date with the completion date established in the original contract or by any extensions thereto. The amount of these damages, to be assessed by reducing the final payment, is calculated simply as the product of the daily liquidated damages amount stated in the contract and the number of days late. If the contract provides for a *bonus for early completion*, that amount will be calculated in the same manner and will be added to the final payment. In some contracts, liquidated damages are assessed against portions of the work and bonus payments may be figured in the same way. The New Zealand *Conditions of Contract for Building and Civil Engineering Construction* (Standards New Zealand Paerewa Aotearoa, 1998) include an important provision regarding liquidated damages: 'Payment or deduction of liquidated damages for late completion shall not relieve the Contractor from any of its other liabilities or obligations under the contract.'

Final payment and release of retainage

If all punch list items have been completed, all certificates, affidavits and other documents submitted and in good order and all other obligations completed, the contractor's request for final payment ought, finally, to be honoured. Included with this payment should be the retainage that has been held by the owner pending satisfactory completion. Note that if 10% has been retained throughout the project the released retainage will be a substantial portion of the project budget. Many contracts provide, however, for reduced amounts of retainage as the project nears completion. When the owner makes final payment and releases retainage to the contractor, both parties waive most of their rights to make further claims against the other. Exceptions include the right of the owner to claim relief due to faulty or defective work or failure of the work to comply with contract requirements and the imposition of the terms of any special guarantees required by the contract documents. The contractor also waives its right to further claims except any that might have been filed previously and remain unsettled (Fisk, 2003).

Final accounting and cost control completion

Was the project a financial success? What profit did the contractor make from this endeavour? Which portions were successful? Which did not meet budget expectations? Only after all of the above steps have been completed can the contractor obtain an accurate picture of the project's final financial status from its point of view. An effective cost control system will be able to compare the actual cost of each work item with its estimate and identify the reasons for any variances, which may be either quantity or rate variances. A quantity variance results from an actual quantity different from that estimated, whereas a rate variance occurs because the cost per hour, per pound or per linear meter is different from the estimate. Note that quantity variances are less important to the contractor in unit-price (measure-and-value) contracts and cost-plus contracts than they are in fixed-price/lump-sum contracts. It is suggested that any variance of 10% or more above or below the estimate ought to incur special analysis (Mincks and Johnston, 1998). Because the actual costs become part of the historic cost database, unusual conditions that led to large variances need to be recorded, so that future tender estimates do not rely too greatly on these abnormally high or low costs. The contractor's organisation ought to have a standard form on which summaries of project costs are reported, so that executive management personnel can make valid comparisons with similar previous projects and use the experience to make informed decisions about future projects.

Certificates

In connection with the procedures described above, several certificates may be issued. We describe five such documents.

Certificate for payment

The *General Conditions of the Contract for Construction* (American Institute of Architects, 1997) provide for the architect to issue this certificate in response to the contractor's request for payment, if the architect believes that 'the work has progressed to the point indicated' and the 'quality of the work is in accordance with the contract documents'. If these conditions are met, such a certificate can be issued for each periodic payment request and for the request for final payment.

Contractor's certificate of completion

Some contracts provide this form as a means for the contractor to notify the owner formally that the project is complete. One such format includes the following wording, to be signed by an executive officer of the contractor's organisation (Fisk, 2003):

> I know of my own personal knowledge, and do hereby certify, that the work of the contract described above has been performed, and materials used and installed in every particular, in accordance with, and in conformity to, the contract drawings and specifications. The contract work is now complete in all parts and requirements, and ready for final inspection.

A producer statement provides the same function; one version allows the contractor to notify the owner that either all of the project or a portion described by appropriate attachments is complete (Standards New Zealand Paerewa Aotearoa, 1998).

Certificate of substantial completion

> Substantial completion is that stage in the progress of the work when the work or designated portion thereof is sufficiently complete in accordance with the contract documents so the owner can occupy or utilize the work for its intended use. (American Institute of Architects, 1997)

By this definition, substantial completion marks the beginning of beneficial occupancy. Issuance of this certificate by the owner or its design professional follows the final inspection. The notion of 'substantial' completion is that some punch list items remain to be completed after the final inspection; thus the project is not yet fully complete, even though the owner can use it for its intended purpose. The certificate of substantial completion includes this list of items yet to be completed. As noted in the previous discussion of beneficial occupancy, the date of substantial completion fixes the point in time at which (1) warranties provided by the contractor commence and (2) no further liquidated damages may be assessed. The term *practical completion* is used in some contracts to refer to this same point in the project closeout and termination phase.

Certificate of completion

When any deficiencies noted on the certificate of substantial completion have been remedied, a certificate of completion may be issued. This certificate is similar to the previous one, except that it contains no list of deficiencies. If the final inspection concludes that no deficiencies exist, the certificate of substantial completion will not be needed. Many owners dispense with the certificate of completion, believing that the certificate of substantial completion, followed eventually by the contractor's request for final payment when deficiencies have been resolved and other documents are in place, provides sufficient protection and notification.

Certificate of occupancy

Some localities or other jurisdictions require inspections by public agency officials of such life safety parts of the work as lifts, fire protection systems, handicapped access and sewage systems. In this case, the regulations require that the appropriate authority issue a certificate of occupancy prior to occupancy by the owner if it finds the facility acceptable.

As-built drawings

Most contracts require that the contractor maintain a set of *record drawings*, commonly known as *as-builts*, as the project progresses. On these documents are recorded actual locations, dimensions and features that are different from the original contract drawings. Usually the original drawings are used and the changes are simply overlain onto them. When the drawings are in electronic format, the contractor may be required to modify the drawing files and highlight the modifications. Examples might be underground utilities located at a depth different from originally designed, a piece of machinery placed in a revised location or a highway project driveway with a modified alignment or location. In each case, the contractor would have been directed, presumably in writing, to make the change, but the design professional would not have changed the relevant contract drawing. Such changes are the contractor's responsibility. It is convenient and useful for the contractor to keep as-built drawings current as the project proceeds; it is a condition of final payment that a complete and accurate set be forwarded to the owner prior to payment.

Operating and maintenance manuals

A typical contract provision requires that the contractor furnish operating and maintenance manuals for all of the project's equipment and operating systems. The main effort involves simply gathering and organising the documents as they are received from the manufacturers. Often they are shipped with the equipment itself and can be easily lost during unpacking. A common practice is to produce bound volumes of these materials, organised by type – pumps, fans and so on – or by system – heating, communication and so on – which then become permanent references for use by the facility's operating and maintenance personnel. Like as-built drawings, it is usually a contract requirement that these manuals be furnished in complete form prior to final payment.

Records archiving and transfer

Project records are essential parts of the project's history and are also important in the operation and maintenance of the finished work. The contractor has two primary obligations in this regard. First, it must organise and store all relevant project records in a secure and convenient location for future use by its own personnel. The list of such records is lengthy, as indicated in our previous section on project documentation. Among other items, the contractor will want to archive cost records, schedule reports, correspondence, various logs and all meeting minutes. The availability of project websites and computer-based document management systems allows the archiving of these materials electronically. One or more CDs can be used to contain all of the project's records exactly as they stood at project completion. Such information will be useful for the planning of future projects and for providing data if questions or legal proceedings arise about the project just concluded.

The contractor's other obligation with respect to project records is to transfer required documents to the owner and/or the design professional in an organised and complete manner. In addition to the as-built drawings, operating and maintenance manuals, keying schedules and certificates discussed above, such records might include progress photographs, material and system test inspection results and any other documents required by the contract.

Furthermore, the contractor will probably have an obligation to furnish the owner with spare parts for various pieces of equipment, extra materials such as paint, carpet, plumbing components and locksets, special tools and in some cases, equipment that has been used to construct the project.

Training sessions

Yet another obligation sometimes required of the contractor is to conduct training sessions for the owner's operating personnel. Imagine a large dam project, with various gate motors, pumps and computerised control, flow monitoring and communications systems. Although the contractor may not be expert on their operations, the contract may require that training be provided. Thus, it is the contractor's responsibility to arrange such sessions, probably led by manufacturers' and/or subcontractors' representatives, after the various components are installed and prior to owner occupancy. A convenient time may be just after the systems and components are tested, when such representatives are already on the site. The operating and maintenance manuals will be an important focus of these instructional sessions.

Warranties, guarantees and defects liability period

The terms *warranty* and *guarantee* tend to be used interchangeably in construction contracts to designate the obligations the contractor assumes for repairing defects in its work for a specified period after the commencement of beneficial occupancy. The general conditions cover the overall obligation and the time period, called the *defects liability period or the maintenance period.* Common defects liability periods are 3 months (Standards New Zealand Paerewa Aotearoa, 1998) and 1 year (American Institute of Architects, 1997) for the entire project, unless certain portions of the work are required to have longer periods. These exceptions are given in the special conditions and often require that the contractor furnish a written guarantee for these elements. For example, a roofing contractor may be required to provide a written guarantee that its work will be free of defects for 15 years, whereas a landscaper may have to warrant its trees and other plantings for 2 years. The New Zealand *Conditions of Contract for Building and Civil Engineering Construction* (Standards New Zealand Paerewa Aotearoa, 1998) provide typical language on this topic:

> 11.1.1 The period of defects liability for the contract works or any separable portion shall commence on the date of practical completion of the contract works or separable portion. The period of defects liability shall be three months unless some other period is stated in the special conditions.
>
> 11.2.1 The contractor shall remedy defects arising before the end of the period of defects liability in the contract works from defective workmanship or materials. The engineer shall give notice in writing to the contractor during the period of defects liability or within five working days thereafter of defects to be remedied.
>
> 11.2.2 The contractor shall not be liable for fair wear and tear during the period of defects liability.

This document also specifies that, if the contractor fails to complete the repairs within a reasonable time, the owner is entitled to have the work done and charge the contractor for its cost.

Post-project analysis, critique and report

Internally, the contractor will want to analyse the entire project to determine what it has learnt that can be applied to the next such endeavour. This step is probably one of the most neglected aspects of construction project management, as there is always pressure to look ahead to the next job rather than backward toward the work just completed. However, a day, or even a few hours, spent with key members of the project team in a post-project critique, followed by the creation of a written report, will pay dividends in the future. Conducted by the project manager, the meeting should cover such topics as personnel and labour relations, construction methods and on-site coordination, safety issues, subcontractor performance, fabrication and delivery matters, cost control, schedule issues, owner and design professional relationships and the quality of the project and its components and systems. Indeed, the table of contents for sections of this book could form the basis for the discussion. The meeting also provides an occasion to thank the members of the team for their efforts and to mark, in some fitting manner, those whose contributions have been especially noteworthy. In addition to the discussion, several members of the team may be asked to contribute written analyses and the project manager will surely prepare their own individual critique. The tenor of the study must not be negative. Questions involving 'What did we do right?' are just as important as 'What could have been improved?' Once the various elements of the discussion and written analyses have been completed, the project manager is responsible for compiling the written report, which becomes a part of the company's historical record, together with the various cost, schedule and other reports.

Owner feedback

Many owners provide their contractors and design professionals with their own project performance evaluations. If they do not, the contractor would do well to seek such feedback from the owner. A friendly visit with the company president, the head of the school board, the mayor or whoever heads the owner's organisation may be sufficient, to offer thanks for the opportunity to construct the project and to assure the owner that any defects will be remedied. Such a visit also serves to remind the owner that the contractor is available for the owner's next project! Beyond that, it is important to solicit comments about the positive and negative aspects of the contractor's performance from the owner's perspective. In that regard, feedback from those closer to the project, such as on-site representatives, will be important.

A closing comment

If this book were to include an epilogue, it would probably suggest that the project is really not complete, because now the real work of using the new facility can begin. Whether that next phase is part of the 'project' may be subject to some disagreement (because by definition a project has a limited life and must be brought to a conclusion), but the importance of facility management, once construction is completed, cannot be denied. It has been estimated that 5% of the UK's gross domestic product is spent on built asset maintenance (Building Research Establishment, 2001). We must have the ability to manage the construction of facilities, which is what this book is all about, but we must also know how to manage them once their construction has been completed. But that is a topic for another time and another place.

Discussion questions

1 The project closeout and termination phase can be approached as a project in itself. Prepare a schedule network depicting the relationships among the various activities that may take place during this phase, based on the information in this chapter and your general knowledge of the construction process. Time estimates are not required.
2 Prepare a punch list of items needing to be remedied for a classroom, laboratory, your place of residence or some other building or room. For each item, note the date, the type of deficiency and the party you believe should be responsible for fixing the defect. Provide a space for recording when the fix is completed and who approved it. Identify any items that might be related to the original construction of the facility or its renovation.
3 The text described a method for determining the final quantity of roadway embankment fill. Name three other work items that might be paid on a unit-price basis and the measurement method for each.
4 Suppose a project consists of three buildings, plus sitework, with stipulated completion dates as follows: Building A, 1 March 2004; Building B, 15 April 2004; Building C, 15 July 2004; Sitework, 1 August 2004. The contract provides for per-calendar-day liquidated damages for the various portions as follows: Building A, US$ 500; Building B, US$ 300; Building C, US$ 750; Sitework, US$ 150. The actual dates of the beginning of beneficial occupancy are as follows: Building A, 1 April 1 2004; Building B, 1 April 2004; Building C, 10 August 2004; Sitework, 10 August 2004.

   (a) Calculate the amount by which the final payment will be modified due to these actual completion dates.
   (b) Now suppose, in addition to the liquidated damages provision, the contract provides for bonus payments for the portions as follows: Building A, US$ 600; Building B, US$ 450; Building C, US$ 1000; Sitework, US$ 200. Revise your answer to (a).
   (c) Suppose a dispute arises over the liquidated damages for sitework, with the contractor claiming unfairness. What might be the arguments of each of the two parties to this dispute? Why might the owner have a stronger position in the case of (b) than (a)?

5 How can modern electronic methods of document management assist with the transfer and archiving of construction project records? What precautions might be taken to assure that archived materials are true and accurate versions of the records as they were at project completion?
6 Interview two of the three parties comprising construction's primary triad: owner, design professional and contractor; find out which three or four end-of-project activities are most difficult from their prospectives and how they recommend making them less difficult. Compare the responses and write a brief report that summarises your findings.

References

American Institute of Architects. 1997. *General Conditions of the Contract for Construction.* AIA Document A201.

Building Research Establishment. 2001. *Centre for Whole Life Performance.* http://www.bre.co.uk/whole_life/service3.html.

Fisk, E.R. 2003. *Construction Project Administration*, 7th edn. Prentice-Hall.

Mincks, W.R. and H. Johnston. 1998. *Construction Jobsite Management.* Delmar.

Standards New Zealand Paerewa Aotearoa. 1998. *Conditions of Contract for Building and Civil Engineering Construction.* New Zealand Standard NZS: 3910: 1998.

Index

Absenteeism, 240
Acceptance period, 114
Access, site, 165
Accident:
  report, 253
  statistics, 222, 223, 224
Accidents:
  fatal, 223
  non-fatal, 224
  reporting, 227
Accounting, 257
  cost, 216
  final, 296
  financial, 216
  managerial, 216
  system, 216
Accounts payable, 216
Accounts receivable, 216
Acoustical consultant, 42
Activity duration, 135
  estimated, 144
Activity table, 140, 152–53
Activity, summary, 145, 149
Activity-on-node method, 134
Actual cost, 162
  of work performed, 215
Actual quantities, 30
Addenda, 83
Add-ons, 98
Adjudication, 267
Adjustment, equitable, 270
Adventurer, joint, 23
Advertisement for tenders, 65
Agency construction management, 16, 17, 20
Agency review, 69
Agency shop, 186
Agreement, 57, 67
  contract, 114–15
  partnering, 128
Air pollution, 232
Alaska, 199, 284, 286
Alternate construction methods, 81
Alternate dispute resolution, 265, 266–69
Alternatives, 54, 55
  identification, 48–9
Alyeska Pipeline Service Company, 199
American Association of State Highway and Transportation Officials, 59
American Institute of Architects, 31, 47, 58, 66, 67, 261, 262, 268, 275
American Society of Civil Engineers, 218
Analysis, post-project, 300
Apparent low tenderer, 107
Application service provider (ASP), 258
Apprentice, 236
  programme, 186, 187
Apprenticeship, 236
Approval, 241
  to construct, 120
Arbitration, 262, 268–69
  advantages and disadvantages, 269
  international, 268
  mediation and final offer, 269
Arbitrator, 268, 269
Archaeology, 235
Architect, 40, 48
Architect-engineer (A/E), 41
Architectural-engineering firm, 41
Archiving, records, 298–99
Arrow method, 134

As-built drawing, 298
Asian Development Bank, 220
Assignment, work, 237
Associated General Contractors, 59, 77, 78, 174, 176–83
Association for the Advancement of Cost Engineering, 87
Association of Consulting Engineers of the UK, 58
At-risk construction management, 16, 17, 26
Australia, 3, 22, 32, 52, 185, 198, 237
Award:
  contract, 110
  subcontract, 174

Backcharge, 294
Backward pass, 138
Balanced tender, 101
Bar chart, 129, 131, 134, 140, 141, 144, 145, 146–48, 206, 208
Bar charting:
  advantages, 131
  disadvantages, 132
Bargaining unit, multi-employer, 185
Barricades, 166
Baseline schedule, 207
Basic network calculations, 206
Belgium, 58
Bench trial, 265
Beneficial occupancy, 292
Benefit package, 238
Best value selection, 111
Bid bond, 122
Bill of quantities, 94, 96, 98
Boilerplate, 60
Bond, 96, 98, 120, 253
  costs, 126
  bid, 122
  labour and material payment, 122, 123, 126
  performance, 122, 123, 125, 126
  proposal, 122, 123, 124
  surety, 122
  tender, 122, 126
Bonding, 73, 122–26
  capacity, 79
Bonus provision in cost plus contract, 32
Borrow areas, 166
Brass tag, 237
Bridging, 19
Brief, 44, 47, 265
British Columbia, 118
British Standards Institution, 65
Brooks Law, 45
Budget, 205
  tracking report, 255
  direct cost, 159
  project, 120, 158, 207, 210
  total, 213, 214
Budgeted cost of work performed, 213, 214, 215
Budgeted cost of work scheduled, 215
Budgeting and cost systems, 158–63
Builder's risk insurance, 127
Building construction, general, 1
Building cost model, parametric, 89
Building permit, 121
Build-operate-transfer (BOT), 21
Build-own-operate (BOO), 22
Build-own-operate-transfer (BOOT), 21–2, 29
Business licence, 122
Business units, 4
Buying, centralised, 169
Buyout, 168

Camara Argentina de Consultores, 58
Canada, 185
Cardinal change, 247
Cash flow, 53
  cumulative, 161
  projection, 160–63
    curve, 160
Categories, construction, 1
Category I and II claim, 271
Centralised buying, 169
Certificate, 289
  for payment, 296
  of completion, 297
    contractor's, 297
  of occupancy, 297
  of substantial completion, 297
Certificates, 296–97
  surety and insurance, 67
Change, 246, 248
Change order, 106, 245, 253, 294
  proposal, 248
  request, 248
Change proposal, value engineering, 102, 251
Change, cardinal, 247
Changed conditions, 270
Changes, 263
  in the work, 240
Characteristics, construction industry, 3
Charter, partnering, 43, 128, 129

Checklist:
construction health and safety, 226
inspection, 221
jobsite, 81–2, 84–5
works tender opening, 109
Civil law, 264
Claim, legal, 262
Claims:
disputed, 293
process, 261–62
Classified advertisement, 78
Clean Water Act, 233
Cleanup, 96
final, 290
Closeout, 253
project, 289–301
Closed shop, 186
Closing out the project, 293–301
Clothing and equipment, personal protective, 227
Code analysis, 50
Code system, cost accounting, 160
Collective bargaining, 120
agreement, 185, 186, 236, 238, 241
unit, multi-employer, 185
Commercial construction, 2
Commissioning, 289
Communication, 205, 245–61, 254
tools, web-based, 260
documentation and, 10
Compacted soil, 231
Company overhead cost, 90, 94
Compensable delay, 272
Competition, 79
Complaint, filing, 264
Completed contract basis, project accounting, 218
Completing the work, 290–93
Completion date, 79
estimated project, 206
Completion:
certificate of, 297
practical, 297
substantial, 297
Complexity, project, 106
Computer applications, network scheduling, 144
Computer simulation, 156
Computer-based document system, 255
Conciliation, 266
Conciliator, 266
Conditions:
changed, 270
general, 60
special, 61, 63
supplementary general, 61
unforeseen, 270
Consent, 120–22
of surety, 294
resource, 121
Constructability analysis, 49–50, 68, 81
Constructability review, 49, 263
Construction camps, 199–204
Construction contract document, 67
quality, 78
Construction cost, 90
Construction document development, 54
Construction estimating, 88
Construction firms, 3
Construction industry characteristics, 3
Construction manager, 16–8, 23, 28, 50, 68, 72, 76, 174
agency, 16, 17, 20
at-risk, 16, 17, 26
Construction office, closing, 293
Construction plan room, 78
Construction Quality Assessment System, 221
Construction Specifications Canada, 62
Construction Specifications Institute, 62, 93
Construction workforce, 3
Consultant selection, 44–7
Consulting Engineers Association of India, 58
Contacts, 253
Contamination, petroleum, 232
Contingency, 90
plan, petroleum spill, 233
Contract, 57, 253
agreement, 114–15
award, 110
document, 57
development, 40, 54, 57–69
drawings, 59
termination, 275–77
terms, 78
type of, 26, 27, 30–3
Contracting licence, 122
Contractor pre-qualification program, 73
Contractor selection phase, 9, 72–115
Contractor's certificate of completion, 297
Contractual relationship, 13
Control, 205, 219, 235, 240
monitoring and, 9
Convenience, termination for, 275
Conversation record, 254
Cornell University, 55, 56

Cost accounting, 216
  code, 209, 210
    system, 160
Cost allowance:
  prime, 93
  provisional, 93
Cost analysis, 154
Cost and budgeting systems, 158–63
Cost control, 154, 207, 209–18, 257
  completion, 296
  curve, 214
  systems, purposes, 158
Cost curve:
  scheduled, 214
  cumulative, 162
Cost database, historic, 94
Cost envelope, 162
Cost estimate, 9, 53, 55, 57, 68, 81, 120, 158, 169
  preliminary, 50, 52, 87
Cost estimating, 83, 87–105, 209, 257
  accuracy, 42
  software, 101–02, 158
Cost estimator, 42
Cost growth, 26
Cost iceberg, 223
Cost insurance freight (CIF), 170
Cost per function, 87
Cost plus contract, 31–2, 33, 72, 76, 248, 261, 294
  variations on, 32–3
Cost plus fixed fee contract, 31
Cost plus percentage contract, 31
Cost records, historical, 91
Cost reports, 210–16
Cost section, cost report, 211
Cost status, 10
  report, 217
Cost sum, prime, 93
Cost summary and forecast, 212
Cost tracking report, 255
Cost variance, 215
Cost:
  company overhead, 90, 94
  equipment, 90, 92, 95
  final, 88
  general overhead, 90, 94, 95
  home office overhead, 90, 94
  indirect, 93
    labour, 91, 93
  job overhead, 90, 93
  labour, 90
  material, 90, 91, 95
  plant, 92
  project overhead, 90, 93
  reimbursable, 31
  site overhead, 90, 93, 95
  subcontract work, 90, 93, 95
Costs:
  bond, 126
  direct, 90
  firm estimate of, 88
  preliminary assessed, 87–8
  reimbursable, 261
  rough order of, 87
Cost-time curve, 214
Court system, 264
Coventurer, 23
  sponsoring, 23
Craft personnel, sources, 235
Craft union, 185
Craftsperson, 236
  lead, 235
Creditors, 216
Crime, jobsite, 238
Criminal law, 264
Criteria, contractor selection, 110–14
Critical activity, 140, 144
Critical path, 140, 206
Critique, post-project, 300
Cross examination, 265
Cumulative cash flow, 161
Cumulative cost curves, 162
Curve:
  cash flow projection, 160
  project progress, 161
CYCLic Operations NETwork (CYCLONE), 156
Czech Republic, 4

Daily report, 254
Damages, liquidated, 273, 295
Data input, 149
Data sources, cost control, 209–10
Debtors, 216
Deciding to tender, considerations, 78
Deep Cove, 193, 195, 196
Default, termination for, 275–76
Defects liability period, 299
Defendant, 264
Delay, 272–275
  compensable, 272, 274
  contractor-caused, 272
  excusable, 273

non-compensable, 274
non-excusable, 273
owner-caused, 272
reasonable, 274
Delivery, 254
material, 165, 285
Demolition products, 233
Denmark, 36
Descriptive specifications, 219
Design:
competition, 46, 76
development, 54, 57, 67
drawings, 59
fee, 45
professional, 13, 40, 44, 47, 50, 54, 75, 80, 83, 105, 121, 163, 185, 241, 245
reputation, 79
stage, 40, 54–7
Design-build, 2, 14–6, 18, 21, 23, 26, 28, 40, 72, 74, 76, 174, 219
Design-Build Institute of America, 14
Design-build-operate (DBO), 22
Design-construct, 14
Design-tender-build, 12–4, 17, 23, 24, 25, 26, 28, 75, 174
Detailed plan and schedule, 129
Differing site conditions, 270–72
Differing soil conditions, 27
Direct cost, 90
budget, 159
Direct examination, 265
Disclaimer clauses, 106
Discovery, legal, 264
Disposal, solid waste, 233
Dispute prevention and resolution, 262–69
Dispute resolution, alternative, 265, 266–69
Dispute review board, 263, 266–67
Dispute, jurisdictional, 236–37
Dispute, legal, 262
Disqualification of bidders, 110
Document and construct, 18–20, 28
Document management, 257
Documentation, 205, 245–61, 289
and communication, 10
Drainage works, temporary, 231
Drawing, 58, 59, 81, 88, 253:
as-built, 298
contract, 59
design, 59
record, 298
shop, 242
Dry shacks, 164
Duration, activity, 135
estimated, 144

Early finish, 135, 140, 162, 206
Early start, 135, 162, 206
Earned value, 213, 214
analysis, 216
e-Builder™, 259, 260
Educational specialist, 42
e-marketplace, 261
Electronic communication, 10, 76, 156, 170, 252
Electronically-enhanced project communications, 257–61
Employee insurance, 128
Endangered species, 234
Enforcement and prosecution powers, safety, 226
Engineer, 40, 41, 48
Engineer's estimate, 43
Engineered construction, 1
Engineering construction, 2
Engineers Joint Contract Documents Committee, 59
Envelope system:
double, 46
single, 46
Environment, 205
Environmental conditions, 188, 240
Environmental management, 228, 231–35
Environmental monitoring, 175
EPC (Engineering, procurement, construction) contract, 21
Equipment, 209, 235, 243
cost, 90, 92, 95
leasing, 244
maintenance, 175, 227, 244
management, 243–45
performance, 188
rental rates, 244
rolling, 243
Equitable adjustment, 270
Error checking, 149
Estimate, 83
at completion, 213
fixed-price, 96
lump-sum, 96
to complete, 213
Estimated activity duration, 144
Estimated cost, 162
Estimated quantities, 30
Estimated variance at completion, 213

Estimating, 172
  construction, 88
  process, 88–101
  software, 158
Eternal triangle, 12
European Committee for Standardisation, 65
Eurotunnel, 22
Everest®, 102
Evidence, 265
  hearsay, 265
  real, 265
Examination:
  cross, 265
  direct, 265
  re-cross, 265
  redirect, 265
Exclusions, subcontract proposal, 173
Excusable delay, 273
Expediting, 241, 242, 254
Expedition®, 256, 257
Experience rating, 127
Expert witness, 265
Expressions of interest and qualification, 44
External conditions, 78

Fabrication, 242, 245
  material, 285
Facilities project, 41
Facilities, temporary, 163–65
Facility management, 301
Fast-tracking, 24
Federal Transit Administration, US, 68
Fédération Internationale des Ingénieurs-Conseils (FIDIC), 21, 44, 45, 47, 58, 67, 219, 269
Fee, design, 45
Fee-bidding, 45, 46
Feedback, 54, 67
  owner, 300
Field memorandum, 254
Final:
  accounting, 296
  cost, 88
  inspection, 291
  judgment, 265
  payment, 295
    request for, 295
  punch list, 291
  quantities, 294
Finances, 289
Financial accounting, 216
Financial feasibility analysis, 52–3
Financial management, 255
Finland, 4
Fire-fighting equipment, 227
Firm estimate of costs, 88
Firms, construction, 3
  number of, 3,4
  sizes, 4
First aid facilities, 164, 227
Fixed price contract, 27, 80, 83, 261, 294
Fixed-price estimate, 96
Float, 138
Food service, 164
Force account, 23–4, 29
Forecast, cost summary and, 212
Foreperson, 235, 239
  general, 235
Forward pass, 138
France, 58
Free on board (FOB), 170, 285
Funding, 53

Gantt chart, 131
Gantt, Henry L., 131
General:
  building construction, 1
  conditions, 60–1
  contractor, 13, 83
  foreperson, 175, 235
  ledger, 216
  overhead cost, 31, 90, 94, 95
  provisions, 60
General Services Administration, US, 69
Geotechnical specialist, 41, 49
Germany, 185
Global Estimating®, 102
Graph, cumulative cash flow, 161
Graphical output, 156
Grievance, 237
Guarantee, 299
Guaranteed maximum cost contract, 32

Handling, material, 243
Hazards, construction site, 224
Health and safety checklist, 226
Hearing, 50, 51
Hearsay evidence, 265
Heavy construction, 2
Helper, 236
Highway construction, 2

Hiring hall, union, 236
Historic cost database, 94
Historical cost records, 91
Home office overhead cost, 90, 94
Hong Kong, 15
  University, 74
Housekeeping, jobsite, 227
Housing and feeding, worker, 188, 200–03
Housing, temporary, 164
Hydrologic specialist, 42

Iceberg, cost, 223
Incentive clause in cost plus contract, 32
Incentive programme, safety, 227, 228
Incentives, 188
Indirect cost, 93
  labour, 91, 93
Indoor storage, 163
Industrial construction, 2
Industrial relations, 175
Industry associations, 65
Information sources, project, 76–8
Information technology, 102, 260
Inspection:
  checklist, 221
  final, 291
  list, 290
  on-site, 221
  pre-final, 291
  public authority, 121
  safety, 227
  site, 271
Inspector, public agency, 185
Installation, material, 242
Institute of Quality Assurance, 218, 219
Institutional construction, 2
Instructions to tenderers, 65–6
Insurance, 67, 120, 127–28, 253
  builder's risk, 127
  employee, 128
  liability, 127
  premium, 127
  property, 127
International arbitration, 268
International Organisation for Standardisation, 219
Internet, 50, 80, 257
  plan room, 77
Inventory:
  material, 254
  tool, 238
Invitation to tender, 65, 75, 76, 80
Invited tender, 75
Invoice, 254
Ireland, 5
Irregular proposal, 110
ISO 9000, 219

Job overhead cost, 90, 93
Job planning, preliminary, 80–3
Jobsite:
  checklist, 81–2, 84–5
  crime, 238
  housekeeping, 227
  layout plan, 168
  office, 163
  safety rules, 228
  visits, 81–2
Joint adventurer, 23
Joint venture, 22–3, 29
Journals, 216
Journeymen, 236
Judgement:
  summary, 264
  final, 265
Judicial precedent, 265
Jurisdictional dispute, 236–37
Jury trial, 265
Just in time (JIT) delivery, 242

Keys, 292
Korea, 284, 286

Labour, 209
  and material payment bond, 122, 123, 126
  conditions and supply, 79
  cost, 90
    indirect, 91, 93
  law, 106
  productivity, 235–41
  skilled, 186
Lag time, 144
Land acquisition, 21, 54
Land surveyor, 42
Late finish, 138, 140, 162, 206
Late start, 138, 162, 206
Late tender, 107, 118–19
Law:
  civil, 264
  criminal, 264

Laws and regulations, environmental, 231
Lay-down areas, 165
Lay-down yard, 285, 287
Layout plan, jobsite, 166, 168
Lead craftsperson, 235
Lease, equipment, 244
Legal issues, 205, 261–77
Legal process, 264
Letter, 254
  of intent, 115
Levels of detail, 87–8
Liability insurance, 127
Licence, 120–22
  business, 122
  contracting, 122
Lien, 294
  final release, 294
  waiver, 294
Life cycle costing, 53
Life cycle, project, 251
Liquidated damages, 273, 295
Litigation, 264
Location, project, 78, 106
Log, submittal, 246
Logistical planning, 193
Louisville and Jefferson County Metropolitan
    Sewer District, 74
Low tenderer, apparent, 107
Lowest qualified tenderer, 110
Lump sum contract, 27, 66, 72, 75, 83, 108, 248,
    249, 250, 294–95
Lump-sum estimate, 96

Machinery, 243
Maintenance period, 299
Maintenance, equipment, 227, 244
Malaysia, 23
Managerial accounting, 216
Manapouri Power Project, 194
Manapouri, Lake, 193
Manual, operating and maintenance, 298
Margin, 90, 95, 105
Markup, 90, 95, 98, 105
  issues, 105
Master builder, 14
Master documents, 59
Master plan, 49
Master programme, 130, 173
Master schedule, 130
Masterformat technical specifications, 63
Material, 209, 235
  acquisition, 120
  cost, 90, 91, 95
  delivery, 165, 285
  fabrication, 285
  handling, 165, 243
  installation, 242
  inventory, 254
  management, 240, 241–43, 254, 284
  procurement process, 169
  quantities, 65
  shipping, 285
  storage, 242
    protection of, 243
  supplier, 185
  take-off, 65
Matrix:
  consultant selection, 46
  scoring, 111–13
Measure and value contract, 27, 30, 31, 66, 80,
    83, 247, 249, 295
Measure-and-value estimate, 98
Measurement, 248–51
Mediation, 266
  and final offer arbitration, 269
Mediator, 266, 269
Medical facilities, 164, 188
Meeting minutes, 254
Meetings, pre-tender, 83
Merit shop, 187
  associations, 187
Method statement, 80–1
Microsoft Project 2000®, 144
Mini-trial, 267
Mobilisation, 96
Modularisation, 188
Monitoring, 205, 219, 235
  and control, 9, 205–35
  schedule, 156
Monthly report, 254
Motivation, employee, 237, 239

National economies, impact of construction, 3
Negotiation, 75–6, 266
  subcontract, 174
Net project cost, 90, 94
  elements of, 90
Network schedule, 134–56
Network, operations simulation, 157
Network, schedule, 136, 137, 139, 142–43,
    150–151

Network-based project schedule, 206
New Zealand, 4, 5, 27, 30, 45, 52, 58, 60, 61, 76, 77, 111, 120, 121, 193, 221, 247, 268, 295, 299
No damages for delay clause, 274
Node method, 134
Noise, 235
Non-contractual relationship, 13
Non-excusable delay, 273
Non-facility projects, 41
Non-productive time, 239
Non-union contracting, 120, 187, 236
Notice to proceed, 105, 114–15
Notice to tenderers, 65
Novation, 19

Obligee, bond, 122, 123
Occupancy:
  beneficial, 292
  certificate of, 297
  partial, 292
Occupational Safety and Health Act (OSHA), 226
Office:
  construction, closing, 293
  manager, 175
  temporary, 163
Offtaker, 22
Oncosts, 90, 93
On-line publications, 76
On-site inspection, 221
On-site testing, 221
Ontario, 118
Open invitation, 9
Open shop, 187
Open tender, 74–5, 76
Opening:
  process, 106
  public, 107
Operating and maintenance manual, 298
Operating personnel, 185
Operations modelling, 156–58
  programme, 158
Operations simulation network, 157
Operative, 185
Opus International Consultants, 45
Order, stop work, 221
Organisation structure, 184, 185
Organising the worksite, 163–68
Over-allocation, 155
Overtime, 240
Owner:
  feedback, 300
  reputation, 79

Panel unit cost, 87
Paperwork, 245
Parameter cost, 87
Parametric building cost model, 89
Partial occupancy, 292
Partial payment, 248
Partial take-off, 88
Particular conditions, 61
Partnering, 43, 128, 263
  agreement, 128
  charter, 43, 128, 129
  workshop, 128
Password, 258
Payment:
  disputed, 293
  final, 295
    request for, 295
  partial, 248
  periodic, 295
  professional services, 47
  request, periodic, 249
  subcontractor, 293–94
Peachtree Complete Accounting 2002®, 158, 216
Penalty provision in cost plus contract, 32
Percent complete, 211
Percentage of completion method, project accounting, 218
Performance bond, 122, 123, 125, 126
Performance specifications, 219
Periodic payment, 295
  request, 249
Permafrost, 271
Permit, 120–22, 231
  building, 121
  plumbing and electrical, 121
Personal protective clothing and equipment, 227
Personnel:
  accommodation, 193
  actions at project termination, 292–93
  operating, 185
  supervision, 235–41
Petroleum:
  contamination, 232
  spill contingency plan, 233
Phased construction, 24–6, 40
Photography, 254
Pipeline workers, 199

Plaintiff, 264
Plan and schedule, detailed, 129
Plan room, 78, 80
Planning, 135, 240
  and design phase, 8–9, 40–69
  and feasibility study, 8, 67
    stage, 40, 44–54
Planning and scheduling, 129–58
Planning, logistical, 193
Plans, 59
Plant, 209, 243
Plant cost, 92
Pollutant, 233
Pollution, air, 232
Post project activities, 300
Post-qualification of tenderers, 72–4
Practical completion, 297
Precedent, judicial, 265
Precision Estimating Collection®, 102
Prefabrication, 81, 188
Pre-final inspection, 291
Preliminary:
  assessed costs, 87–8
  cost estimate, 50, 52, 87
  design, 76
  job planning, 80–3
  project schedule, 129
  schedule, 82–3
Premium, insurance, 127
Pre-project phase, 8, 12–33
Pre-qualification, 9
  criteria, 74
  of tenderers, 72–4
  process, 73, 75
Pre-tender meetings, 83
Pre-trail activity, 264
Prevention, dispute, 263–64
Price alone, 110
Prime contractor, 20
Prime contracts, separate, 20, 29
Prime cost allowance, 93
Prime cost sum, 93
Principal, bond, 122, 123
Procurement process, materials, 169
Product data, 246
Productivity, 14, 90, 91, 96
  definition, 238–39
  labour, 235, 238–41
  trends, 239
Professional construction management, 16
Professional licensing, 41
Profit, 90
Programme, 47, 253, 263
  development, 47–8
  manager, 43
  master, 130, 173
  project, 206
  safety, 222, 238
    incentive, 227, 228
  substance abuse, 238
  training, 238
Programming, 129–58, 156, 240, 257
Progress curves, 161
Progress payment, 248–51
  request, 250
Project:
  brief, 8
  budget, 9, 120, 158, 207, 210
  characteristics, 78
  closeout and termination, 289–301
    phase, 10
  communications, electronically-enhanced, 257–61
  completion date, estimated, 206
  complexity, 106
  delivery options, 40
  delivery system, 8, 12, 26, 174
    advantages, 28, 29
    features, 28, 29
    limitations, 28, 29
  development schedule, 56
  duration, total, 138
  engineer, 175
  engineering, 175
  information sources, 76–8
  Labour Agreement (PLA), 186
  life cycle, 7, 87, 251
  location, 78, 106
  Management Institute, 149
  manager, 18, 23, 28, 43, 47, 72, 75, 83, 175
    contractor's, 175
  manual, 58
  mobilisation phase, 9, 120–88
  operations phase, 9–10, 205–77
  organisation, 36, 38
  overhead cost, 90, 93
  programme, 206
  progress curves, 161
  recommendation, 53
  records, 263, 289
  schedule, 205
    detailed, 120
    preliminary, 129
  scope, 247

size, 78, 106
sponsor, 22
staffing, 174, 175, 184–87
superintendent, 175
type, 78
website, 50
Projection, cash flow, 160–63
Project-specific quality requirements, 220
Project-specific websites, 257, 261
Promotion, 238
Property insurance, 127
Proposal:
bond, 122, 123, 124
design, 44, 45
form, 108
irregular, 110
material, 92
opening, 105–07
preparation, 105–06
subcontract, 172
submittal, 105–07
Protection of stored materials, 243
Provisional cost allowance, 93
Public:
agencies, involvement, 69
agency inspector, 185
authority inspection, 121
Buildings Service, US, 252
input, to design, 50
tender opening, 107
Publications, on-line, 76
Punch list, 253, 290
final, 291
preliminary, 290–91
Purchase order, 169, 170, 171, 254
Purchasing policy, company, 168

Qualification, in tender, 114
Qualifications-based selection, 45
Qualified workforce, 236
Quality, 205
assurance, 218
control, 218
report, 254
definition, 218
management, 218–22
total, 219
programme, company-wide, 220
Quantities:
actual, 30
estimated, 30
final, 294–95
schedule of, 65
Quantity section, cost report, 211
Quantity survey, 43, 158
Quantity surveyor, 43, 94, 249
Quarries, 166

R.S. Means Company, Inc., 82
Real evidence, 265
Reasonable delay, 274
Recommendation, project, 53
Reconciliation, 267
Record drawing, 298
Records:
archiving and transfer, 298–99
project, 289
Re-cross examination, 265
Recruiting, 187
Redirect examination, 265
Reimbursable cost, 31, 261
Relationship:
contractual, 13
non-contractual, 13
Release of retainage, 295–96
Remote projects, mobilisation, 187
Remote regions, 120
Renovation products, 233
Rental rates, equipment, 244
Report:
post-project, 300
tabular, 149
Request:
for information (RFI), 254, 259
for payment, 255
Residential construction, 1
Resource:
analysis, 154
consent, 121
curve, 155
levelling, 154, 155
management, 9–10, 205, 235–45
Responsible tender, 75
Responsible tenderer, 110
Responsive tender, 75
Responsive tenderer, 110
Retainage, 248–51
release of, 295–96
Retention, 251
Return on investment, 52
Review board, dispute, 266–67
Review, agency, 69

Risk, 90, 263
Roles, of planning and design phase parties, 40–4
Rolling equipment, 243
Rough order of costs, 87
Royal Incorporation of Architects of Scotland, 41

Safe workplace laws, 225
Safety, 79, 106, 166, 175, 188, 205
  health and, checklist, 226
  inspection, 227
  management, 222–28
  programme, 222, 38
    and policy, company, 226
    site, 226, 227
  rules, jobsite, 228
  signage, 230
Sales tax, 98
Samples, 246
Sanitary facilities, 164
Sanitation, worker, 234
Scandinavia, 23, 36
Schedule, 253
  control, 240, 257
  detailed project, 120
  growth, 26
  monitoring, 156
  network, 136, 137, 139, 142–43, 150–51
  of prices, 30
  of quantities, 43, 65
  of values, 250
  progress, 10, 206
  updating, 156, 206
  variance, 215
  master, 130
  network-based project, 206
  preliminary, 82–3, 86
  project development, 56
Scheduled cost curve, 214
Scheduling, 129–58
Schematic design, 8, 54–5, 57, 67
Scope, project, 247
Scoring matrix, 111–13
Security:
  site, 166
  tool, 238
Segments, construction industry, 1
Separate prime contracts, 20, 29
Services, temporary, 163–65
Shacks, dry, 164
Shipping, material, 285
Shop drawing, 242, 245, 246, 285
Signage, 166–67
  safety, 230
Silt fence, 231, 232
Simulation, computer, 156
Singapore, 14, 221, 222, 239
Site:
  access, 165
  conditions, differing, 270–72
  inspection, 271
  investigation, 49
  layout plan, 166
  overhead cost, 90, 93, 95
  safety, 166
    programme, 226, 227
  security, 166
  selection, 54
Size, project, 78, 106
Skilled labour, 186
Slack, 138, 140, 206
Slovenia, 107
Snagging list, 290
Social charter, 185
Software, 255
  cost estimating, 101–02, 158
  features, network scheduling, 149, 154–56
Soil conditions, 49
Solid waste disposal, 233
Spain, 36
Spare parts, 289
Special conditions, 61, 63, 231
Special provisions, 61, 63
Specialists, 42
Specialty trade contractors, 1,6
Species, endangered, 234
Specifications, 60, 62, 81, 253
  descriptive, 219
  performance, 219
Sponsoring coventurer, 23
Square metre of contact area (smca), 96
Standards New Zealand, 58
Startup, 290
Steward, union, 236
Stipulated price contract, 27
Stop work order, 221
Storage:
  areas, 165
  indoor, 163
  material, 242
Strategic positioning, 78
Subconsultant, 13, 41
Subcontract, 170, 209, 253
  award, 174

negotiation, 174
proposal, 172
sample, 176–83
work cost, 90, 93, 95
Subcontracting, 170, 172–74
advantages and disadvantages, 172
Subcontractor, 13, 83, 93, 106, 120,163, 165, 175, 228, 251, 274, 291
payment, 293–94
Submittal, 241, 245–46, 251
detail, 257
log, 246, 256
process, 106
Substance abuse programme, 238
Substantial completion, certificate of, 297
Successful contractor, selecting, 107, 110–14
Summary activity, 145, 149
Summary judgement, 264
Summons, 264
Superintendent, project, 175
Supplementary general conditions, 61
Surety, 73, 122
Association of Canada, 122
bond, 67, 122, 251
consent of, 294
Suretyship, 122
Sweden, 36, 39
Switzerland, 58
System, cost accounting code, 160

Table, activity, 140, 152–53
Tabular report, 149, 154
Tag:
brass, 237
in tender, 114
Take-off, 94
material, 65
partial, 88
Tardiness, 240
Taxes, 96
Taxonomy, cost estimate, 90
Technical specification, 57, 58, 61, 62–5, 96, 169, 241, 246
Telephone record, 254
Temporary:
facilities, 163–65
housing, 164
quarters, 199
services, 163–65
utilities, 164
Tender, 9
balanced, 101
bond, 122, 126
decision, contractor's, 76–9
form, 66
invitation to, 65, 75
open, 76
price, total, 90, 100
proposal form, 107
responsible, 75
responsive, 75
unbalanced, 101
Tenderer:
apparent low, 107
lowest qualified, 110
responsible, 110
responsive, 110
Tenderers, instructions to, 65–6
Tendering, 105, 172, 251
Termination:
contract, 275–77
for convenience, 275, 276
for default, 275–76
project, 289–301
Testimony, 265
expert, 265
opinion, 265
Testing, 290
on-site, 221
Texas Department of Transportation, 110
Time:
and materials contract, 33
calculations, 154
card, 209, 237
non-productive, 239
of the essence clause, 273
record, 237
Timing of expenses and revenues, 160
Tools, 238
inventory, 238
security, 238
Topography, 49
Total:
budget, 213, 214
project duration, 138
quality management, 219
tender price, 90, 100
variance, 215
Trade:
contract, 32
magazines, 78
union, 185

Tradespeople, 175, 185
Traffic consultant, 42
Training, 186, 187, 239, 289
  owner personnel, 299
  programme, 238
Trans Alaska Oil Pipeline, 23, 186, 188, 199–204
Transfer, records, 298–99
Transit New Zealand, 221
Transmittal, 254
Triad of parties, 1
Trial:
  activity, 264
  bench, 265
  jury, 265
Triple constraint, 218
Turnkey, 21
  contract, 20–1, 29
Type of contract, 8, 26, 27, 30–3
Type, project, 78

Unbalanced tender, 101
Unforeseen conditions, 270
Union:
  agreement, 91
  hiring hall, 236
  labour, 185–86
  shop, 186
  steward, 236
Unit area cost, 87
Unit cost section, cost report, 213
Unit price, 103
  contract, 27, 30, 31, 33, 66, 72, 75, 80, 83, 247, 249, 295
  estimate, 98
  schedule, 101
Unit volume cost, 87
United Kingdom, 4, 21, 26, 221, 222, 223, 225, 226, 267, 268
United States of America, 4, 6, 19, 44, 45, 58, 59, 107, 114, 121, 185, 186, 226, 233, 234, 237, 239, 268
Updating, schedule, 156, 205
Up-front documents, 57–8
Utilities, temporary, 164

Value engineering, 68, 102,105, 251–52, 245
  change proposal, 102, 251
Value, earned, 213, 214
Variance:
  at completion, estimated, 213
  cost, 215
  schedule, 215
  total, 215
Variation, 30, 106, 245, 246–48, 253, 255, 294
  proposal, 248, 255
  request, 248, 255
Verdict, 265
Virginia Department of Transportation, 73
Visits, jobsite, 81–2

Wage rate, 236
Waiver of lien, 294
Wanganella, M.V., 193, 197–98
Warranted maximum cost contract, 32
Warranty, 299
Water drainage and runoff, 231
Web-based:
  collaboration, 245
  communication tools, 260
Website, 76, 102, 259, 260
  project-specific, 257
Weekly report, 254
Wetlands, 121
Wildlife protection, 234
Wilmot Pass, 193, 196
  Road, 195, 197
Winest®, 102, 158
Witness, expert, 265
Work:
  assignment, 237
  breakdown structure (WBS), 130–33, 144, 145
  performed, actual cost of, 215
  performed, budgeted cost of, 213, 214, 215
  scheduled, budgeted cost of, 215
  site, 120
Worker sanitation, 234
Workforce:
  construction, 3
  qualified, 236
Working conditions, 239
Works tender opening checklist, 109
Workshop, 163
  partnering, 128
Worksite, 163–168
  organization structure, 174, 175, 184–85
World Bank, 76, 266

Yellowknife, 65
Yukon River, 199